ENGINEERING MECHANICS AND ELEMENTS OF CIVIL ENGINEERING

S.S. Bhavikatti

Emeritus Fellow
B.V.B. College of Engineering & Technology
Hubli, Karnataka
(Formerly Principal, RYMEC, Bellary
Dean and Professor at NITK, Surathkal
and SDMCET, Dharwad)

S.K. Duggal

Professor and Head
Civil Engineering Department
Motilal Nehru Institute of Technology
Allahabad (U.P.)

PUBLISHING FOR ONE WORLD

NEW AGE INTERNATIONAL (P) LIMITED, PUBLISHERS

New Delhi • Bangalore • Chennai • Cochin • Guwahati • Hyderabad
Jalandhar • Kolkata • Lucknow • Mumbai • Ranchi

Visit us at www.newagepublishers.com

Published by New Age International (P) Ltd., Publishers
First Edition: 2010

Branches:

- 36, Malikarjuna Temple Street, Opp. ICWA, Basavanagudi, **Bangalore**. ✆ (080) 26677815
- 26, Damodaran Street, T. Nagar, **Chennai**. ✆ (044) 24353401
- CE-39/1016, Carrier Station Road, Near Paulson Park Hotel, Ernakulam South, **Cochin**. ✆ (0484) 4051303
- Hemsen Complex, Mohd. Shah Road, Paltan Bazar, Near Starline Hotel, **Guwahati**. ✆ (0361) 2543669
- No. 105, 1st Floor, Madhiray Kaveri Tower, 3-2-19, Azam Jahi Road, Nimboliadda, **Hyderabad**. ✆ (040) 24652456
- RDB Chambers (Formerly Lotus Cinema) 106A, Ist Floor, S.N. Banerjee Road, **Kolkata**. ✆ (033) 22275247
- 16, Jopling Road, **Lucknow**. ✆ (0522) 2209578
- 142C, Victor House, Ground Floor, N.M. Joshi Marg, Lower Parel, **Mumbai**. ✆ (022) 24927869
- 22, Golden House, Daryaganj, **New Delhi**. ✆ (011) 23262368, 23262370

ISBN: 81-224-1440-0

Rs. 125.00

C-09-08-3911

Printed in India at Mohanlal Printers, Delhi

PUBLISHING FOR ONE WORLD

NEW AGE INTERNATIONAL (P) LIMITED, PUBLISHERS
4835/24, Ansari Road, Daryaganj, New Delhi - 110002
Visit us at **www.newagepublishers.com**

PREFACE

Every engineer should know elements of civil engineering and engineering mechanics. This book is written for the benefit of all students of engineering. First part deals with elements of civil engineering comprising of role and functions of civil engineers, basics of construction management, introduction to traditional, alternative, composite and smart materials and also to surveying. Second part deals with the principles of mechanics and their applications to engineering problems. Coplanar force systems, friction, centroid and moment of inertia problems are dealt with in detail. Illustrative examples and practice problems are included throughout the book to provide a thorough understanding of the subject.

S.S. Bhavikatti

S.K. Duggal

CONTENTS

PART-A : ELEMENTS OF CIVIL ENGINEERING

UNIT - I

UNIT - II

ENGINEERING MECHANICS AND ELEMENTS OF CIVIL ENGINEERING

UNIT–I

CHAPTER 1

Civil Engineering and Infrastructural Development

1.1 INTRODUCTION

Engineering is the profession that puts scientific knowledge to practical use. The word engineering comes from the Latin word ingeniare, which means to design or to create. Engineers use principles of science to design structures, machines, and products of all kinds. They look for better ways to use existing resources and often develop new materials. Engineers have had a direct role in the creation of most of modern technology—the tools, materials, techniques, and power sources that make our lives easier.

The field of engineering includes a wide variety of activities. For example, engineering projects range from the construction of huge dams to the design of tiny electronic circuits. Engineers may help produce guided missiles, industrial robots, or artificial limbs for the physically handicapped. They develop complex scientific equipment to explore the reaches of outer space and the depths of the oceans. Engineers also plan our electric power and water supply systems, and do research to improve automobiles, television sets, and other consumer products. They may work to reduce environmental pollution, increase the world's food supply, and make transportation faster and safer.

The key theme in the civil engineering is the ''protection of the welfare and safety of the public''. Indeed, the *civil* in civil engineering refers to the discipline's envolvment in public works, including government buildings, military bases, mass transit systems—highways, railways, airports, shipping ports, water treatment works, waste management etc. In ancient times, there was no formal engineering education. The earliest engineers built structures and developed tools by trial and error. Today, special college training prepares engineers to work in a certain branch or field of engineering and standards of quality and performance guide them on the job.

Civil engineering, the oldest of the main disciplines of engineering, involves planning and supervision of such large construction projects as bridges, canals, dams, tunnels, and water supply systems. Civil engineers also cooperate with architects to design and erect all types of buildings. Other civil-engineering projects include airports, highways, levees, irrigation, sewerage systems and landfills, pipelines, and railways.

Civil engineers work to build strong and safe structures that meet requirements of building codes and other regulations and are well suited to their surroundings. They are responsible for surveying and preparing construction sites and for selecting appropriate materials. Civil engineers must also understand the use of bulldozers, cranes, power shovels, and other construction equipment.

Some civil engineers specialize in the study of the physical characteristics of soils and rocks and the design of foundations. Others concentrate on the management of water resources, including the construction of flood control and irrigation systems, hydroelectric power plants, water supply and sewerage systems and environment protection works. Still others are concerned with designing transportation systems and methods of traffic control. Many civil engineers are involved in city planning, urban renewal programmes, ground improvement and land reclamation works.

1.2 ROLE OF CIVIL ENGINEER

Construction is one of humanity's earliest organized activity. Civil engineering affects many of our daily activities: the buildings we live in and work in, the transportation facilities we use, the water we drink, and the drainage and sewage systems that are necessary to our health and well-being. Civil engineers:

- Measure and map the earth's surface.
- Design and supervise the construction of bridges, tunnels, large buildings, dams, and coastal structures.
- Plan, layout, construct, and maintain railroads, highways, and airports.
- Devise systems for the control and efficient flow of traffic.
- Plan and build river navigation and flood control projects.
- Provide plants and systems for water supply and sewage and refuse disposal.

To build may be a primal urge. Our constructions, while they may be simply for shelter or transportation, often include aesthetic touches that are there to make us feel good about what we have built. Thus, bridges have geometrical designs intended to support weight, but they also have an artistic detailing and a ''look'' that defines the era in which they were built. In constructing buildings, highways, and bridges, civil engineers work with architects to develop the appearance of the structure. Ugly buildings represent a failed communication between the two professionals; a building that falls down, or cannot be maintained, also represents a failure, but one that the civil engineer could possibly have prevented.

Civil engineering is much more than erecting skyscrapers or bridges. Civil engineers are trained in the interactions among structures, the earth, and water, with applications ranging from highways to dams and water reservoirs. Deeply involved with specifying appropriate construction materials, many civil engineers and others are employed by the manufacturers of those materials. Since constructing a large building or public-works project can involve elaborate planning, civil engineers can be outstanding project managers. They sometimes oversee thousands of workers and develop advanced computerization and planning policies.

Most significantly, many civil engineers are involved with preserving, protecting, or restoring the environment. Most water treatment and water purification projects are designed and constructed by civil engineers (in these two areas, many of them are known as environmental engineers). A growing number of civil engineers are involved in billion-dollar projects to clean up toxic industrial or municipal wastes at abandoned dump sites. Civil engineers engage in such diverse projects as preserving wetlands or beaches, maintaining national forest parks, and restoring the land around mines, oil wells, or factories.

1.3 INFRASTRUCTURAL DEVELOPMENT

A key word that arose in 1980's and will remain important for civil engineers for many years to come is *infrastructure*. This term refers to the facilities that local, state, and central governments provide in order for private industry to expand, or for improving the services for private citizens.

The economic development of a country depends upon the developments in agricultural and industrial fields. Both, the agriculture and industry, require power, machinery, transport, planning, management and banking. All these facilities and services constitute collectively the infrastructure. The development and expansion of these facilities are essential for the increase in production and have great bearing over the economy of the nation. The infrastructural facilities, referred to as economic and social overheads, consist of energy (coal, electricity, oil, etc.), transport (railway, highway, airway, shipping, etc.), science and technology, communication, social (health and hygiene), and banking.

Indian planners were fully aware of the link between infrastructural facilities and general economic development. Accordingly, from the very first five year plan, the expansion of these facilities was given top priority. Large scale industrialization led to the migration of people from rural areas to the cities. Civil engineers also comes to the fore when social changes faster new development. The population swing also require new construction for roads, housing, water systems, and schools.

Construction is the back bone of the progress of a country. It is impossible to think of development of infrastructural facilities without civil engineering. From getting into small projects—construction of buildings, etc. for say Post and Telegraph—to big projects such as dams, river valley projects etc., it is the civil engineers who take the lead. Depending upon the field of specialisation a civil engineer is classed as structural engineer, irrigation engineer, etc. as follows.

1. **Structural engineer:** This is the classic civil engineer, concerned with designing walls, towers, bridge spans, dams, or foundations. A knowledge of

construction materials and methods is combined with analytical techniques that determine how much weight or mass a structure is carrying, what forces it must withstand (such as wind or water) and, in cases where an architect is involved, how best to accomplish the architect's vision.

2. **Construction engineer:** This engineer works at the construction site, transforming blueprints and drawings into concrete and steel reality. Besides understanding the principles by which a structure was designed, the construction engineer must manage the actual work. This can involve elaborate scheduling and planning so that materials and workers are brought to the site to complete their purpose in the proper order. Time pressures and an awareness of the financial elements of a project are constant objectives.

3. **Surveying and mapping engineer:** Even before a design is completed, and as construction progresses, teams of surveying and mapping engineers are at work. They use electronic instruments and even satellites (which provide detailed overhead views) to measure the dimensions of the project. Some construction projects can cover dozens of square miles of territory. Elevations must be determined, calculations made regarding how much earth needs to be moved, and the *exact* location of structure(s) must be determined.

4. **Irrigation engineer:** India has the largest irrigated area in the world. Canal irrigation and well irrigation have been developed around the country for agricultural purposes, tubewells account for 30% of toal area irrigated. Agriculture which is one of the key indices of economic growth is directly dependent upon the irrigation facilities. Though the government has placed emphasis on irrigation in its each five year plans, it has been possible to irrigate only 60% of the agricultural land available. Civil engineers are part and parcel of the irrigation system from wells, canals, dams to river valley projects.

   Utility companies, factories, farms, and river barges depend on a steady source of water. These engineers perform the planning, design, construction, and maintenance to keep supplies available. Dam design and construction, flood control, and the design and construction of reservoirs, wells and aqueducts are all common projects.

5. **Geotechnical engineer:** Along with geological engineers, these engineers help determine the underlying rock strata and soil conditions that affect roadways, water reservoirs, bridges, and other large structures. Earthquake planning and preparation also fall into this category.

6. **Transportation engineer:** Transportation is the backbone of development. The movement of raw materials, fuel, machinery, manpower to the place of production/construction depend on the transportation facilities. Transport development helps to open up remote regions and resources for construction/production purposes.

   The transport system includes highways, railways, air and water navigation facilities, of which rail and road transport dominate. The role of civil engineer is construction and maintenance of the rail and road networks. Strengthening and improving the crucial sections and multilaning of highways, improvement of roads in the remote and rural areas are some of the tasks expected from civil engineers. The slow pace of development has forced the government to allow the private sector to participate in construction. Transportation engineer also oversees the design and construction of mass transit systems, such as subways, which require tunneling, and research on commuting plans with regards to water navigation there are 11 ports in India. A civil engineer are responsible for the construction, maintenance and upgradation of docks and harbours. Air navigation is another area where a civil engineer is called upon to construct airports, air fields, terminal buildings etc.

7. **Environmental (sanitary) engineer:** These engineers specialize in water and wastewater projects, land remediation, and solid waste disposal. This field is currently one of the fastest growing of all engineering specialities. Billions of rupees are being allocated for water and wastewater treatment, for methods of processing solid wastes, and for cleaning up hazardous waste dumping sites for better environment.

1.4 IMPACT OF INFRASTRUCTURAL DEVELOPMENT

Construction is the key part of the overall Indian economy. Infrastructure demands will be a big element of civil engineering for the next several years. The pace of construction activity is directly related to the central, state, and city budgets for construction. Following are some of the direct and indirect impacts of infrastructural developments.

1. The heavy investments by the government on infrastructural facilities have provided the necessary impetus for rapid industrial expansion and agricultural development. This has led to multifold rise in agricultural and industrial productions and percapita income.
2. The overwhelming emphasis has been on the provision of infrastructural facilities mainly for the urban areas. The concentration of the infrastructure in the urban areas and the relative neglect of the rural areas resulted in inadequate development and inadequate employement opportunities in rural areas. Consequently there has been exodus of the rural people to urban areas for the search of their livelyhood. This has caused problems such as urban congestion, acute housing shortage, transport bottle necks, slums, pollution, etc.
3. Infrastructural development has not only shown an urban-bias, but also a bias in favour of rich. It is the higher income groups who could take full advantages of power, transport, health and other infrastructural facilities.
4. Irrigation projects also have been for the benefit of rich farmers only — the small and marginal farmers though large in number could not derive any benefit.

Overall, however, the civil engineering field in India, at present, is not as dynamic as it was two to three decades ago, when new industries were coming up and communities were popping up all over the land. Today, the reduced demand for civil engineers can be seen in the slightly lower salaries that civil engineers earn coming out of college. However, the scene is changing again. The financial aids from World Bank for various infrastructural projects has opened up gates for civil engineers to envolve more in the development and construction programs. At the same time, saying that there is less growth in the Indian economy is not the same as saying there is no growth. New industries are coming up, new skyscrapers and bridges are going up across the land, and more environmental work is being scheduled. The need of civil engineers in the economic growth of the country is non ending and unparalleled.

EXERCISE

1. What is the literal meaning of civil engineering? What are the activities a civil engineer is expected to perform?
2. Describe briefly the role of civil engineer. How do they contribute to the society?
3. How does infrastructural development affect the economy of a country? Discuss the role of civil engineer in infrastructural development.
4. Describe briefly the impact of infrastructural development in India.

CHAPTER 2

CONSTRUCTION MANAGEMENT

2.1 INTRODUCTION

The efficiency of an organization depends largely on the quality of management — its ability to organize and direct the various functions. Management is concerned with the economy and timely execution of the project. A project is composed of jobs, activities, and functions. All the three are interrelated and hence influence the completion of the project to a great extent.

The basic objective of managing any task is achieving the same in the most economical way. This is accomplished by

- planning for optimum utilisation resources
- planning the resources in the least expensive way in case new resources are to be added
- achieving the task in least possible time
- adhering to the specifications and maintaining the quality
- monitoring the programme
- having good controlling mechanism to control the deviations, if any, found.

To achieve the above objectives a proper and effective project management is desirable. It is not only contractor alone involved in executing the project. The purchaser and consultant also play a vital role in minimising the completion time and cost of the project. The management has to be such that all the three work in tandem in the interest of the project.

The project management involves planning, scheduling, and monitoring or controlling. The first two phases are accomplished before the project starts whereas the third phase (controlling) operates during the execution of the project.

2.2 PLANNING

The design, construction techniques and quality of materials while influence the economics of construction on one end, how effectively the execution is planned affects it to a larger extent on the other hand. Planning is not one time affair but is a continuous process. It is not just enough to have the programme charts framed up on the walls. It will deliver fruitful results only when it is updated from time to time by monitoring it by having a good controlling mechanism.

Planning has been practised in some form or other through ages; the structures of the past are examples of the skill and fore-thought of builders of the past. However, with the growing sophestication of the technology, the projects have tended to be more and more complex, necessitating newer and better methods of planning. Techniques such as CPM (Critical Path Method) and PERT (Program Evaluation and Review Technique), the network analysis techniques, provide a tool to plan rationally and to manage effectively.

Planning consists in defining the objectives of the project; listing of tasks to be performed to achieve the objectives; estimating the requirements of material, equipment, and manpower; and preparation of estimates of costs and durations for the various activities. Following are some of the advantages of planning:

1. It provides direction to achieve the set goals in stipulated time.
2. It provides unifying frame work.
3. It helps to reveal future opportunities and exposes complexities.
4. It provides performance standards.

Planning is of particular importance in execution of projects. Considerable fore-thought and preparatory work is required before taking up the actual execution.

2.2.1 Pre-planning

While undertaking any new project (bridge, dam, highway etc.) it is necessary to think about how the work will be done, how much time will be required, and what will be the probable cost. The project should be well defined and the steps to achieve the objectives must be clearly visualized.

After deciding the objective the possible alternatives likely to help in achieving the objective are explored and a cost analysis is done to identify the economic viability. The project may involve private or public investment. In case of private investment it will have to be ascertained that whether the project will yield the minimum return on the capital or not in case of public investment the cost analysis has to be much more elaborate. Here, the projects may be intended to improve social conditions of a community, to initiate economic development by some backward area or investment on education etc. In such cases *Economic Analysis* is carried out and it involves calculation of socio-economic benefits to the country.

Steps

1. Detailing the objectives of the project.
2. Formulation of the frame-work of the project.
3. Justification for undertaking the project
4. Analysis of investment alternatives
5. Cost - benefit analysis
6. Decision — amongst the various investment alternatives.

2.2.2 Detailed Planning

Whether the project is to be taken up or not, is decided after preplanning. Once it is accepted for execution a detailed planning is required. This requires a considerable amount of planning. The first step in this is site selection with the help of field surveys. Then a detailed design and estimate is prepared. Then, arrangement of working drawings is made. The need of plant and equipments is identified. Specifications of the works, plant and machinery are made. Also, quantities of work of different types and costs involved are worked out. In the last it is planned as to how thc job will be done and the method of execution is detailed. Following are the steps involved.

1. Preparation of detailed designs and working drawings.
2. Deciding specification of works.
3. Preparation of bill of quantities and estimates.
4. Working out detailed plan for execution of work.

The information, a planner needs is about:

1. Material resources
2. Financial resources
3. Equipment resources
4. Space
5. Manpower resources
6. Time

2.3 SCHEDULING

The arrangement of required resource also called *allocation of resources* is known as scheduling. It may be defined as the laying out of different activities of the project in time order in which they will be executed; calculating the man power, material requirements, equipments etc.; along with the expected completion time of each of the activity, and hence the project duration.

Scheduling is a mechanical process of formalising the planned functions, assigning the starting and completion dates to each activity so that the whole project proceeds in a logical sequence and in a systematic way.

From the schedule so prepared, it is possible to workout the requirement of materials, machinery, manpower, and funds at various stages of the project.

2.4 MONITORING

Once the time chart of various activities is evolved, the project can be taken up for execution. The schedule prepared is based on the estimates of duration of various activities. Assumptions are made about the site conditions and availability of materials equipments, manpower etc. These assumptions may or may not prove to be true in practice. This leads to variations in the completion time of various activities, from the original time chart. This requires updating the original schedule on the basis of feed-back of information about the actual progress time estimates of the availability of resources.

Thus, project monitoring is the formal mechanism established to determine deviations from the basic schedule, to determine the effect of these deviations on the project duration, and to replan and reschedule to compensate for the deviations.

A prerequisite for any project to be successful, serious efforts should be made for effective monitoring and controlling the programs. There has to be coordinated thinking and action amongst the client, contractor and consultant if any project has to be completed economically and in time.

4.5 ROLE OF AN ENGINEER

It is engineer, who is envolved in the project from the very first stage of planning, executes it till the completion, and is called Project Manager. He has lot of responsibilities and is answerable to all the odds. He has to decide as to which plan is to be implemented, what is the target, etc. A poor or erroneous decision may lead to the failure of the project.

EXERCISE

1. Describe briefly the role of management in project execution.
2. How is planning important for a project? What are the disadvantages of improper planning.
3. What are the three phases of project management? Discuss them alongwith the role of civil engineer.
4. Write notes on following:
   (a) Planning
   (b) Scheduling
   (c) Monitoring

UNIT–II

CHAPTER 1

Materials of Construction

1.1 INTRODUCTION

The importance of building materials industry in our national economy is enormous as its output governs both the rate and the quality of construction work. Although their most important use is in construction activities, no field of engineering is conceivable without their use.

Every construction project is intended to result in a finished product which will perform certain functions in conformance with, and sometimes in spite of, the effects of nature. Whether or not satisfactory results are achieved depends upon the materials selected and how they are used. They must perform under specific conditions of usage. The designer, the builder, and the user must all understand construction materials to produce the finished facility and to use it to best advantage. Knowledge of design procedures, construction methods, and maintenance practices is needed. Underlying all these qualifications is a knowledge of materials. In order to completely satisfactory, each material used must perform its function well over a sufficiently long time, and both original cost and maintenance expense must be reasonable.

The wide span of service conditions of buildings and installations, the great diversity of production processes, etc., place a great variety of requirements upon building materials, and this calls for a very wide range of their properties; strength at normal and high temperatures; water resistance; resistance to various salts, acids and alkalis; resistance to slag (of particular importance in metallurgy), etc. No less important in building and engineering work is the permeability (or the impermeability) of materials to liquids, gases, heat, cold, electric field, radioactive radiation. And finally, materials for decorating interiors of residential and public buildings, gardens and parks, etc., should be, by their very purpose, pleasant to the eye, durable and strong. A material with universal properties is as yet a challenge of the future.

The chief properties of building materials predetermine their applications. Only a deep and comprehensive knowledge of the properties of materials allows a rational—both from the engineering and economic standpoints—choice of materials for specific service conditions.

All construction materials are the subject of the Bureau of Indian Standard specifications which name the materials, define its major properties and requirements placed upon it and specify, first, the testing procedures to be used to establish and quantify its properties; second, the rules of acceptance, transportation and storage. These technical specifications are complemented by the codes which are not permanently fixed. As production processes and construction practice improve, the requirements upon building materials also grow more stringent and their scope is enlarge. The above documents are subject to revision as required.

The importance of standardization is very great. It requires the quality of materials and manufactured items to be not below a specified standard level, this allowing to design dependable and durable construction irrespective of the techniques by which the constituent materials and the manufactured items are produced. However, the importance of standardization is not limited to this factor alone, since each new or revised. Standard places higher requirements upon the products than the preceding one, with the effect that the industry concerned has to keep up with the standards and improve production techniques.

Thus, the industry of building materials gains both in quantity and quality, so that new, more efficient products are manufactured and the output of conventional materials is increased.

A task of great importance is to compare the performance (under specified service conditions) of similar kinds of materials with a view to develop products of greater economic efficiency.

It is common knowledge that today the manufacture of building materials lags behind the demands for them, and, therefore, the main current tasks are to speed up the development of the building materials industry and to decrease steadily the cost price of manufacture and the specific capital investments, i.e., investments per unit of products at newly commissioned plants.

The types of materials used in smaller buildings have become somewhat standardized. However, new materials are constantly being proposed, and the use of the better ones results in lowered costs or improved living conditions. Their development and proper use require an understanding of materials. As man's desires expand, the need is created for materials with new qualities. In order to explore space, lightweight materials were needed that could resist heat of a higher degree than ever before. Necessary qualities may be obtained by developing special treatments for common materials or by developing entirely new materials.

1.2 SELECTING MATERIALS

We constantly encounter man-made objects built of materials carefully selected to be the most satisfactory ones for that particular use. Any satisfactory choice always requires a knowledge of construction materials and an adequate selection procedure.

A construction project originates in the mind of the owner. The owner may be a city administrator determined to build a sewer system and a plant for treating sewage; a landowner who wants to build an office building and lease office space; or a government body planning to build a dam or bridge. The owner is concerned with the cost of the project and the service it will provide.

A designer is selected who, among other things, is responsible for selection of all construction materials to achieve the desired performance within the budget cost. He considers the service each component must perform, appearance, original cost, maintenance expense, and useful life expectancy. Maintenance includes such operations as cleaning, preventing and repairing corrosion damage, and repairing or replacing damaged material.

Original cost and maintenance expense must be weighed together against useful life expectancy. Original cost and maintenance expense must also be balanced against each other. Often a low first cost means high maintenance expense and vice versa. However, this is not always so. Expensive material may be expensive to maintain. Even though it is the total cost that must be considered, it is important to remember that the original cost must be paid during construction and immediately thereafter, while the maintenance expense is paid through the life of the facility.

The designer may select the material or assembly he wants, or prepare specifications describing the performance required and let the builder do the selecting within the requirements of the specifications subject to the approval of the person supervising construction.

If the designer specifies exactly what materials and assemblies are to be incorporated into the project, he knows, either from past experience or from investigation, that they will be satisfactory. He avoids the risk of using something new or unfamiliar. He also misses the opportunity of using something that is more economical or performs better. Specifications prepared this way are called *material specifications*.

If the designer specifies performance in terms of appearance, strength, corrosion resistance, and other features, he has the benefit of the builder's experience in selecting the most economical materials. Specifications prepared this way are called *performance specifications*. They must be very carefully written to prevent any inferior products from satisfying the specification requirements; and the builder's selections must be carefully investigated to be sure they are acceptable according to the specifications. Both types of specification and various combinations of the two types are used. Each type is suitable for certain cases.

The process of selection includes the following steps:

1. Analysis of the problem (performance required, useful life required, allowable cost and maintenance expense);
2. Comparison of available materials or products with the criteria of step 1;
3. Design or selection of type of material, size, shape, finish, method of presenting, and method of fastening in place.

1.3 PROPERTIES OF MATERIALS

Density The mass of a unit volume of homogeneous material expressed as g/cm^3. For example, density of brick is 2.5–2.8 g/cm^3 and that of wood is 1.5–1.6 g/cm^3.

Bulk Density is defined as the mass of a unit volume of material in its natural state. For convenience it is expressed as kg/m^3. For example the bulk density for bricks is 1600-1800 kg/m^3 and that for sand is 1450-1650 kg/m^3.

Porosity is defined as the degree to which volume of the material is interspersed with pores. It is expressed as a ratio of the volume of pores to that of the specimen. Porosity is indicative of other major properties of material, such as bulk density, heat conductivity, durability, etc. Dense materials, which have low porosity, are used for constructions requiring high mechanical strength on the other hand, walls of buildings are commonly built of materials, featuring considerable porosity.

Hygroscopicity is the property of a material to absorb water vapour from air. It is influenced by air-temperature and relative humidity; pores—their types, number and size, and by the nature of substance involved.

Water Absorption denotes the ability of the material to absorb and retain water. It is expressed as percentage in weight or of the volume of dry material:

$$W_w = \frac{M_1 - M}{M} \times 100 \qquad W_v = \frac{M_1 - M}{V} \times 100$$

where M_1 = mass of saturated material (g)
M = mass of dry material (g)
V = volume of material including the pores (mm^3)

Water absorption by volume is always less than 100 per cent, whereas that by weight of porous material may exceed 100 per cent.

Fire Resistance is the ability of a material to resist the action of high temperature without any appreciable deformation and substantial loss of strength. Fire resistive materials are those which char, smoulder, and ignite with difficulty when subjected to fire or high temperatures for long period but continue to burn or smoulder only in the presence of flame, e.g. wood impregnated with fire proofing chemicals. Non-combustible materials neither smoulder nor char under the action of temperature. Some of the materials neither crack nor lose shape such as clay bricks, whereas some others like steel suffer considerable deformation under the action of high temperature.

Refractoriness denotes the ability to withstand prolonged action of high temperature without melting or losing shape. Materials resisting prolonged temperatures of 1580°C or more are known as refractory.

High-melting materials can withstand temperature from 1350–1580°C, whereas low melting materials withstand temperature below 1350°C.

Chemical Resistance is the ability to withstand the action of acids, alkalis, sea water and gases. Natural stone materials, e.g. limestone, marble and dolomite are eroded even by weak acids, wood has low resistance to acids and alkalis, bitumen disintegrates under the action of alkali liquors.

Durability is the ability to resist the combined effects of atmospheric and other factors.

1.4 STRENGTH AND STRESS

All construction materials must resist force. A *force* is a push or pull that has a value and a direction. The pull of gravity is responsible for most of the forces dealt with in construction. However, there are other causes such as wind and water currents.

A force exerted on the surface of an object is assumed to spread uniformly over the internal area of the object. A force does actually spread within the material, but not uniformly. *Stress* is internal force per unit area over which the force acts. It is obtained by dividing the force by the area on which it acts and is expressed as N/mm^2 or MPa.

Strength of a material, in general terms, is the ability to resist a force. That ability depends on the size and shape of the object as well as on the material of which it is made. An object with a large area is able to resist more force than an object of the same material but with a smaller resisting area.

In order that strength may be considered as a property of the material, it is necessary to relate strength to the material itself regardless of its size or shape. Therefore, the *strength* of a material in technical terms is equal to the stress that the material can resist. Strength has the same units as stress, that is, MPa.

The useful strength of a material is equal to the stress at failure. Failure takes place when an object can no longer serve its purpose. The material may fail by breaking or by excessive deformation. *Deformation* means a change in the outside dimensions of an object caused by a force. The amount of deformation depends on the size and shape of the object as well as on the material of which it is made. As in the case of strength it is desirable to relate deformation to the material itself regardless of its size or shape.

The term strain means the total change in dimension divided by the original dimension. Strain is the effect caused by stress. Strain is a ratio and therefore has no units. The amount of deformation and the original length must be measured in the same units to provide a correct ratio. They are usually measured in mm.

Materials differ in their response to stress. A *ductile* material can be drawn into a thin, long wire by a tensile force. A *malleable* material can be flattened into a thin, wide sheet by a compressive force. A *brittle* material breaks with very little deformation. It appears to fail suddenly because there is no noticeable deformation to serve as a warning.

In this discussion, forces have been assumed to be applied once for a brief period of time. Ordinary tests made to determine strength consist of subjecting a sample of the material to a force that increases steadily until the material breaks. These tests take a few minutes and indicate static stress at failure. However, in a structure, forces may be applied for extended periods of time, they may be applied and removed many times, and they may be applied suddenly with impact or shock.

A material, even if brittle, deforms slowly when a force is applied to it for an extended period of years, even though the force is too small to cause failure in a short time. This deformation is called *creep*. The creep may be great enough to constitute failure.

Although a force of a certain amount may not cause breaking no matter how long it is applied it may be large enough to cause breaking if it is applied and removed many times (tens of thousands of times), even if over a shorter time. For example, structural members of a bridge are subjected to application and removal of stress each time a vehicle crosses. Failure from this cause is called *fatigue*, and it occurs with very little deformation.

Because there is so little deformation, there is no warning and the break seems to be sudden. However, it begins as a tiny crack and becomes larger over many cycles until the structure fails by breaking. The smaller the stress, the more times it must be repeated to cause failure. There is a stress below which the material will not fail at any number of cycles, called the *endurance limit*. Any stress above this limit will cause failure if repeated enough times. Specimens tested for endurance are generally subjected to bending first one way and then the opposite way. This reversal of stress produces failure at fewer cycles than the simple application and removal of force. The endurance limit found this way may be less than half the static stress at failure.

Toughness is the capacity of a material to absorb energy while a force is applied to it. Energy is expended by a force acting over a distance and is absorbed by a material being forced to deform through a distance. Toughness is the product of stress and strain up to the point of fracture. It is computed by determining the area under the stress-strain curve (see Figure 1-4), which is equivalent to multiplying the average stress by the total strain (force times distance). The result is called the *modulus of toughness*. Strength and ductility are both involved. The toughness of a material indicates its ability to withstand a sudden force, known as an *impact load* or *shock load*.

Resilience is the ability of a material to recover its original size and shape after being deformed by an impact load. The *modulus of resilience* is the product of stress and strain up to the elastic limit. It is a measure of the useful toughness, since beyond the elastic limit permanent deformation ordinarily renders the material unfit for further use. It is computed by determining the area under the stress-strain curve from zero to the elastic limit.

Hardness is the ability of a material to resist penetration by a harder body. Mohs scale is used to find the hardness of materials. It is a list of ten minerals arranged in the order of increasing hardness. Hardness of metals and plastics is found by indentation of a steel ball.

Elasticity is the ability of the material to restore its initial form and dimensions after the load is removed. Within the limits of elasticity of solid bodies, the deformation is proportional to the stress. Ratio of unit stress to unit deformation is termed as modulus of elasticity. A large value of it represents a material with very small deformation.

Plasticity is the ability of material to change its shape under load without cracking and to retain this shape after the load is removed. Some of the examples of plastic materials are steel, copper and hot bitumen.

1.5 THERMAL EXPANSION

All building materials change size with a change in temperature, becoming smaller when colder and larger when hotter. A piece of material, if heated uniformly, expands, with each unit length becoming a certain percentage longer. This elongation takes place in all directions and is somewhat different for each material. In order to predict amounts of expansion and contraction to be expected, a *coefficient of expansion* is determined for each material. It is a decimal representing the increase in length per unit per degree increase in temperature. The coefficient varies somewhat at different temperatures but is nearly constant for the range of temperatures involved in most cases so that one coefficient can be used for each material.

Materials to be used together in an assembly must have approximately the same coefficients of expansion, or else some provision must be made for their different expansions. The use of steel reinforcement in concrete is a good example of combining materials with approximately the same linear coefficients of thermal expansion. Long structural members may expand and contract so much that expansion room must be provided at the ends.

1.6 THERMAL CONDUCTIVITY

A building used by people must be kept warmer than the surrounding air in cold climates and cooler than the surrounding air in hot climates. Heat flows to a cooler area much like water flows to a lower level. The flow continues until outside and inside temperatures are equal. Heat movement takes place by conduction through any solid object separating areas of different temperatures.

It costs money to heat or cool a building, and the movement of heat in the wrong direction is expensive. The rate of movement varies with the material through which the heat passes. For large areas such as walls and roofs this rate is an important consideration. The rate is measured as thermal conductivity (U) of heat transmitted per square meter of cross section per hour per °C difference in temperature between the two sides of the material.

Insulation, which is material with a very low U, is used to line large surfaces to lessen the rate of heat flow. The U of a material varies directly with its density. Dead air spaces in a material are effective in reducing the U factor. The best insulation, expanded plastic foam, consists of bubbles with the proportion of solid material less than 1 per cent of the volume and the rest consisting of air or gas. Insulation is also made of fibers, ground particles, or other porous material. However, some structural materials have a low U factor and therefore serve as insulation also. Wood and certain types of lightweight concrete are two such materials that are covered in this book.

1.7 STANDARDS

The designer or builder may desire any number of properties in the material he is going to use. He must be able to specify the degree of each property in terms that he and the supplier understand, and he and the supplier must have some mutually acceptable means of determining whether or not the materials posses each property in sufficient amounts. The supplier must be able to prove to the buyer that the material possesses the properties desired to the degree desired.

EXERCISE

1. Why is it important to study the properties of building materials.
2. List and define the physical properties of materials.
3. Define and differentiate the following:
   (a) Density and bulk density

(b) Stress and strength
(c) Porosity and voids
(d) Elasticity and plasticity

4. Write short notes on the following:
(a) Refractoriness
(b) Heat conductivity
(c) Fire resistance
(d) Thermal expansion

5. What factors should be borne in mind while selecting a building material?

CHAPTER 2

Stones

2.1 INTRODUCTION

The history of mankind is supposed to have begun with the stone age marked by the use of implements and weapons made of stone. Prior to that, the difference between animals and homosapiens was largely physical. But once human beings started using stones, the world of both changed entirely.

Stone has been defined as the natural, hard substance formed from minerals and earth material which are present in rocks. Rock may be defined as the portion of the earth's crust having no definite shape and structure. Almost all rocks have a definite chemical composition and are made up of minerals and organic matter. Some of the rock-forming minerals are quartz, felspar, mica, dolomite, etc. The various types of rocks from which building stones are usually derived are granite, basalt, trap, marble, slate, sandstone and limestone.

Use of stone in building construction is traditional in the places where it is produced, although even there its high cost imposes limitations on its use. The conditions which govern selection of stone for structural purposes are cost, fashion, ornamental value and durability. As building material, however, it has lost importance with the advent of cement and steel. Secondly the strength of the structural elements built with stones cannot be rationally analysed. Also, it is difficult to transport, dress and polish consuming lot of time and slowing the pace of construction.

The natural stone is produced by quarrying natural rocks. After quarrying it is dressed and left, for 6 to 12 months, exposed to atmosphere, for seasoning—a freshly cut stone carries some natural moisture known as *quarry sap* (a mineral solution), the mineral constituents of which react with atmosphere and make stone harder.

2.2 DRESSING OF STONE

A quarried stone has rough surfaces, which are dressed to obtain a definite and regular shape. Dressing of stones is done immediately after quarrying and before seasoning to achieve less weight for transportation, pleasing appearance, proper bedding with good mortar joints, special shapes for arches, copings, pillars, etc. The various types of dressed stones are shown in Fig. 2.1.

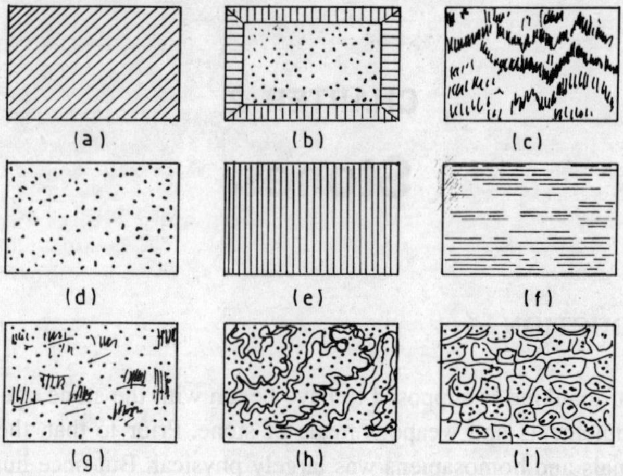

(a) Stroked
(b) Punched
(c) Rock faced
(d) Sparrow picked
(e) Tooled
(f) Sawn
(g) Combed
(h) Vermiculated
(i) Reticulated

Fig. 2.1 Dressed Stone Surfaces

2.3 DURABILITY OF STONES

Quarrying and cutting have a great bearing on the weathering properties of stones. Stone from top ledges of limestone, granite and slate and from the exposed faces of the rock bed is likely to be less hard and durable. Highly absorbent stone should not be quarried in freezing whether since the rock is likely to split. The method of blasting and cutting also influences the strength of the stone and its resistance to freezing and temperature changes. Small, uniformly distributed charge of blasting powder has a lesser weakening effect than large concentrations of explosives. A porous stone is less durable than a dense stone, since the former is less resistant to freezing. Also, rocks with tortuous pores and tubes are more apt to be injured by freezing than those of equal porosity having straight pores and tubes. Repeated hammering in cutting is likely to injure the stone. Polished stone is more enduring than rough surfaced work, since the rain slides off the former more easily. Stones from stratified rocks should be placed along the natural bed in order to secure maximum weathering resistance. Pyrite, magnetite and iron carbonate oxidize in weathering and cause discolouration of the stone in which they are present. Since oxidation is accompanied by a change in volume, the surrounding structure is weakened.

2.4 PRESERVATION OF STONES

Preservation of stone is essential to prevent its decay. Different types of stones require different treatments. But in general stones should be made dry with the help of blow lamp and then a coating of paraffin, linseed oil, light paint, etc. is applied over the stone. However, this treatment is periodic and not permanent. When treatment is done with the linseed oil, it is boiled and applied in three coats over the stone. Thereafter, a coat of dilute ammonia in warm water is applied.

The structure to be preserved should be maintained by washing stones frequently with water and steam so that dirt and salts deposited are removed from time to time. The best way is to apply preservatives. Stones are washed with thin solution of silicate of soda or potash. Then, on drying a solution of $CaCl_2$ is applied over it. These two solutions called Szerelmy's liquid, combine to form silicate of lime which fills the pores in stones. The common salt formed in this process is washed afterwards. The silicate of lime forms an insoluble film which helps to protect the stones.

Sometimes lead plant is also used to preserve the stones, but the natural colour of the stone is spoilt. Painting stone with coal tar also helps in the preservation but it spoils the beauty of the stone. Use of chemicals should be avoided as far as possible, especially the caustic alkalis. Although cleaning is easy with chemicals, there is the risk of introducing slats which may subsequently cause damage to the stone.

In industrial towns, stones are preserved by application of solution of baryta, $Ba(OH)_2$—Barium hydrate. The sulphur dioxide present in acid reacts on the calcium contents of stones to form calcium sulphate. Soot and dust present in the atmosphere adhere to the calcium sulphate and form a hard skin. In due course of time, the calcium sulphate so formed flakes off and exposes fresh stone surface for further attack. This is known as *sulphate attack.* Bartya reacts with calcium sulphate deposited on the stones and forms insoluble barium sulphate and calcium hydroxide. The calcium hydroxide absorbs carbon dioxide from the air to form calcium carbonate.

$$Ba(OH)_2 + CaSO_4 \longrightarrow \underset{\text{(Barium sulphate)}}{BaSO_4} + \underset{\text{(Calcium hydroxide)}}{Ca(OH)_2}$$

$$Ca(OH)_2 + CO_2 \longrightarrow \underset{\text{(Calcium carbonate)}}{CaCO_3} + H_2O$$

The question whether or not stone preservatives should be used on old and decayed stone is a difficult one. Real evidence of the value of various treatments is most difficult to assess. The treatments, if carefully applied under favourable circumstances, may result in an apparent slowing down of the rate of decay. However, the rate of decay of stone is so slow that a short period experience is of very little value in establishing the effectiveness of the treatment. Also, there is some evidence that treatments which appear to be successful for few years, fail to maintain the improvement. In fact, the value of preservatives is not yet proved, and they may actually be detrimental if judged over a long period.

2.5 CHARACTERISTICS OF GOOD BUILDING STONE

A good building stone should have the following qualities.

Appearance: For face work it should have fine, compact texture, light-coloured stone is preferred as it is more durable.

Structure: A broken stone should not be dull in appearance and should have uniform texture free from cavities, cracks, and patches of loose or soft material. Stratifications should not be visible to naked eye.

Strength: A stone should be strong and durable to withstand the disintegrating action of weather.

Weight: It is an indication of the porosity and density. For stability of structures such as dams, retaining walls, etc. heavier stone are required, whereas for arches, vaults, domes, etc. light stone may be the choice.

Hardness: This property is important for floors, pavements, aprons of bridges, etc. The hardness is determined by the Mohs scale (Section 3.2).

Toughness: The measure of impact that a stone can withstand is defined as toughness. The stone used should be tough when vibratory or moving loads are anticipated.

Porosity and Absorption: Porosity depends on the mineral constituents, cooling time and structural formation. A porous stone disintegrates as the absorbed rain water freezes, expands, and causes cracking. Permissible water absorption for some of the stones is given in Table 2.1.

Table 2.1 24 Hours Water Absorption of Stones by Volume

S.No.	*Types of stone*	*Water absorption (% not greater than)*
1.	Sandstone	10
2.	Limestone	10
3.	Granite	1
4.	Trap	6
5.	Shale	10
6.	Gneiss	1
7.	Slate	1
8.	Quartzite	3

Seasoning: The stone should be well seasoned.

Weathering: The resistance of stone against the wear and tear due to natural agencies should be high.

Workability: Stone should be workable so that cutting, dressing and bringing it out in the required shape and size may not be uneconomical.

Fire Resistance: Stones should be free from calcium corbonate, oxides of iron, and minerals having different coefficients of thermal expansion. Igneous rock show marked disintegration principally because of quartz which disintegrates into small particles at a temperature of about 575°C. Limestone, however, can withstnad a little higher temperature; i.e. up to 800°C after which they disintegrate.

Specific Gravity: The specific gravity of most of the stones lies between 2.3 to 2.5.

Thermal Movement: Thermal movements alone are usually not troublesome. However, joints in coping and parapets open-out inletting the rain water causing trouble. Marble slabs show a distinct distortion when subjected to heat. An exposure of one side of marble slab to heat may cause that side to expand and the slab warps. On cooling, the slab does not go back to its original shape.

2.6 SELECTION OF STONES

The conditions which govern the selection of stone for structural purposes are cost, fashion, ornamental value and durability, although the latter property is frequently overlooked or disregarded. Cost is largely influenced by transportation charges, difficulties in quarrying and cutting, the ornamental features, and the durability of stone. The type of dressing of stone may make a difference to the cost, particularly with the stones derived from igneous rocks. When the cost of quarried stone to cost of finished stone is considered, it will be found that the labour cost is far greater than the price of the stone. Thus, a difference in the price between two alternative stones is unimportant and it would be unwise to reject a more durable stone on the grounds that it was costly.

Another factor which should be considered is the suitability of the stone for the type of design, for example, for a highly carved design if, by mistake, a harder stone such as granite is selected the cost will be affected. Colour arrangement and shape of mineral constituents greatly influence fashion and ornamental value. One of the first factors influencing the selection of stone for a particular work will be colour. It is important that the designer is aware about how the colour is likely to change after long exposure and in particular how it may vary in polluted atmosphere. As an example limestone, being slightly soluble in water, will remain clean in portions facing rain but retain a film of soot in sheltered areas. This results in strong colour contrast.

Resistance to fire and weathering—factors which are largely influenced by the mineral constitution of the rock—are the most important determinators of durability. It is very important to select a stone according to its exposure conditions. Limestones when used in areas not exposed to rain but acted upon by sulphur gases of polluted atmosphere, from a hard and impermeable surface skin which subsequently blisters and flakes off. It must be noted that flaking of this kind occurs mainly on external work only, although the air inside the building is almost equally polluted, probably due to the damper conditions inside.

Limestones, sandstones and granites all tend to crack and spall when exposed to fire, and there is really little to choose between them in this respect. The type of stones usually selected for specific engineering works are given in Table 2.2.

APPLICATIONS AND USES OF STONES

Use of stone as building material depends upon the nature of the work, type of the structural element in which it is to be used and its quality, availability and transportation cost. For structural purpose, granite, gneiss, trap, sandstone, limestone, marble, quartzite and slate are most useful. A list of building stones commonly used along with the classification and characteristics is given in Table 2.3.

Table 2.2 Stones for the Specific Uses

S.No.	Type of work	Recommended Stone Type	Reasons
1.	Heavy engineering works such as bridge, piers, and abutments, break waters, docks and light houses, retaining walls	Granite (of its three varieties, viz. biotite-granite, homblende-granite and tourmaline-granite; biotite-granite is most widely used	It is heavy strong, durable and is capable of resisting large thrust
2.	Building facing the sea	Granite, fine grained sandstone	These are not affected by the weathering action of sand particles blown by wind
3.	Buildings in industrial areas	Granite, compact sandstone	These are acid fumes and smoke proof
4.	Arches	Fine grained sandstone	Strong, durable
5.	Building face work, carved works, ornamental works and stattutes	(a) Marble, close grained sand stone (b) Fine grained granite	These are light weight, soft and easy to work and have pleasing colour and appearance It takes high polish
6.	Fire resisting structure	Compact sandstone	Fireproof
7.	Road metal and aggregate for concrete	Granite, Basalt, quartizite	Hard, tough and has high abrasion resistance
8.	Railway ballast	Coarse grained sandstone, quartizite	These are hard and compact
9.	Electrical switch board	Slate, marble	Poor conductor of electricity

On the basis of the method of manufacture, items and materials from natural stones are classified as Sawn—obtained either from massive rocks by stone-cutting and stone-splitting machines (large stones) or from semi-product blocks by appropriate working (facing slabs, windows sill slabs, etc.); Split—obtained by splitting and finishing blocks (curb stones, paving blocks, etc.); Roughly split—manufactured by oriented splitting of blocks (bedded stone); Fractured—produced by blasting rocks and separating finer sizes (quarry stone); Crushed—produced by crushing and screening (crushed stone, artificial sand) and; Ground—obtained by grinding rocks (ground mineral powder, stone powder).

Foundation and Wall Items: Quarry, Split and sawn stones from rocks are used to erect the substructure of buildings. Piece stones sawn and split from limestone, sandstone, dolomite and volcanic tuff are used for walls, piers, abutments, etc.

Foundation and Architectural Items: Facing slabs and stones, stairs and landings, parapets, etc. are made of slabs sawn or split from semi-finished product blocks with glossy, dull, ground, sawn, pointed, fluted or rock finish. Facing slabs of granite, gabbro, basalt, marble, breccia, limestone, sandstone and volcanic tuff are generally used.

Building Items: Elements of stairs, landings, parapets and guard rails are manufactured from granite, marble, limestone, tuff, etc. Pedestal slabs and stones for farming doorways, cornices and window-sill slabs are made from the same material as the facing slabs.

Road Construction Items: Curb stones—intended to separate roadways from sidewalks; Paving blocks—used for pavements; Cobble stone—used to reinforce slopes of earth works and banks of water basins; Crushed stone—a mixture of jagged stone fragments (< 70 mm); Gravel—loose agglomeration of rock fragments (70 mm) and Sand—loose mass of mineral and rock particles (0.14-5 mm) obtained from natural stone are used in road construction.

Underground Structures and Bridges: are built of slabs and stones from igneous and sedimentary rocks. Tunnels and above-water elements of bridges are built of granite, diorite, gabbro and basalt. Face stones and facing slabs for tunnels and bridges are given rock face, grooved or fluted finishes.

Heat and Chemically Resistant Items: are manufactured from non-weathered rocks. For high temperature working conditions, they are made from chromite, basalt, andesite and tuffs. Building elements are protected against acid (except hydrofluoric acid and fluosilicic acids) by using slabs made of granite, syenite, and silicious stones. Limestones, dolomites, marble and magnesite show excellent resistance against alkalis. When high temperature and chemical attack is expected, crushed stone and sand for concrete and mortar are used.

Table 2.3 Classification and Uses of Building Stones

S.No.	Types	Classification	Characteristics	Suitability
1.	Granite	Igneous	1. Sp. gr. 2.63-2.75 2. Water absorption < 1% 3. Compressive strength 77-130 N/mm^2 4. Difficult to work with 5. Fine grained variety takes high polish 6. Colour depends upon colour of felspar 7. Excess of felspar causes early decay	Most suitable for important engineering works such as bridge abutments, piers, dams, sea walls, light houses, etc.
2.	Trap and Basalt (green stone, white stone, blue basalt)	Igneous	1. Sp. gr. 2.6-3 2. Compressive strength 150-190 N/mm^2 3. Not easily to work with	Suitable for road metal and concrete aggregate. Its red and yellow varieties are used for decorative features in structures.
3.	Serpentine	Igneous	1. Compact, but soft and easy to work with 2. Affected by smoke and fumes	Suitable for ornamental works for high quality building works.
4.	Syenite	Sedimentary	Similar to granite	Most suitable for road metal.

(*Contd.*)

Table 2.3 Classification and Uses of Building Stones (Contd.)

S.No.	Types	Classification	Characteristics	Suitability
5.	Sandstone	Sedimentary	1. Sp. gr. 2.65-2.95 2. Compressive strength 65 N/mm^2 3. Weathers well when free from lime and iron	In the form of flag stone for paving, tile stone for roofing, natural stone for ornamental work and grit for heavy engineering works.
6.	Limestone	Sedimentary	1. Sp. gr. 2.0-2.75 2. Compressive strength 55 N/mm^2 3. Affected by frost and atmosphere 4. Tough but soft enough to be cut	Suitable for flooring, paving and roofing and in the manufacture of lime and cement.
7.	Kankar (impute limestone)	Sedimentary	1. Irregular in shape 2. Porous structure 3. Nodular kankar variety is hard and tough	Black kankar is hard and is used as building material. Nodular kankar is used to produced hydraulic lime.
8.	Mooram (Decompos ed Laterite)	Sedimentary	1. Strong and hard	Most suitable for surfacing fancy paths in gardens and bunglows.
9.	Gneiss (Stratified granite)	Metamorphic	1. Strong and durable 2. Can be split into thin slabs for railway tracks.	Most suitable for surfacing fancy paths in gardens and bunglows.
10.	Laterite (Sandy clay stone)	Metamorphic	1. Porous or cellular structure 2. Soft and easy to work with 3. Affected by the action of water	Suitable for rough stone masonry work. The nodular variety yields road metal
11.	Marble	Metamorphic	1. Specific gravity 2.65 2. Crushing strength 70 N/mm^2 3. Hard and compact 4. Takes fine polish 5. Sufficiently hard and compact	Most suitable for monuments, statutes flooring, decorative and ornamental works.
12.	Slate	Metamorphic	1. Specific gravity 2.89 2. Compressive strength 77-210 N/mm^2 3. Hard and tough 4. Splits into thin slabs	Most suitable for roof coverings floorings, damp proofing and partitions.

EXERCISE

1. Write short notes on following, giving their uses in civil engineering works: Granite, lime stone, marble, slate

2. Name the three geological classes into which rocks can be divided. What are important building stones which are derived from these rocks?
3. Briefly describe the following:
   (a) Dressing of stone
   (b) Qualities of good building stones
   (c) Classification of stones for engineering works
   (d) Preservation of stones
4. What considerations would guide you in selecting stone for use in the situations mentioned below:
   (a) Face work of a building
   (b) Railway ballast
   (c) Road metal
   (d) Dams
5. Name the stones you would select for the following works. Justify your choice.
   (a) Coarse aggregate for reinforced cement concrete
   (b) building in industrial area
   (c) Railway goods platform
6. Give the characteristics and uses of following stones.
   (a) Granite
   (b) Marble
   (c) Slate
   (d) Sand stone

CHAPTER 3

Bricks

3.1 INTRODUCTION

One of the oldest building material brick continues to be a most popular and leading construction material because of being cheap, durable and easy to handle and work with. Clay bricks are used for building-up exterior and interior walls, partitions piers, footings and other load bearing structures.

A brick is rectangular in shape and of size that can be conveniently handled with one hand. Brick may be made of burnt clay or mixture of sand and lime or of Portland cement concrete. Clay bricks are commonly used since these are economical and easily available. The length, width and height of a brick are interrelated as below:

Length of brick = 2 × width of brick + thickness of mortar

Height of brick = width of brick

Size of a standard brick should be 19 × 9 × 9 cm and 19 × 9 × 4 cm. When placed in masonry the 19 × 9 × 9 cm brick with mortar becomes 20 × 10 × 10 cm. Weight of such a brick is 3.0 kg. An indent called frog, 1-2 cm deep, as shown in Fig. 3.1, is provided for 9 cm high bricks. Frog is not provided in 4 cm high bricks and extruded bricks. The size of frog should be 10 × 4 × 1 cm. The purpose of providing frog is to form a key for holding the mortar and therefore, the bricks are laid with frogs on top.

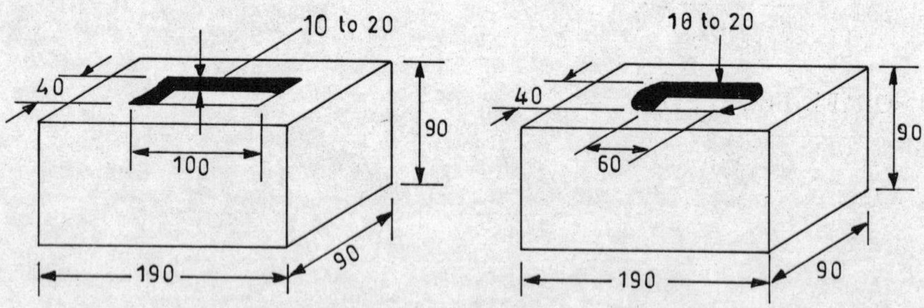

Fig. 3.1 Bricks with Frog. (all dimensions in mm)

Clay bricks though most popular and used extensively in all the construction projects at some level or other are responsible for the loss of useful agricultural land. Such lands from which clay is dug have to be reclaimed latter and cost.

Specifications are being formulated to develop alternate materials (e.g. fly ash) as substitute of clay bricks. Irrespective of the fact, the clay bricks will remain, in near future, as important a building material as it is today.

3.2 INGREDIENTS OF GOOD BRICK EARTH

For the preparation of bricks, clay or other suitable earth is moulded to the desired shape after subjecting it to several processes. After drying, it should not shrink and no crack should develop. The clay used for brick making consists mainly of silica and alumina mixed in such a proportion that the clay becomes plastic when water is added to it. It also consists of small proportions of lime, iron, manganese, sulphur, etc. The proportions of various ingredients are as follows:

Silica	50-60%	
Alumina	20-30%	
Lime	10%	
Magnesia	< 1%	Less than 20%
Ferric oxide	< 7%	
Alkalis	< 10%	
Carbon dioxide		Very small percentage
Sulphur trioxide		
Waters		

Functions of Various Ingredients

Silica: It enables the brick to retain its shape and imparts durability, prevents shrinkage and warping. Excess of silica makes the brick brittle and weak on burning. A large percentage of sand or uncombined silica in clay is undesirable. However, it is added to decrease shrinkage in burning and to increase the refractoriness of low alumina clays.

Alumina: absorbs water and renders the clay plastic. If alumina is present in excess of the specified quantity, it produces cracks in brick on drying. Clays having exceedingly high alumina content are likely to be very refractory.

Lime normally constitutes less than 10 per cent of clay. Lime in brick clay has the following effects:

1. Reduces the shrinkage on drying.
2. Causes silica in clay to melt on burning and thus helps to bind it.
3. In carbonated form, lime lowers the fusion point.
4. Excess of lime causes the brick to melt and the brick looses its shape.
5. Red bricks are obtained on burning at considerably high temperature (more than 800°C) and buff-burning bricks are made by increasing the lime content.

Magnesia: rarely exceeding 1 per cent, affects the colour and makes the brick yellow. In burning; it causes the clay to soften at slower rate than in most case is lime and reduces warping.

Iron: Iron oxide constituting less than 7 per cent of clay, imparts the following properties.

1. Gives red colour on burning when excess of oxygen is available and dark brown or even black colour when oxygen available is insufficient, however, excess of ferric oxide makes the brick dark blue.
2. Improves impermeability and durability.
3. Tends to lower the fusion point of the clay, especially if present as ferrous oxide,
4. Gives strength and hardness.

3.3 HARMFUL SUBSTANCES IN BRICK EARTH

Lime: When a desirable amount of lime is present in the clay, it results in good bricks, but if in excess, it changes the colour of the brick from red to yellow. When lime is present in lumps, it absorbs moisture, swells and causes disintegration of the bricks. Therefore, lime should be present in finely divided state and lumps, if any, should be removed in the beginning itself.

Pebbles, Gravels, Grits: do not allow the clay to be mixed thoroughly and spoil the appearance of the brick. Bricks with pebbles and gravels may crack while working.

Iron Pyrites: tend to oxidise and decompose the brick during burning. The brick may split into pieces. Pyrites discolourises the bricks.

Alkalis: forming less than 10 per cent of the raw clay, are of great value as fluxes, especially when combined with silicates of alumina. However, when present in excess, alkali makes the clay unsuitable for bricks. They melt the clay on burning and make the bricks unsymmetrical. When bricks come in contact with moisture, water is absorbed and the alkalis crystalise. On drying, the moisture evaporates, leaving behind grey or white powder deposits on the brick which spoil the appearance. This phenomenon is called *efflorescence*.

Organic Matter: On burning green bricks, the organic matter gets charred and leave pores making the bricks porous; the water absorption is increased and the strength is reduced.

Carbonaceous Materials: in the form of bituminous matter or carbon greatly affects the colour of raw clay. Unless proper precaution is taken to effect complete removal of such matter by oxidation, the brick is likely to have a black core.

Sulphur: is usually found in clay as the sulphate of calcium, magnesium, sodium, potassium or iron, or as iron sulphide. Generally, the proportion is small. If, however, there is carbon in the clay and insufficient time is given during burning for proper oxidation of carbon and sulphur, the later will cause the formation of a spongy, swollen structure in the brick and the brick will be decoloured by white blotches.

Water: A large proportion of free water generally causes clay to shrink considerably during drying, whereas combined water causes shrinkage during burning. The use of water containing small quantities of magnesium or calcium carbonates, together with a sulphurous fuel often causes similar effects as those by sulphur.

3.4 CHARACTERISTICS OF GOOD BRICK

Size and Shape: The bricks should have uniform size and plane, rectangular surfaces with parallel sides and sharp straight edges. The surfaces should not be too smooth to cause slipping of mortar.

Colour: The brick should have a uniform deep red or cherry colour as indicative of uniformity in chemical composition and thoroughness in the burning of the brick.

Texture and Compactness: The brick should have precompact and uniform texture. A fractured surface should not show fissures, holes grits or lumps of lime.

Hardness and Soundness: The brick should be so hard that when scratched by a finger nail no impression is made. When two bricks are struck together, a metallic sound should be produced.

Water Absorption: should not exceed 20 per cent of its dry weight when kept immersed in water for 24 hours.

Crushing Strength: should not be less than 10.5 N/mm^2.

Brick Earth: should be free from stones, kankars, organic matter, saltpeter, etc.

3.5 CLASSIFICATION

I.S. Classification

I.S. Code has classified bricks on the basis of strength and dimensional tolerance. Refer to Table 3.1.

Table 3.1 Classification of Bricks based on Compressive Strength

Class	*Average compressive strength not less than (N/mm^2)*
350	35.0
300	30.0
250	25.0
200	20.0
175	17.5
150	15.0
125	12.5
100	10.0
75	7.5
50	5.0
35	3.5

Notes: (i) The burnt clay bricks having compressive strength more than 40.0 N/mm^2 are known as heavy duty bricks and are used for heavy duty structures such as bridges, foundations for industrial buildings, multistory buildings, etc.

(ii) Each class of bricks as specified above is further divided into sub classes A and B based on tolerances and shape. Subclass-A bricks should have smooth rectangular faces with sharp corners and uniform colour, Subclass-B bricks may have slightly distorted and round edges.

	Subclass-A		Subclass-B	
	Dimension (cm)	Tolerance (mm)	Dimension (cm)	Tolerance (mm)
Length	380	± 12	380	± 30
Width	180	± 6	180	± 15
Height				
(i) 9 cm	180	± 6	180	± 15
(ii) 4 cm	80	± 3	80	± 6

Field Classification

Clay bricks are classified as first class, second class, third class and fourth class as per prevalent practice.

First Class Bricks

1. These are thoroughly burnt with deep red, cherry or copper colours.
2. The surface should be smooth and rectangular, with parallel, sharp and straight edges and square corners.
3. They should be free from flaws, cracks and stones.
4. They should have uniform texture.
5. No impression should be left on the brick when a scratch is made by a finger nail.
6. Water absorption should be 12-15% of its dry weight when immersed in cold water for 24 hours.
7. The fractured surface of the brick should not show lumps of lime.
8. A metallic or ringing sound should come when two bricks are struck against each other.
9. The crushing strength of the brick should not be less than 10.5 N/mm^2.

Uses: First class bricks are recommended for pointing, exposed face work in masonary structures, flooring and reinforced brick (R.B.) work.

Second Class Bricks are supposed to have the same requirements as the first class ones except that

1. Small cracks and distortions are permitted.
2. A little higher water absorption of about 16-20% of its dry weight is allowed.
3. The crushing strength should not be less than 7 N/mm^2.

Uses: Second class bricks are recommended for all important or unimportant hidden masonry works and centering of reinforced brick and reinforced cement concrete (R.C.C.) structures.

Third Class Bricks are underburnt. They are soft and light-coloured producing a dull sound when struck against each other. Water absorption is about 25 per cent of dry weight.

3.6 FORMS OF BRICK

The various forms of clay bricks in common use are shown in Fig. 2.2.

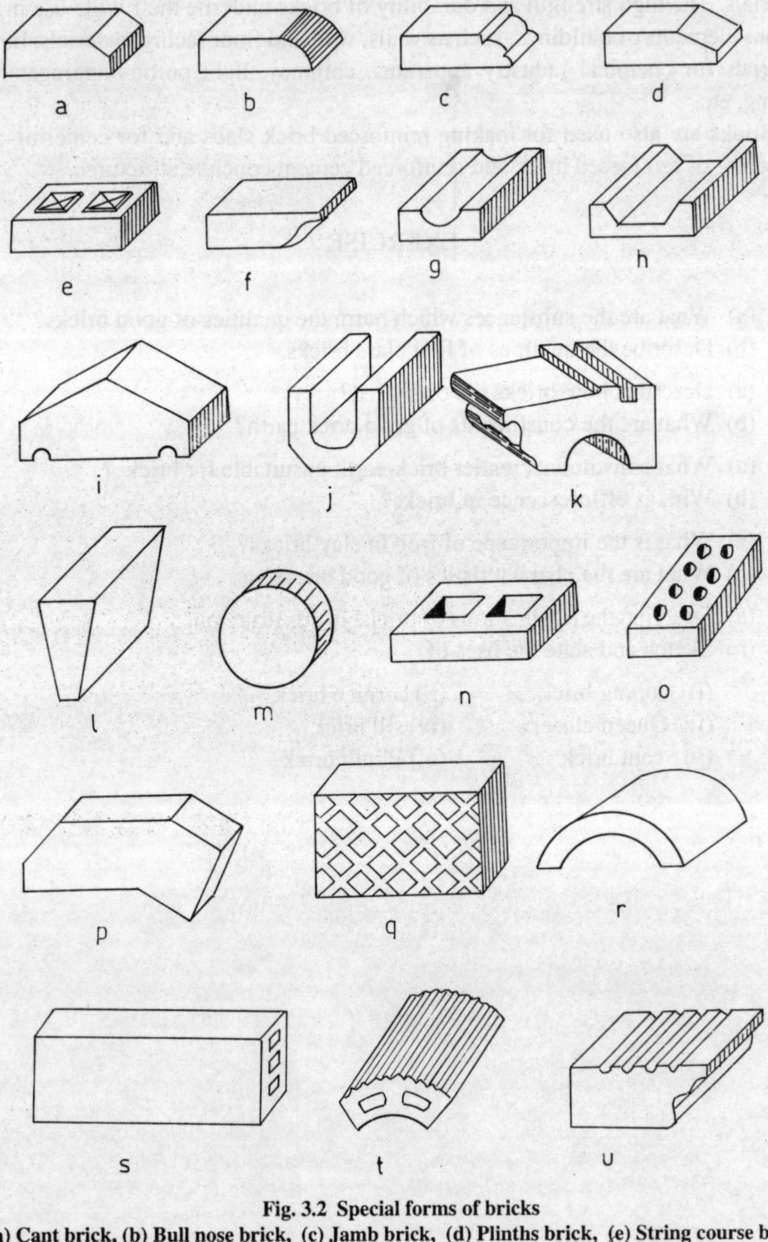

Fig. 3.2 Special forms of bricks
(a) Cant brick, (b) Bull nose brick, (c) Jamb brick, (d) Plinths brick, (e) String course brick, (f) Cornice brick, (g) Cutter brick, (h) Channel brick, (i) Coping brick, (j) Drain brick, (k) Sill brick, (l) Key brick, (m) Round brick, (n) Hollow brick, (o) Perforated brick, (p) Angle mullion brick, (q) Chequered brick, (r) Coping brick, (s) Hourdi brick, (t) Tubular brick and (u) Corbel brick.

3.7 APPLICATIONS

The availability of clay in abundance, comparative simplicity of manufacture and excellent durability has put clay bricks in the forefront among other construction materials. The high strength and durability of bricks underlie their wide use in the various elements of buildings, such as walls, wall and floor facing materials, lining materials for chemical industry apparatus, chimney, light porous aggregate for roofing, etc.

Bricks are also used for making reinforced brick slabs and for centering and shuttering of reinforced brick and reinforced cement concrete structures.

EXERCISE

1. (a) What are the substances which harm the qualities of good bricks?
   (b) Describe the qualities of first class bricks.
2. (a) Describe, how bricks are classified?
   (b) What are the constituents of good brick earth?
3. (a) What constituents render brick-earth unsuitable for bricks?
   (b) What is efflorescence in bricks?
4. (a) What is the importance of frog in clay bricks?
   (b) What are the characteristics of good bricks?
5. (a) Describe the applications of bricks in construction.
   (b) Sketch and state the uses of

   (i) coping brick (ii) cornice brick
   (iii) Queen closer (iv) sill brick
   (v) cant brick (vi) Plinth brick

CHAPTER 4

Timber

4.1 INTRODUCTION

The terms timber and wood are often used synonymously, but they have different meanings in the building industry. Wood is the hard, fibrous material that makes up the tree under the bark, whereas timber may be defined as a wood which retains its natural physical structure and chemical composition and is suitable for various engineering works.

Wood is a hard and fibrous substance which forms a major part of the trunk and branches of a tree. It can also be defined as a natural polymeric material which practically does not age. Wood as a building material falls in two major classes—natural and man made. With the advances in science and technology, wood in its natural form as timber, lumber, etc. is being rapidly replaced by composite wood materials in which natural wood is just a basic ingredient of matrix or a laminate. The latter are found to be more useful and adaptable as they may be treated chemically, thermally or otherwise as per requirements. Some examples are plywood, fibreboards, chipboards, compressed wood, impregnated wood, etc.

Wood has many advantages due to which it is preferred over many other building materials. It is easily available (this won't be true after some years!) and easy to transport and handle, has more thermal insulation, sound absorption and electrical resistance as compared to steel and concrete. It is the ideal material to be used in sea water. Wood is a good absorber of shocks and so is suitable for construction work in hilly areas which are more prone to earthquakes. Finally, since wood can be easily worked, repairs and alterations to wood work can also be done easily.

Owing to the above mentioned advantages, wood is very widely used in buildings as doors, windows, frames, temporary partition walls, etc. and in roof trusses and ceilings apart from form work.

4.2 STRUCTURE OF TIMBER

A tree can be divided into three portions, crown composed of branches and leaves, trunk and roots. The trunk accounts for about 80 per cent of the total bulk of wood.

Figure 4.1 shows the structure of well grown timber from trunk of the exogenous tree. The structure of timber visible to naked eye or at a small magnification is called *macro structure*, and that apparent only at great magnifications, the *micro structure*. Macro structure of the timber can be studied by cutting the trunk in three directions (Fig. 4.1(a)). In the cross-sectional and radial ducts, the following main parts of a tree, e.g. bark, cambium, sap wood, heart wood and pith, become readily apparent (Fig. 4.1(b)). Each of the components has a specific function. The bark protects the wood against mechanical damage. Its inner layer, called bast conveys the nutrients from the crown downwards and stores them. The function of cambium is to grow wood cells on the inside and smaller bast cells on the outside. The sapwood assists in the life process of tree by storing up starch and conducting sap. The cells in the sap wood are active. The heart wood gives a strong and firm support to the tree. With the growth of tree, the cells in the inner older portion of trunk gradually become inactive and lifeless, but do not decay. This portion of the trunk is called heart wood. At the centre of the cross-section is the pith, a small area occupied by friable tissues consisting of thin walled, loosely connected cells called pith. In a felled tree, it easily crumbles and rots. In the cross-sectional direction, nutrients pass from bast to the heart through groups of cells running at right angles to the cambium layers and are referred to as medullary rays.

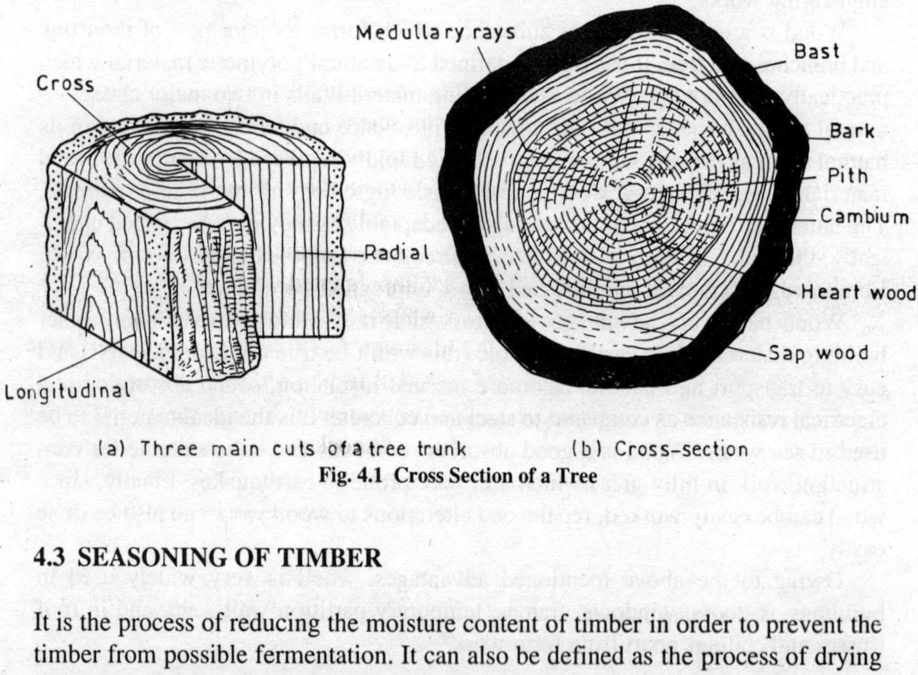

Fig. 4.1 Cross Section of a Tree

4.3 SEASONING OF TIMBER

It is the process of reducing the moisture content of timber in order to prevent the timber from possible fermentation. It can also be defined as the process of drying the wood to a moisture content approximately equal to the average humidity of the surroundings, where it is to be permanently fixed. Very rapid seasoning after removal of bark should be avoided since it causes case hardening and thus increases resistance to penetration of preservatives. Some of the objects of seasoning wood are as follows:

1. Reduce the shrinkage and warping after placement in structure.
2. Increase strength, durability and workability.
3. Reduce its tendency to split and decay.
4. Make it suitable for painting.
5. Reduce its weight.

Methods of Seasoning

Timber can be seasoned naturally or artificially

Natural or Air Seasoning: The log of wood is sawn into planks of convenient sizes and stacked under a covered shed in cross-wise direction in alternate layers (Fig. 4.2) so as to permit free circulation of air. The duration for drying depends upon the type of wood and the size of planks. The rate of drying is however very slow. Air seasoning reduces the moisture content of the wood to 12-15 per cent.

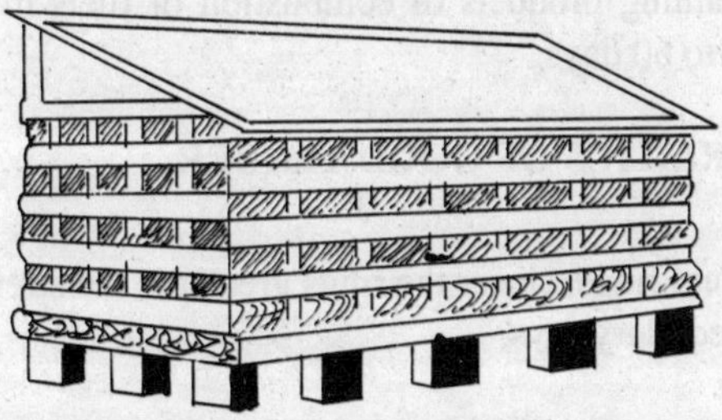

Fig. 4.2 Shed for Air Seasoning of Timber

Artificial Seasoning: The prevalent methods are as follows:

Water Seasoning: The logs of wood are kept completely immersed in running stream of water, with their larger ends pointing upstream. Consequently the sap, sugar, and gum are leached out and are replaced by water. The log is then kept out in air to dry. It is a quick process but the elastic properties and strength of the wood are reduced.

Boiling: in water or exposing to the action of steam spray is a very quick but expensive process.

Kiln Seasoning: is adopted for rapid seasoning on large scale to any moisture content. The scantlings are arranged for free circulation of heated air with some moisture or superheated steam. The circulating air takes up moisture required from wood and seasons it. Two types of kilns, the progressive and the compartment are in use. For most successful kiln-seasoning the timber should be brought to as high a temperature as it will stand without injury before drying is begun; otherwise the moisture in the hot outer fibers of the wood will tend to flow towards the cooler interior.

Chemical or Salt Seasoning: An aqueous solution of certain chemicals have lower vapour pressures than that of pure water. If the outer layers of timber are treated with such chemicals the vapour pressure will reduce and a vapour pressure gradient is setup. The interior of timber, containing no salts, retains its original vapour pressure and, therefore, tends to dry as rapidly as if there had been no treatment. The result is to flatten the moisture gradient curves, to reduce the slope of the curves, and consequently to reduce the internal stresses induced during drying. Since it is these stresses which are responsible for defects such as checks, etc.

a chemically treated timber will exhibit fewer defects. Common salt or urea are generally used; the latter is preferred. The corrosive action of common salt is a drawback.

Electric Seasoning: The logs are placed in such a way that their two ends touch the electrodes. Current is passed through the setup, being a bad conductor, wood resists the flow of current, generating heat in the process, which results in its drying. The drawback is that the wood may split.

Mc. Neill's Process: has no adverse effects; it is the best method although most expensive. The timber is stacked in a chamber with free air space ($^1/_3$rd of its capacity) and containing products of combustion of fuels in the fire place. The time required is 15 to 60 days.

4.4 CHARACTERISTICS OF GOOD TIMBER

1. Narrow annual rings, closer the rings greater is the strength.
2. Compact medullary rays.
3. Dark colour.
4. Uniform texture.
5. Sweet smell and a shining fresh cut surface.
6. When struck sonorous sound is produced.
7. Free from the defects in timber.
8. Heavy weight.
9. No woolliness at fresh cut surface.

4.5 FIRE RESISTANCE OF TIMBER

Timber is very inflammable. The fire hazard of timber structures is, however, often overemphasized. When wood burns, the first step is the vapourisation of moisture (118°C) and the next is the volatilisation of extraneous materials (110-165°C). At temperatures (165-220°C) well below the ignition point, destructive distillation or decomposition begins with scorching and evolution of inflammable gases which ignite and form the first flames around the heated wood. As the temperature builds up, the evolution of gases is more rapid and the surface of the timber begins to char. Finally, the point is reached at which the wood itself begins to glow and to ignite (220-390°C). Until this point is reached, the woody portion does not support its own combustion. Quick ignition of inflammable gases and glowing of charcoal occur at about 390-530°C.

Fire proofing makes timber resistant to fire to a degree that it is difficult to ignite and support its own combustion. The fire resistance of wood can be enhanced either by impregnating it with chemicals like phosphates of ammonia, mixture of ammonium phosphate and ammonium sulphate, borax and boric acid, sodium arsenate, sodium tetra-borate or by designing wood to provide slow burning construction.

Chemical Methods

These include the impregnation of timber with effective chemicals, or by coating the surface of timber with a layer of noncombustible paint.

Impregnation with Chemicals: As chemicals are water soluble, they are leached out due to rain when applied on exposed structures. Therefore, a second shallow impregnation with a fire-retardant, water-repellent sealer or paints like flamex and bitulac fire-retardant paint is applied which substantially retards leaching. Fire-retardant salts are impregnated by pressure process. Able's process for making wood fire resistant is as follows:

The surface of wood is painted by a weak solution of sodium silicate. Thereafter slaked lime solution of the consistency of a paste is applied followed by the application of concentrated solution of sodium silicate in two coats, the second being applied after 6 hours of the first coat. The composition of the solution is:

Sodium silicate	56 g
Water	50 g
Kaolin	75 g

Surface Coating Method: The most commonly used flame retardant coatings are cement grouts, clay-sulphate paste; paints such as silicate, chloride, phosphate paints and; emulsions like chloro paraffin. Some fire retardants (chemicals) form a film on the surface of wood and decompose under heat yielding non-inflammable gases that dilute the inflammable gases and consequently retard the ignition of the latter. Some of the other fire retardants have low melting point. These melt under heat and form a barrier to the supply of oxygen to the inside.

Structural Method

There is considerable difference between the burning of thin wooden members and that of large timbers. Timber of substantial dimension offers high resistance to fire. Heavy timber on burning form a protective coating of charcoal, which being a heat insulator retards the penetration of heat to the interior. On the contrary, thin members quickly reach the ignition point and burn rapidly. Once the fire is started, flames rise upward in case of wood and hence, wood which is in vertical direction catches fire easily. That is why windows and doors burn more rapidly as compared to beams and floors.

4.6 PROPERTIES OF WOOD

Physical Properties

Density: of wood is approximately equal for all species and averages 1540 kg/m^3.

Bulk density: depends on the volume of pores and moisture content of the wood. For most wood species, the bulk density is less than unity. Bulk density value is used to determine the quality factor which is the ratio of compressive strength to the bulk density. It is 0.6 for pine and 0.57 for oak.

Moisture movement: Timber is liable to shrink or swell with the movement of moisture. This movement is not the same in all the directions. Fig. 4.3 shows two pieces A and B cut from a log. In piece A, the layers producing the annual rings run roughly parallel to the face of the timber, while in piece B, they run roughly at right angles to the face. Movement of the wood due to variations in moisture content is greatest in the plane of the annual rings. The movement a_1 will be greater than movement b. Also, as a_2 is nearer to the condition of b, it will move rather less than a_1. It is clear, therefore, that if a piece of timber is to be used as a board or panel, it will move less if cut like *B* than like *A*. It should also be obvious that the slightly varying tendency to movements on the faces of piece may lead to stresses which will cause it to warp. Recommended moisture content for structural elements is 12-20 per cent for doors and 10-16 per cent for windows.

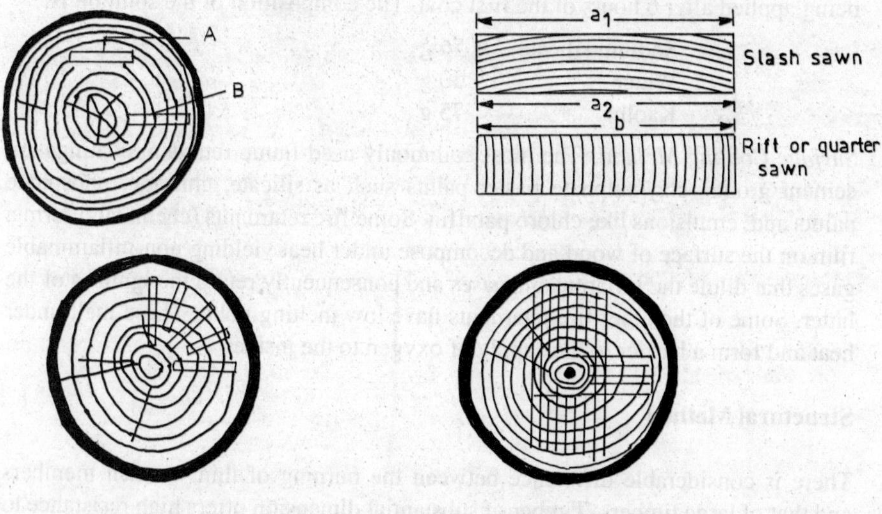

Fig. 4.3 Methods of Converting Timber

Shrinkage: is the reduction in linear and volumetric dimensions in drying of wood. Evaporation of capillary water is not accompanied by shrinkage, the latter taking place only when hygroscopic moisture evaporates. Because of structural non-uniformity, wood shrinks or swells irregularly in various directions. Linear shrinkage along the fibres lies between 0.1 and 0.3 per cent, in radial direction between 3 and 6 per cent and in tangential direction between 7 and 12 per cent. The volume shrinkage factor for various species ranges from 0.2 to 0.75 per cent.

Swelling: is the capacity of wood to increase both its linear and volumetric dimensions when it absorbs water. Swelling of wood along the length of fibres ranges from 0.1 and 0.8 per cent, 3 to 5 per cent in the radial direction and 6 to 12 per cent in the tangential direction.

Heat Conductivity: is quite low. The coefficient of heat conductivity along the fibres is 1.8 times greater than that across the fibres and averages 0.15 to

0.27 K cal/mh°C. as the bulk density of wood increases and its moisture content decreases, the amount of air entrapped inside cavities decreases, the effect being greater heat conductivity of wood.

Sound Conductivity: The velocity of sound in wood is 2 to 17 times greater than in air and as such wood may be considered to have high sound conductivity.

Resistance to Action of Acids and Alkalis: Wood is not affected by weak alkali solution but decays in an acid medium (pH< 4).

Mechanical Properties

Wood has three principal axes—longitudinal, radial and tangential—along which properties are fairly constant. Since wood is a nonisotropic material, it has three values of modulus of elasticity varying by as much as 150 to 1, three shear moduli varying by 20 to 1, and six Poisson's ratios varying by 40 to 1. There is no sharply defined elastic limit in wood but there is a proportional limit. However, the stress-strain diagram in any direction is fairly straight over a considerable range before it gradually curves off.

The relative stress-strain curves for direct tension, direct compression and bending stress intensities parallel to the grain in Fig. 4.4 show that in both, direct compression and bending, the proportional limit is in the vicinity of 65 to 75 per cent of the ultimate strength. For all practical purposes, there is no proportional limit in direct tension.

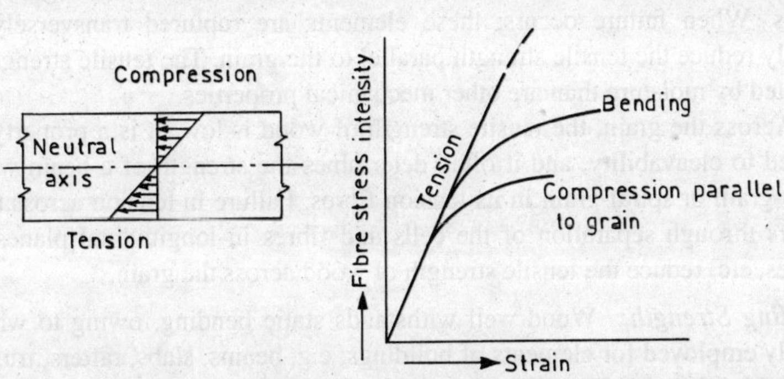

Fig. 4.4 Stress Strain Curves Parallel to the Grain

Modulus of elasticity of the grain is practically the same in direct tension, direct compression and bending, if shear deformation in bending is eliminated. Because modulus of rigidity of wood is low (approximately 1/15 to 1/20 the modulus of elasticity), the apparent bending modulus of elasticity varies somewhat with the type of loading and span, but in no case it is less than 90 to 95 per cent of the true bending modulus of elasticity.

Compressive strength: Compressive strength parallel to fibres, at 15 per cent moisture content, varies from 30.0 to 77.5 N/mm^2. When subjected to compressive force acting parallel to the axis of growth, wood is found to be one of the strongest structural material. However, compressive strength perpendicular to fibres of wood is much lower than that parallel to fibres of wood. Furthermore, a knowledge of the compressive strength is of value in estimating strength in bending, since experiments have demonstrated that the yield point of wooden beams is determined by the compressive strength of the wood.

When wood is subjected to compression parallel to the grain, it may fail through collapsing of the cell walls or through lateral bending of the cells and fibres. In wet wood and in the hardwoods, which are composed to thick-walled fibres and vessels, incipient failure is due to bending of the individual fibres. In cross-grained pieces, the failure is likely to take place through shear parallel to the grain.

The strength of timber compressed across the grain is brought into play whenever a concentrated load is imposed on a beam. Since the compressive strength across the grain is only a small fraction of the compressive strength parallel to the grain, proper allowance for this discrepancy must be provided with a footing to distribute the pressure.

Tensile Strength: Parallel to the fibres is of the order 80.0 to 190.0 N/mm^2. However, wooden parts restrained at their ends suffer from shearing stresses and crushing which wood resists poorly, and cannot be extensively used in structures working under tension. Moreover, since the tensile strength parallel to the grain is two to four times the compressive strength, the latter governs the strength of beams. The tensile strength parallel to the grain is influenced to some extent by the nature of the wood elements and their arrangement, but principally by the straightness of the grain and the thickness of the walls of the longitudinal elements. When failure occurs, these elements are ruptured transversely. Knots greatly reduce the tensile strength parallel to the grain. The tensile strength is less affected by moisture than are other mechanical properties.

Across the grain, the tensile strength of wood is low. It is a property closely related to cleavability, and it often determines the strength of a beam which has cross-grain or spiral-grain in its tension fibres. Failure in tension across the grain occurs through separation of the cells and fibres in longitudinal planes. Knots, shakes, etc. reduce the tensile strength of wood across the grain.

Bending Strength: Wood well withstands static bending, owing to which it is widely employed for elements of buildings, e.g. beams, slabs, rafters, trusses, etc. The initial failure of long beams of uniform width is indicated by a wrinkling of the overstressed compression fibres, much like the failures which occur in compression prisms. Final failure of such beams is generally in tension. It is accompanied more or less by snapping as the individual fibres begin to break when the maximum load is reached. Very dry specimens sometimes fail very suddenly in tension before any wrinkling of the compression fibres is noticeable. However, green test pieces fail silently in compression without rupturing of the tensile fibres. Short deep beams fail by horizontal shear suddenly, and this is more common in well seasoned timber of structural sizes than in green timbers or in small beams. Very often shear failures result from defects.

Long narrow beams must be restrained laterally and supported in such manner that the wide sides are vertical. If the first condition is not fulfilled, the member will fail from column action in the compression fibres. Unless the second condition is met, strength of the beam may be very seriously reduced through the inclination of the neutral axis. These requirements are satisfied by diagonal bracing known as *bridging*. The bending strength of some of the timbers is given in Table 4.1.

Table 4.1 Bending Strength of Timbers

S.No.	*Trade name*	*Unit weight N/m³ at 12% moisture content*	*Bending strength (N/mm²) Location*		*Durability*
			Inside	*Outside*	
1.	Babul	8350	18.2	15.4	Low
2.	Chir	5750	8.4	7.0	Low
3.	Deodar	5600	10.2	8.8	High
4.	Fir	4650	7.8	6.6	Low
5.	Jaman	8500	15.2	12.6	Moderate
6.	Mango	6550	12.4	10.2	Low
7.	Oak	8650	14.8	12.4	Moderate
8.	Sal	8000	16.8	14.0	High
9.	Teak	6250	14.0	11.6	High
10.	Neam	8360			–
11.	Rosewood	7550			High
12.	Shishum	7850			Moderates
13.	Call	5150			Low

Shearing Strength: Wood has low shearing strength of 6.5–14.5 N/mm² along the fibres. Resistance of wood to cutting across the fibres is 3 to 4 times greater than that along the fibres, but pure shear generally does not take place since the fibres are also subjected to crushing and bending.

Stiffness: In general, denser woods are more stiff. A green timber is less stiff than when seasoned. The structural sizes of timber are about as stiff as the clear small sticks.

Toughness: A wood which has a large capacity to resist shock or blows is called tough. Hard wood as a class excels in toughness. Long leaf pine is the only one of the conifers possessing much toughness. In general, green wood is tougher than seasoned wood.

Cleavability: is the measure of the ease with which wood may split. Most hard woods split more easily along radial planes than along tangential surfaces. Among the conifers, with an exception of longleaf pine, the difference in cleavage strength in the two directions is not great.

Hardness: is defined and measured as resistance to indentation and resistance to scratching. Both are important properties in woods used for finishing and for furniture. These properties, together with the ability to wear without splintering, determine the wearing resistance of wood for floors and pavements.

Effect of Moisture on Mechanical Properties of Wood

Variations in the moisture content of the cell walls are accompanied by large changes in the strength and stiffness of wood. After years of seasoning, large timbers may lose enough water to effect an increase in tensile and compressive strength and in stiffness, but defects arising from shrinkage stresses often decrease the resistance to horizontal shear stresses. In kiln-seasoning, the normal increase in strength due to loss of moisture is often nullified by case-hardening, a condition which prevents complete drying of the piece and produces internal stresses.

The mechanical properties of wood are not materially affected by a reduction of the moisture content until the point of fibre-saturation is reached. Further drying causes a large, proportionate increase in strength and stiffness.

Effect of Temperature on Strength of Wood

The effects of temperature on wood are dependent upon the moisture content. Dry wood expands slightly when heated, while wet wood shrinks owing to evaporation of moisture. Very high temperatures, such as those used in vulcanizing, slightly weaken dry wood. Freezing somewhat increases both the strength and stiffness of wood. If wood is kept moist during the heating process, it is rendered very pliable and is weakened.

4.7 SUITABILITY OF TIMBER FOR SPECIFIC USES

The characteristics and suitability of timbers is given in Table 4.2.

Table 4.2 Characteristics and Uses of Timber from Various Trees

S.No.	*Purpose*	*Requirements*	*Nature of tree*
1.	Agricultural implements	1. Hard and durable 2. Should take good polish	Bel, Arjun, Babul, Black wood
2.	Houses	1. Sufficiently close grained 2. Takes good polish 3. Toughness and durability 4. Pleasing colour 5. Easy to work with	Sisoo, Teak, Ablus, Babul, Haritaki, Jiyal, Kath bel, Mahua, Nirmali, Red wood, Walnut
3.	Bridge	1. Strength, hardness 2. Resistance to salt water action 3. Durability in moist places and under water	Babul, Red cedar, Iron wood, Jarul, Nageshwar, Sal, Satin, Sisoo
4.	Carts and wheels	1. Hardness and durability 2. Close grained	Jiyal, Arjun, Babul, Tamarind

(Contd.)

Table 4.2 (*Contd.*)

S.No.	Purpose	Requirements	Nature of tree
5.	Columns, beams door frames, etc.	1. Hardness and durability 2. Should take good polish 3. Flexibility 4. Light in weight 5. Easy to work 6. Lasts under water	Arjun, Gamhar, Bamboo, Coconut, palm, Jam, Mango, Pial, Palm
6.	Furniture	1. Light, soft and durable 2. Should take good polish 3. Close grains 4. Easy to work with	Teak, Champa, Deodar, Rakta-ghandan, Walnut, Shishum
7.	Fancy goods and toys	1. Easily workable 2. Strength and durability 3. Fairly hard and light weight 4. Scented	Simul, Sandal, Bamboo, Mahogony
8.	Music instruments	1. Pleasing brick red colour 2. Soft and easy to work 3. Beautiful motteling	Toon, Walnut
9.	Scientific instruments	1. Moderately hard and tough 2. Light 3. Easy to work with	Guava
10.	Packing boxes	1. Soft and light in weight 2. Lasts under water 3. Cheap	Simul, Palasa, Mango, Deodar
11.	Pegs	1. Hardness and durability 2. Cheap	Arjun, Coconut, Palm, Kher
12.	Piles	1. Strength, hardness and durability 2. Close grains 3. Durable in moist places and under water	Red cedar, Sisoo, Sal, Nageswar, Iron wood
13.	Railway carriage	1. Close grains, hardness and durability 2. Should take good polish	Black wood, Teak, Iron wood, Red wood
14.	Railway sleepers	1. Hardness, toughness and strength 2. Durability under moist conditions 3. Close grains	Sundari, Red cedar, Sal, Kath bel, Deodar
15.	Scaffolding	1. Strength and durability 2. Flexible 3. Easy to work	Bamboo
16.	Shuttering	1. Hardness and durability 2. Easy to work 3. Durability in moist conditions	Nageshwar, Gamhar, Haritaki
17.	Ships, Boats	1. Hardness, toughness and strength 2. Durability under satly water	Jarul, Babul, Teak, Nageshwar, Bakul, Mahua, Khair Red cedar, Palm, Banyan, Babul, Gamhar Mango

(*Contd.*)

Table 4.2 (*Contd.*)

S.No.	Purpose	Requirements	Nature of tree
18.	Well curbs	1. Lasts long under water 2. Softness and light weight 3. Easy to work	Kail, Simul
19.	Match Box	1. Moderately hard or soft	

4.8 WOOD PRODUCTS

Veneers

The primary process in the manufacture of wood based products is veneering which produces thin sheets of wood known as *veneers*. The thickness of veneers varies from 0.4 to 0.6 mm. The most suitable wood for this purpose is walnut. However, other species like teak, sissoo, rose wood, etc. are also used. The logs to be used for this purpose are kept in wet storage to avoid end splitting and are softened by heating with hot water or steam and the bark is removed. The log is then cut to veneers. Depending on the cutting process, the veneers are classified as rotary veneers (Fig. 4.5) and sliced veneers (Fig. 4.6). These are used in the manufacture of plywood and other laminated boards.

Plywood

A wood panel glued under pressure from an odd number (usually 3 to 13) of layers/piles of veneers is known as plywood (Fig. 4.7). The outer most veneer sheets in a plywood panel are called *faces*. The interior ply/plies which have their grain directions parallel to that of the faces are termed as *core/centre*. Other piles which have grain directions perpendicular to that in the face are termed as *cross bands*.

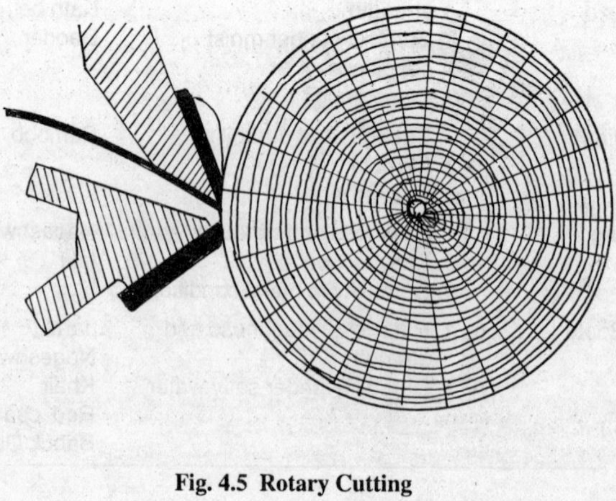

Fig. 4.5 Rotary Cutting

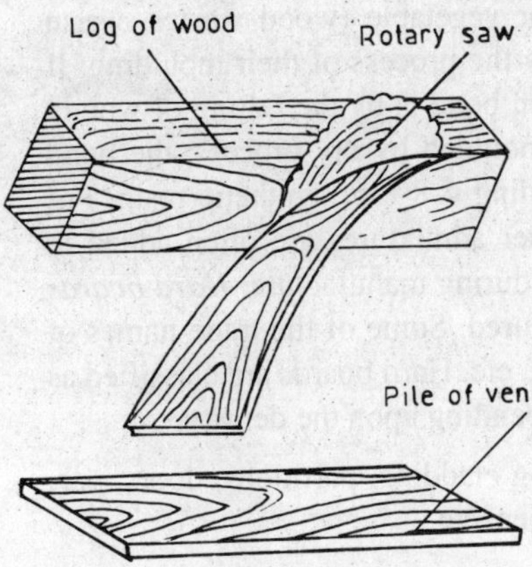

Fig. 4.6 Sliced Veneer

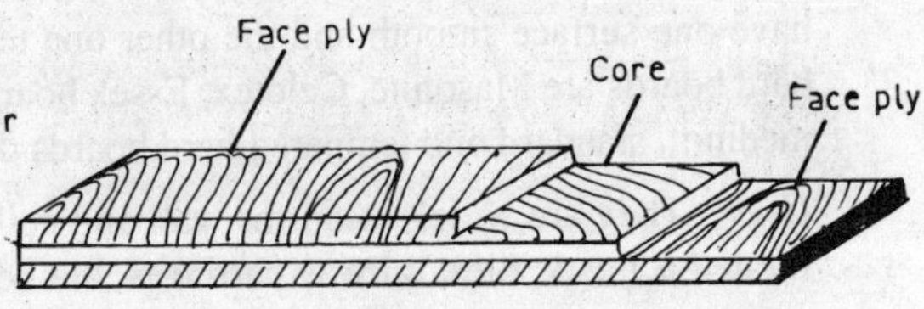

Fig. 4.7 Plywood

Plywood may be classified upon direction of grains in the piles and on the type of adhesive used. Normally the alternate piles are oriented at 30° or 60° in *star* plywood. The faces are arranged with the grain at 45° to that of the centres in *diagonal* plywood. When the piles are bonded together with water-soluble glues such as casein glue, *interior grade* plywood is obtained and when bonded with phenol formaldehyde adhesive it is identified as *exterior grade* plywood which is completely water proof.

Plywood Grades: Plywood is graded as boiling water proof (BWP) grade, boiling water resistant (BWR) grade, warm water resistant (WWR) grade and, cold water resistant (CWR) grade, depending upon the type of adhesive used for bonding of the veneers. These are further classified as AA, AB, AC, BB, BC and CC based on the quality of the two faces, each face being of the kind A, B, or C. After pressing, the finished plywood is recommended to a moisture content not less than 8 per cent and not more than 16 per cent.

Advantages

1. It has good strength both along as well as across the grains.
2. The wood shrinks or swells more across the grains. Since plywood has cross-grained construction, the tendency to shrink or swell is reduced.
3. It has better splitting resistance due to the grains in adjacent veneers in cross direction as such nailing can be done very safely even near the edges.
4. Plywood can be curved into desired shapes.

Uses: These are extensively used for partitions, ceilings, doors, concrete form work, plywood boards, lamin boards (built-up boards with core strips up to 7 mm wide and 7 mm thick) and block boards (built-up boards) etc.

Fibre Boards

These boards built up of felting from wood or vegetable (wood wastes, waste paper, agricultural wastes, etc.) are classified by the process of their moulding. If the boards are moulded by wet process, the main bond is by the felting of woody fibres and not by added glue. For the boards moulded by dry process, the bond between the predried fibres is improved by adding 4-8% of synthetic resin. For better performance wood preservatives and other admixtures are often added to the pulp. *Insulating boards* are not compressed during manufacture. *Hard boards* have one surface smooth and the other one textured. Some of the trade names of hard boards are Masonite, Celotex, Essex boards, etc. Hard boards are classified as medium, standard and tempered hard boards depending upon the density.

Uses: They are widely used for wall and ceiling cladding, partitions, doors, perforated acoustic tiles, railway carriages, but bodies, etc.

Particle Boards or Chip Boards

They are manufactured from particles of wood or other ligno cellulose materials which are agglomerated, formed and pressed together by the use of an organic binder together in the presence of heat, pressure or moisture. They are manufactured from small timber pieces and wood wastes. The latter is first converted into small chips. The moisture content of chips is reduced to 15 per cent and then some gluing material, usually phenol formaldehyde, is sprayed. The chips are then spread to form a mat and then pressed in a hydraulic press in presence of heat and moisture.

Dimensions and Tolerances: The size of particle boards in mm should be as follows:

Length (mm)	4850, 3650, 3000, 2750, 2400, 2100, 1800, 1500, 1200, 100, 900
Width (mm)	1850, 1800, 1500, 1200, 1000,900, 600, 450
Thickness (mm)	6, 9, 12, 15, 19, 22, 15, 27, 30, 35, 40
Tolerances	
Length	± 8 mm
Width	± 8 mm
Thickness	± 2.5%
Above 25 mm	± 2.5%
Up to 25 mm	± 5%
Density	Mean density of particle boards shall be between 500.0-900.0 kg/m^3

Uses: These are widely used in buildings, partitions, ceilings, floor slabs, doors, furniture, etc.

Block Boards: The core is made up of strips of wood each not exceeding 25 mm in width, forming a slab, glued between at least two surface veneers (Fig. 4.8). The

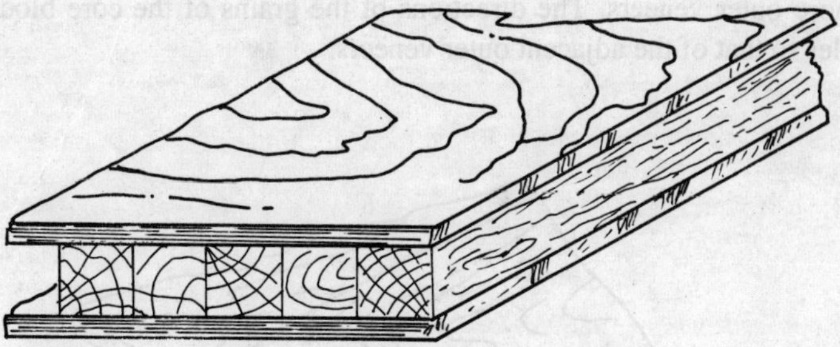

Fig. 4.8 Block Board

directions of the grains of the core blocks run at right angles to that of the adjacent outer veneers.

The grades and types of block boards are represented as:

ICOM Interior grade commercial type
IDEC Interior grade decorative type
XCOM Exterior grade commercial type
XDEC Exterior grade commercial type

These are further subgrades as Grade 1 and Grade 2. Grade 1 is *exterior grade* used for bus bodies, railways coaches, prefabricated houses, etc. and Grade 2 is *interior grade* used for furniture, partition, panelling, ceiling, etc.

Uses These are extensively used for construction of railways carriages, bodies of buses, marine and river crafts, partitions, furniture, etc.

Batten Boards have core made up of 80 mm wide wood pieces as shown in Fig. 4.9, forming a slab glued between at least two surface veneers. The directions of the grains of the core block runs at right angles to that of adjacent outer veneers.

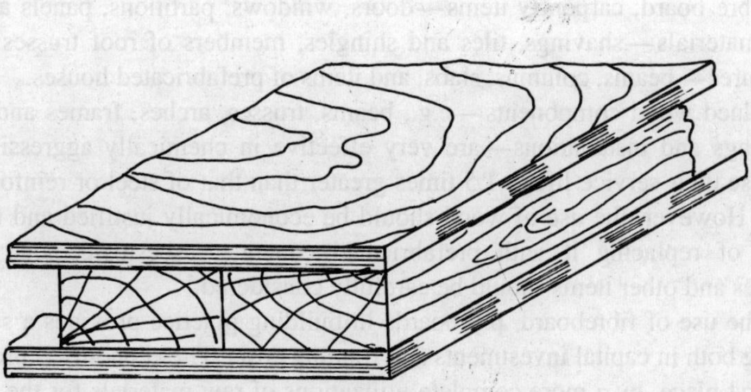

Fig. 4.9 Batten Board

Lamin Boards have a core of strips, each not exceeding 7 mm in thickness as shown in Fig. 4.10, glued together to form a slab which in turn is glued between two or more outer veneers. The directions of the grains of the core block run at right angles to that of the adjacent outer veneers.

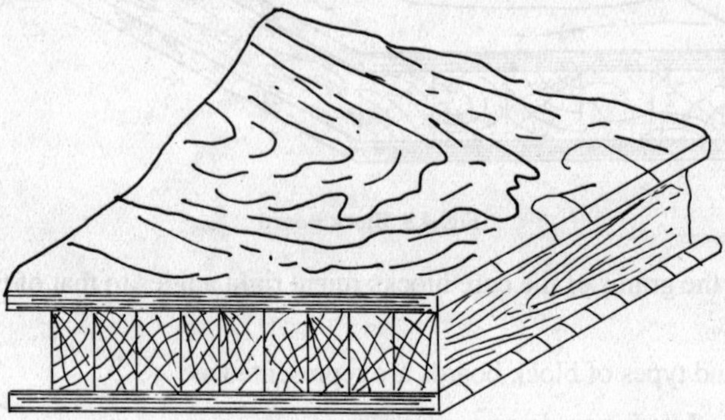

Fig. 4.10 Lamin Board

4.9 APPLICATION OF WOOD AND WOOD-PRODUCTS

In modern building practice, wood is extensively used for walls and floors of buildings, carpentry and graded plank items, as well as prefabricated standard wooden cottages. A great quantity of wood is consumed in building and installation work for making piles, poles, various load-bearing components formworks, scaffolds, etc.

Wood materials and items used in building practice include round timber (log); sawn timber and planks; floor materials—flooring boards, end-grain blocks and fibre board; carpentry items—doors, windows, partitions, panels and gates; roof materials—shavings, tiles and shingles, members of roof trusses; wooden structures—beams, columns, slabs, and items of prefabricated houses.

Glued wood components—e.g., beams, trusses, arches, frames and roofs of buildings and installations—are very effective in chemically aggressive media because their service life is 1.5 times greater than that of steel or reinforced concrete. However, the use of wood should be economically justified and the possibility of replacing it with prefabricated concrete, asbestos-cement, gypsum, plastics and other items should be carefully considered.

The use of fibreboard, ply-boards in building practice provides a substantial saving both in capital investments and running costs. The economy is provided, in the first place, by a more complete utilizations of raw materials for the manufacture of building material and items.

The use of boards made of pressed wood shavings in dwelling house construction has a great economical effect. Currently, wood waste is utilized to manufacture polymer and cement based fibreboard and wood shavings board. This also allows manufacturing materials of better physical, mechanical and decorative properties than wood.

EXERCISE

1. Draw a cross-section of a matured tree.
2. What is seasoning of timber and why is it done?
3. State the qualities you will consider in selecting timber for construction purposes.
4. (a) What are the requirements of a good timber?
   (b) Describe the methods of seasoning timber.
5. Write short notes on the following:
   (a) Plywood
   (b) Hard-board
   (c) Fire-resistance
6. Describe the physical and mechanical properties of timber.
7. Suggest suitable timber for following purposes. Give reasons for your choice
   (a) Doors (b) Shuttering
   (c) Piles (d) Railway sleepers
   (e) Pegs (f) Scaffolding
8. Describe the chief characteristics and uses of following in the construction industry.
   (a) Bamboo (b) Teak
   (c) Sal (d) Deodar

CHAPTER 5

Mortar

5.1 INTRODUCTION

Building mortars are used for jointing of bricks, stones etc. It may be defined as a paste (capable of setting and hardening) obtained by adding water to a mixture of fine aggregate such as sand and binding material, e.g., cement. The mortar composition is designated by the volume or weight of material in 1 m^3 of mortar or by the relative amount of materials with the amount of binding material taken as unity. For simple mortars composed of one kind of binding material (e.g. cement mortar), the composition will be designated, say 1:4, i.e. one part (by weight or volume) of cement and 4 parts of sand. Combined mortar composed of two binding materials and mineral admixtures are identified by three figures, e.g. cement-clay mortar 1:0.4:5 (cement:clay:sand). Some of the important uses of mortars are as follows:

1. In brick and stone masonry—it is used in the vertical joints and is spread over each layer to give an even bed for successive layers of masonry.
2. In plastering and pointing—to cover exposed walls and joints to protect against weathering besides better appearance.
3. As matrix in concrete.

In order to ensure a good quality of masonry, a mortar must be equally well suited for both brick laying and load carrying when hardened. The workability of the fresh mortar must be such that the mason can fill all joints easily. When a course of units has been laid, the mortar and brick system must attain a reasonable rigidity before the next course is laid, in order to prevent excessive racking movements. On the other hand, if the mortar and brick system stiffens too fast, it can be impossible for the mason to make the necessary corrective movements of the newly laid unit.

The requirements mentioned above are not easily fulfilled with one single type of mortar since the properties of the units, especially the suction, also play an important role for the total result. It is possible that the ratio between the suction of the masonry units and the mortar's water retentivity may be such that only a thin mortar layer close to the units dries out—and, more specifically, dries out too fast. In such cases the bond between units and the mortar will suffer, and the cracks will appear between the units and the hardened mortar joints. It is also possible that the water retentivity may be so small that the mortar is dried out by the lower unit, so that the new unit is laid in a partly dried and set mortar bed. In such cases the strength of the mortar joint will usually be low, and cracks will soon appear.

The governing factors in deciding a particular type of mortar for a specific structure depend upon the desired strength of masonry, resistance to penetration of rain water, immediate and long term appearance, hardening temperature, expected working conditions of the building and cost.

5.2 INGREDIENTS

Aggregate

The aggregate used for mortar is mostly sand as found in nature, with small adjustments. The sand must be well graded so that different sizes of particles are present. A graded sand will give a dense mass of aggregate, requiring a minimum amount of cementitious material for a given strength. At the same time, creep and shrinkage are minimized, and the mortar is economical in use.

Aggregates should be spherical or rounded, since these shapes produce a type of mortar which is easy to work with. Strength and workability of the mortar are usually better when the mortar sand contains comparatively large particles. If the largest particles are too large, however, they may cause stress concentrations and, at worst, they simply may not be accommodated in the joints. As a general rule, the maximum particle size should not be more than 1/3 or 1/2 of the thickness of the joint.

Cementitious material

There are many cementitious materials used for making mortars. *Lime* is produced from calcite or limestone or from magnesium limestone ($CaCO_3$ and $MgCO_3$). When calcium carbonate is heated, it decomposes into calcium oxide (CaO) and carbon dioxide at approximately 900 to 1000°C. When a proper amount of water is added, the burned and crushed lime (quicklime, CaO) is converted to calcium hydroxide [$Ca(OH)_2$]. This hydrated lime is a fine-grained product which is screened and sometimes further ground to a very fine powder. When water is later added to the slaked lime on the building site, a paste mainly of crystalline calcium hydroxide plus a colloid is formed. As the water evaporates, the paste sets. When the paste is exposed to the air, the preliminary setting is followed by gradual hardening, during which the hydroxyl (OH) is replaced by carbon dioxide from the air, forming calcium carbonate again. Lime mortar depends on the carbon dioxide from the air for its hardening and can therefore not harden in a wet environment.

The hydrated lime which is marketed in powder form has a unit weight of approximately 650 kg/m^3 and contains up to 98% magnesium and calcium hydroxide. Quicklime (about 75% calcium oxide and 20% magnesium oxide) must be slaked and aged into lime putty before use. The lime putty weighs about 1300 kg/m^3.

Portland cement is made of a mixture of materials: of calcium oxide, silica, aluminium, and iron oxide plus some impurities. When water is added to Portland cement a plastic paste is obtained. The properties depend upon the amount of cement and water. The paste begins to set within a few hours and sets finally within about 10 hours.

After setting, the paste will continue to harden for a very long period of time.

The main hardening, however, takes place in the first month, so that after 1 month about 80 to 90% of the final strength is obtained. Portland cement hardens even in water, and the rate of hardening can be varied within rather wide limits by adjusting the proportions of the different constituents and by different curing processes.

In order to add hydraulic properties (ability to harden in a wet environment) to lime mortar (without magnesium oxide), Portland cement may be added to the mortar. Most mortars used for masonry are cement-lime-sand mortars.

Portland cement gives a fast setting, high strength mortar, whereas lime give a more workable mortar.

Sometimes glue is added to the lime-cement-sand mortar to make a rapidly setting and strong mortar. The costs are high, however, and this type of mortar is only used for very thin joints, such as in light weight cellular concrete masonry.

Lately *plastic binders* have been marketed. They add greatly to the tensile strength of masonry, but again the costs are high, and at the present time this type of binder can only be used where it is economically possible or technically necessary.

Admixtures

Accelerating additives such as calcium chloride have sometimes been used in mortar. There is a risk of steel corrosion, and staining, however, and the use of calcium chloride is normally not allowed. Retarding agents are sometimes used to delay the setting of the mortar, for example at building sites with very high temperatures.

Air-entraining admixtures are used to make the mortar more workable and to improve, in some cases, the resistance against freezing. To improve the resistance against early freezing, alcohol is sometimes used. The freezing point can be decreased by 2 to 3°C by adding 1 to 2% alcohol. Water-repelling additives are sometimes used in mortar. These additives are usually wetting agents or silicon powders.

Most of the additives must be added to the mortar in very small amounts, and the recommendation of the manufacturer, the designer, and the codes must be followed strictly. While a correct dose of additive may improve the mortar, it is almost certain that an overdose will harm it. A very strict control over the manufacturing process of such mortar is therefore recommended. The use of admixtures is normally not permitted.

Functions of Ingredients

Cement, lime and clay used as binding materials impart adhesive power and strength.

Sand increases the crushing strength of mortar and reduces shrinkage. When used in lime mortar, it assists the hardening of fat lime by allowing air to penetrate providing carbon dioxide for carbonisation.

Surkhi is used for economy and for furnishing hydraulic properties to lime mortar.

Flyash and cinders are used in lime mortar as fine aggregate in place of surkhi.

Molasses or gur is mixed with fat lime mortar; solubility of lime is increased and it readily crystallises. Consequently the mortar solidifies easily. One part of molasses is used with 80 parts by weight of water used for mixing that fat lime.

Water in mortar lubricates the surfaces of aggregate, spreads the binding material uniformly so that it can fill the pores in the fine aggregate and cause hydration of cement and hydraulic lime. The pH value of water used should not be less than 6.

5.3 CLASSIFICATION

Mortars are classified on the basis of their bulk density, kind of binding material, applications and physical and mechanical properties.

On the Basis of Bulk Density

Type of Mortar	Bulk Density (kg/m^3)	Aggregate
Heavy weight	> 1500	Heavy quartz or sand
Light weight	< 1500	Light porous sand from pumice, tuffa, slags, etc.

On the Basis of Binding Material

For most practical purposes a building mortar will fall in one of the following classes:

Cement Mortars are prepared from Portland cement or its varieties, sand and water.

Lime Mortars are mixtures of air hardening lime or hydraulic lime, sand and water.

Gypsum Mortars are prepared from gypsum or anhydride binding materials.

Mud Mortars are prepared from clay nodules and are used in construction of houses for poor and temporary construction works.

Composite Mortars may be surkhi-mortar (surkhi, lime and water), lime surkhi-sand mortar, cement-lime mortar and cement-clay mortar.

On the Basis of Application

Brick Lying Mortars are intended for brick work.

Finishing Mortars are intended for architectural or ornamental parts, application of decorative layers on walls and panels.

Special Mortars are intended for specific purposes for acoustics, x-ray shielding, plugging concrete at oil fields, etc.

On the Basis of Physical and Mechanical Properties

The basis of this classification is the strength of concrete which underlies the durability of concrete. Building mortars are subdivided into nine grades on the basis of compressive strength from 0.4 to 30.0 N/mm^2.

5.4 CEMENT MORTAR

Cement mortar can be prepared by mixing cement, sand and water in desired proportions. Portland cement and blast furnace slag cement from excellent mortars for walls from bricks, stones and large blocks. Puzzolana Portland cement and sulphate-resisting cement form mortar which are used for constructions exposed to aggressive and waste waters. Cement mortars are used for plastering, rendering smooth finishes and damp proof courses. The mix proportions are given in Table 5.1.

Table 5.1 Mix Proportions

S.No.	*Type of Work*	*Cement*	*Sand*
1.	Masonry	1	4–5
2.	Plastering (a) Interior	1	4
	(b) Exterior		5–6
3.	Pointing	1	1–3
4.	Reinforced brick work	1	3
5.	Foundation	1	3–4

Effect of Alakli Waters and Sea Water on Cement Mortar: Neat cement may disintegrate by the combined chemical and mechanical action of waters containing salts. The sulphates and chlorides are chemically active and remove lime from the cement. The carbonate of soda alone or in solution with sodium chloride or sodium sulphate withdraws silica. Under the conditions of alternate wetting and drying this process is accelerated. Crystals of large sizes are formed and expansive forces are produced which disintegrate the neat cement paste. This effect is less pronounced in lean mortars.

Effect of Oil and Acids on Cement Mortar: Well-cured cement mortars are not affected by oils. Lean mortars may develop less strength after 7 days when partially immersed in oil than when moist-cured for a month prior to immersion. The mortar surfaces soaked with oil show a marked reduction in abrasive resistance. When mineral oils are incorporated in the mixing water they retard the set of cement and reduce strength.

Animal and vegetable oils attack the lime compounds in the cement and form lime soap causing disintegration. Therefore, only mineral oils should be used for lubricating moulds.

For pH value less than 7 water removes the lime from cement in proportion to the decrease in the pH number and consequently the strength is reduced.

Effect of Sugar on Cement Mortar: Up to 0.15 per cent of sugar added to cement delays the setting time and destroys the early strength. However, when added up to 2 per cent, it increases the strength at an age of 2 to 3 months. The action of sugar is attributed to the formation of a soluble calcium saccharate ($C_{12}H_{22}O_{11}$. CaO + $2H_2O$).

Effect of Low and High Temperatures on Cement Mortar: The rate of setting of cement falls for temperature falling below 4.5°C. When the temperature falls below freezing the particles of cement in unset cement paste separate by the expansion of water. Alterations in freezing and thawing before set break the bond between cement particles and consequently there is loss of power. If binding cement freezes before setting but thaws without refreezing, it achieves half the normal strength under proper curing.

Considerable chemical activity is noticed in neat cement paste when setting at –18°C. At such low temperatures neat cement paste gains strength at a very slow rate but develops a high proportion of its normal value after several years. Cement paste hardening at room temperature attains higher strength than when allowed to harden for a like period after exposure to freezing temperatures.

Effect of Premixing and Retempering Cement Mortar: Only half of the cement grains are hydrated by water in ordinary cement paste. The powder obtained by crushing and grinding neat cement briquettes has cementitious properties, and briquettes made after a second regrinding possess a low strength. The strength is found to reduce in proportion to the increase in water/cement ratio caused by retempering.

5.5 LIME MORTAR

Lime mortar is made by mixing lime, sand and water. Lime used for mortar may be fat lime (quick or hydrated lime) or hydraulic lime. Fat lime has high calcium oxide content. Its hardening depends on loss of water and absorption of carbon dioxide from the atmosphere and possible recrystallisation in due course. Hydrated lime is a dry powder obtained by treating quick lime with just sufficient water to satisfy the chemical affinity of lime for water under the conditions of its hydration. Hydraulic lime contains silica, alumina and iron oxide in small quantities. When mixed with water it forms putty or mortar having the property of setting and hardening under water.

Slaked fat lime is used to prepare mortar for plastering, while hydraulic lime is used for masonry construction and are most suitable for construction of chimneys and lightly loaded superstructure of buildings. The mix proportions of lime mortar for various types of works are given in Table 5.2.

Lime mortars have high plasticity and placability, good cohesion with other surfacing and little shrinkage. They harden and develop strength very slowly continuously gaining strength over long period. Lime mortars do not set but stiffen only as water is lost by absorption (by masonry units) and evaporation. The gain in strength is a very slow reaction of lime with carbon dioxide absorbed from air.

Table 5.2 Mix Proportions

S.No.	Type of lime	Lime	Sand	Fineness modulus of sand	Type of work
1.	Fat Lime	1–2	2–3	2–3	Plastering
		1.5	2–3	2–3	Pointing
2.	Hydraulic	2–3	1.5–2.5	1.5–2.5	Masonry

Note: Sand in lime mortar is an adulterant, and reduces its shrinkage. Lime mortar becomes porous allowing air to penetrate and helps the mortar in hardening.

5.6 SURKHI MORTAR

It is prepared in the same way as lime mortar, with surkhi replacing sand. Surkhi should be ground to pass through 4.75 mm sieve and about less than 15 per cent through 150 micron sieve. Sand is also mixed sometimes. The mix proportions are given in Table 5.3.

Table 5.3 Mix Proportions

S.No.	Type of Mortar	Lime	Surkhi	Sand	Type of work
1.	Surkhi	1	2	–	Masonry and foundation
2.	Lime-surkhi	1	1	1	Masonry and foundation

5.7 LIME-CEMENT MORTAR

Also known as *guarded mortar* it is made from cement and lime. The advantages of lime-cement mortar are increased water retentivity, workability, bonding properties and frost resistance. The mortar gives good and smooth plaster finish and is used in buildings. The mix proportions are given in Table 5.4.

Table 5.4 Mix Proportions

S.No.	Location	Ratio (by volume)		
		Cement	Lime	Sand
1.	Outside wall	1	1	6
		1	2	9
2.	Inside wall	1	2	9
		1	3	12

5.8 MUD MORTAR

They are the cheapest type of mortar prepared with locally available ingredients and are used for masonry works, surfacing floors and plastering wall surfaces in low cost houses. To improve resistance to rain water, the plastered surfaces are sometimes sprayed with bituminous material.

5.9 CHARACTERISTICS OF GOOD MORTAR

The chief properties of hardened mortar are strength, resistance to weathering and those of green mortar mixes are mobility, placability and water retention.

Strength

The strength of masonry depends upon both the mortar and the building unit (brick, stone or block). A very strong mortar with weak building units will be of little use. It is also important to consider whether full strength is required within a short time. In cold weather, when the strength of lime or cement mixes develops slowly, this is likely to affect the choice of mix. Strong cement mortars are most likely to lead to shrinkage cracks, and should, therefore be avoided except where high strength is an essential requirement. On the other hand the use of much weaker mortar say, 1:10 cement mortar is not satisfactory since reduction in cement content leads to less workability, less cohesion and will produce porous joints of low frost resistance. Strength of hardened mortar depends on the activity of binding materials, the water/cement ratio, consumption of binding material and the quality of sand. It has been found that:

1. The density and strength of mortars made of the same class of aggregate decrease as the proportion of fine aggregate is increased.
2. It requires about twice as much cement to produce a mortar of given strength when fine sand is used as it does with coarse sand.
3. When the percentage of mixing water is increased beyond that required to form a placeable mix, the density and strength of mortar reduces. The proportionate effect is greatest at the early ages.
4. Even small percentage of mica if present considerably lowers the tensile strength and adversely affects the compressive strength.
5. A replacement of less than 15 per cent of cement by hydrated lime does not affect the tensile strength. There is a loss of compressive strength by the replacement of less than 25 per cent of cement by hydrated lime.
6. The strength of concrete increases with the cement content. A richer concrete requires less water for a definite consistency which may offset the shrinkage. For constant water/cement ratio, an increase in the cement content improves the workability without affecting the strength.

Resistance to Penetration of Rain

The mortar for plastering should protect the masonry joints and units by forming an impermeable sheet. A satisfactory bond between the building units, mortar and plaster should be ensured.

Mobility/Workability

Depending on its composition a mortar may have a consistency ranging from stiff to fluid. Mortars for masonry, finishes and other works are made sufficiently mobile. The mobility of mortar mix determines its placability. Mortars prepared from Portland cement alone are frequently deficient in cement paste, stiff and non-placable and plasticizers are added.

Water Retention

It is characterized by the ability of mortar not to stratify during transportation and to retain adequate humidity in a thin layer spread over a porous bed. A mortar mix of low water retention will show the defects after hardening. Mortar may lose so much water that the amount left may be insufficient for its hardening and required strength. Mineral and organic plasticizing agents may be added to enhance water retention.

5.10 SPECIAL MORTARS AND APPLICATIONS

Some of the mortars intended for use under special circumstances are as follows:

Mortars for Filling Joints between Prefabricated Reinforced

Concrete Components: Prepared with Portland cement and quartz sand with a mobility of 70 to 80 mm the grade of the mortar should be same as that of the concrete and in no case less than M-10.

Cement-sand and cement paste injection mortars intended for filling grooves inside prestressed concrete components should have a grade not less than M-30, usually M-40 grade or more is preferred. Sulphite-alcohol vinasse or naphetene soap is added in amounts up to 0.2 per cent of the weight of cement to reduce viscosity.

Packing Mortars: Used for packing oil wells these mortars may be of cement-sand, cement-loam and cement-sand-loam. Slag Portland cement, puzzolana and sulphate resisting cements are used for aggressive water and packing Portland cement when water pressure is expected. These mortars should have high homogeneity, water resistance, predetermined setting time, adequate water yield under pressure and ability to form soil water-proof plugs in cracks and voids of rocks. Cement-sand-loam mortar with 5 per cent calcium chloride is especially suitable for tunnelling.

Damp-proofing Mortars: are prepared using high grade sulphate-resisting Portland cement or sulphate-resisting puzzolana cement as binding material and quartz sand or sand from crushed solid rock. An approximate composition of the mortar is 1:2:5 or 1:3:5. Water proof seams and joints are made from damp-proofing mortars prepared with expanding cement.

Sound-absorbing Mortars: are prepared with Portland cement, slag cement, lime or gypsum as binding material, caustic magnesite and single-size fraction sand (3.5 mm) from light weight porous materials such as pumice, cinders, ceramsite, etc. They have a bulk density of 6-12 kN/m^3 and are used as sound absorbing plaster to reduce the noise level.

Fire-shielding Mortars: are used for setting refractory bricks in the furnace linings where the temperature is too high for ordinary mortars. Aluminous cements and finely powdered fire bricks in the ratio 1:2 give excellent fire resisting mortars. Its trade name is Accoset 50.

X-ray Shielding Mortars: Heavy mortars of bulk density over 22 kN/m^3 are required for plastering walls and ceiling of X-ray cabinets. The binding materials are Portland cement and slag cement, and the aggregates are from heavy rocks in the form of sand (up to 1.25 mm) and dust. Admixtures containing light weight elements (hydrogen, lithium, cadmium) are added to enhance the protective properties.

EXERCISE

1. What is mortar? Briefly describe various types of mortars.
2. State the functions of following in mortar.
   (a) sand
   (b) cement
   (c) water
   (d) lime
3. (a) Under what conditions will you recommend cement mortar over lime mortar for masonry.
   (b) Write a note on special mortars.
4. What are the different types of mortars used in construction? State the composition and use of each.
5. What are the advantages of adding surbhi to lime mortar.
6. Give reasons:
   (a) Why are bricks soaked with water before they are laid in mortar?
   (b) Why is sand added to mortar?
   (c) Why is puzzolana added to mortar?
7. Write short notes on:
   (a) Lime mortar
   (b) Lime-surkhi mortar
8. What types of mortars will you recommend under the following situations? Give reasons.
   (a) masonry in foundation
   (b) Plastering interior of your house
   (c) 20 cm thick walls

CHAPTER 6

Concrete

6.1 INTRODUCTION

Concrete a composite man-made material, is the most widely used building material in the construction industry. It consists of a rationally chosen mixture of binding material such as lime or cement, well graded fine and coarse aggregates, water and admixtures (to produce concrete with special properties). In a concrete mix, cement and water form a paste or *matrix* which fills the voids of the fine aggregate and binds them (fine and coarse) together. The matrix is usually 22-34% of the total volume. Freshly mixed concrete before set is known as *wet* or *green* concrete whereas after setting and hardening it is known as *set* or *hardened* concrete. The moulded concrete mix after sufficient curing becomes hard like stone due to chemical action between the water and binding material. It would be impossible to discuss all the aspects of this material in few pages and the discussion is confined to the characteristics and properties of concrete only.

The popularity of the concrete is due to the fact that from the common ingredients, it is possible to tailor the properties of concrete to meet the demands of any particular situation. The concept of treating the concrete in its entity as a building material rather than ingredients is gaining popularity since the user is now interested in the concrete itself having desired properties without bothering about the ingredients. This concept has paved way for ready mixed concrete industry.

The concrete to be used under a particular situation should be workable and should exhibit the desired strength and durability after hardening. These characteristics depend upon the properties of ingredients, on the mix proportions, and the controls during various stages of concrete production. Among the various properties of concrete, its compressive strength is considered to be the most important and is taken as an index of its overall quality.

6.2 ADVANTAGES AND LIMITATIONS OF CONCRETE

Concrete as a construction material has the following advantages.

(i) Concrete is economical in the long run as compared to other engineering materials. Except cement, it can be made from locally available coarse and fine aggregates.

(ii) Concrete possesses a high compressive strength, and the corrosive and weathering effects are minimal. When properly prepared its strength is equal to that of a hard natural stone.

(iii) The green concrete can be easily handled and moulded into any shape or size according to specifications. The form work can be reused a number of times for similar jobs resulting in economy.

(iv) It is strong in compression and has unlimited structural applications in combination with steel reinforcement. The concrete and steel have approximately equal coefficients of thermal expansion. The concrete is extensively used in the construction of foundations, walls, roads, airfields, buildings, water retaining structures, docks and harbours, dams, bridges, bunkers and silos, etc.

(v) Concrete can even be sprayed on and filled into fine cracks for repairs by the guniting process.

(vi) The concrete can be pumped and hence it can be laid in the difficult position also.

(vii) It is durable and fire resistant and requires very little maintenance.

Following are the disadvantages of concrete.

(i) Concrete has low tensile strength and hence cracks easily. Therefore, concrete is to be reinforced with steel bars or meshes.

(ii) Fresh concrete shrinks on drying and hardened concrete expands on wetting. Provision for contraction joints has to be made to avoid the development of cracks due to drying shrinkage and moisture movement.

(iii) Concrete expands and contracts with the changes in temperature. Hence expansion joints have to be provided to avoid the formation of cracks due to thermal movement.

(iv) Concrete under sustained loading undergoes creep resulting in the reduction of prestress in the prestressed concrete construction.

(v) Concrete is not entirely impervious to moisture and contains soluble salts which may cause efflorescence.

(vi) Concrete is liable to disintegrated by alkali and sulphate attack.

(vii) The lack of ductility inherent in concrete as a material is disadvantageous with respect to earthquake resistant design.

6.3 GRADES OF CONCRETE

The concrete is generally graded according to its compressive strength. The various grades of concrete as stipulated in IS: 456-2000 and IS: 1343-1980 are given in Table 1.1. In the designation of concrete mix, the letter M refers to the mix and the number to the specified characteristic strength of 150 mm work cubes at 28 days, expressed in N/mm^2. The concrete of grades M5 and M7.5 is suitable for lean concrete bases and simple foundations of masonary walls. These need not be designed. The concrete of grades lower than M15 is not suitable for reinforced concrete works and grades of concrete lower than M 30 are not to be used in the prestressed concrete works.

Table 1.1 Grades of Concrete

Grade Designations	M5	M7.5	M10	M15	M20	M25	M30	M35	M40	M45	M50	M55
Specified characteristic strength at 28 days, N/mm^2	5	7.5	10	15	20	25	30	35	40	45	50	55

6.4 CURING

Cement gains strength and hardness because of the chemical reaction between cement and water. This chemical reaction requires moisture, favourable temperatures and time referred to as the curing period. Curing freshly placed concrete is very important for optimum strength and durability. The major part of the strength in the initial period is contributed by the clinker compound C_3S and partly by C_2S is completed in about three weeks. The latter strength contributed by C_2S is gradual and takes long time. As such sufficient water should be made available to concrete so that it gains full strength. The process of keeping concrete damp for this purpose is known as curing. The object is to prevent the loss of moisture from concrete due to evaporation or any other reason, supply additional moisture or heat and moisture to accelerate the gain of strength. Curing must be done for at least three weeks and in no case for less than ten days.

Approximately 14 litres of water is required to hydrate each bag of cement. Soon after the concrete is placed, the increase in strength is very rapid (3 to 7 days) and continues slowly thereafter for an indefinite period. Concrete moist cured for 7 days is about 50 per cent stronger than that which is exposed to dry air for the entire period. If the concrete is kept damp for one month, the strength is about double that of concrete exposed to dry air.

The most common method of curing is water curing-concrete surface is kept covered with gunny bags and water is regularly sprinkled over it. Some of the other methods of curing concrete are by steam, infra-red radiations, electricity, and chemicals.

6.5 WORKABILITY

In fresh concrete—concrete in the plastic state, which can be moulded into desired shape—the theoretical w/c ratio to meet the requirements of water for chemical combination with cement, and to occupy the gel-space is about 0.4 for maximum strength. The w/c ratio used at site may vary because of:

(a) the presence of free surface moisture in the aggregates, and (b) the absorption of moisture of by the dry or porous aggregates.

Because of the above limitations another characteristic workability, which is again a reflection of w/c ratio, becomes important.

The theoretical w/c ratio used will not give the maximum strength because of the reasons stated above. 100 per cent compaction of concrete will give maximum strength and this can be obtained by increasing the w/c ratio. The water lubricates the concrete which can be compacted at site with the specified efforts. The lubrication required for handling concrete without segregation, for placing without loss of homogeneity, for compacting with specified effort and for easy finish are indication of workable concrete.

As per Road Research Laboratory U.K. workability is defined as the property of concrete which determines the amount of useful internal work necessary to produce full compaction. It can also be defined as the ease with which concrete can be compacted 100 per cent with regard to mode of compaction and place of deposition.

Workability is different than consistency. The latter indicates degree of fluidity or mobility. A concrete with high consistency need not be workable for a particular job. For example a concrete workable for foundation may not be workable for slab. Even for slab different workabilities will be required for compaction by hand and that by vibration for the requirement of workability is less.

Factors Affecting Workability

Water Content: The fluidity of concrete increases with water content. At site the normal practice is to increase the water content to make the concrete workable which lowers strength. In controlled concrete this cannot be resorted and even in uncontrolled concrete this should be the last choice. More water can be added provided cement content is proportionately increased.

Mix Proportions: Aggregate-cement ratio influences the workability to a large extent. The higher the ratio leaner will be the concrete. In a lean concrete, paste available for lubrication of per unit surface area of aggregates will be less and hence the workability is reduced.

Aggregate Size: For big size aggregate the total surface area to be wetted is less, also less paste is required for lubricating the surface to reduce internal friction. For a given water content big size aggregate give high workability.

Shape of Aggregates: For a given water content, round and cubical shape aggregates are more workable than rough, angular or flaky aggregates, because the former requires less cement paste for lubrication as these have less surface area and lesser voids. In case of round aggregates frictional resistance is also small so less lubrication is required. For this reason river sand and gravel provide greater workability than crushed sand and aggregates.

Surface Texture: A rough surface aggregate will have more surface area than a smooth round textured aggregate. Hence, latter will be more workable for the reasons discussed above.

Grading of Aggregates: Properly graded aggregates are more workable. It is so because such a mix will have least voids and thus excess cement paste will be available as lubricant. This also prevents segregation.

Admixtures: Air entrained concrete is more workable. It is so because air forms bubbles, on which the aggregates slide past each other increasing the workability.

Another factor is that air entraining agents are surface active and they reduce the internal friction between the aggregates.

6.6 STRENGTH OF CONCRETE

The most useful property of concrete is its compressive strength. However, it is weak in tension. Till date no relation exists between compressive, tensile, bending, and shear strengths of concrete.

The indicated compressive strength increases as the specimen size is decreased whereas modulus of elasticity decreases. Concrete containing about 6 per cent of entrained air which is weaker in strength is found to be more durable than dense and strong concrete.

Factors Influencing Strength of Concrete

Factors affecting the strength of concrete can be broadly grouped into those depending upon the testing methods and the others independent of the testing methods.

Factors depending on testing methods are size of test specimen, size of specimen relative to maximum size of aggregate, moisture condition of specimen, rate of loading adopted, and type of testing machine used; and those independent of testing method are type of cement and age of cement, type and size of aggregates, degree of compaction, water/cement ratio, aggregate/cement ratio, air-voids, curing method and curing temperature, and type of stress situation that may exist (uniaxial, biaxial and triaxial).

Size of Test Specimen: The concrete cubes of side 150 mm, cured for 28 days are tested in the most saturated condition in compression. For materials which rupture on planes inclined more than 45° with the horizontal, the cube is not suitable; since the strength is increased by frictional restraint acting at the surfaces under pressure. Frictional restraint from the platens opposes the lateral expansion of the specimen, and subjects ends to inward compressive forces. These diminish with distance from the ends, but a cube is so short that their effect extends throughout its volume. A cube is therefore tested under nonuniform triaxial compression. Prisms or cylinders with a height equal to twice the least lateral dimension are more suitable for such materials. End effects have less influence in these specimens, and it is generally agreed that the cylinder strength of concrete is a good estimate of the monoaxial compressive strength. A cube of concrete is expected to have a strength 15 per cent greater than a cylindrical specimen. If height to side ratio is changed the compressive strength of the prisms relative to cube strength changes and is given in Table 6.2. Beyond a height to side ratio of 4 it stabilises. The effect of height to lateral dimension ratio of specimen on compressive strength is shown in Fig. 6.1.

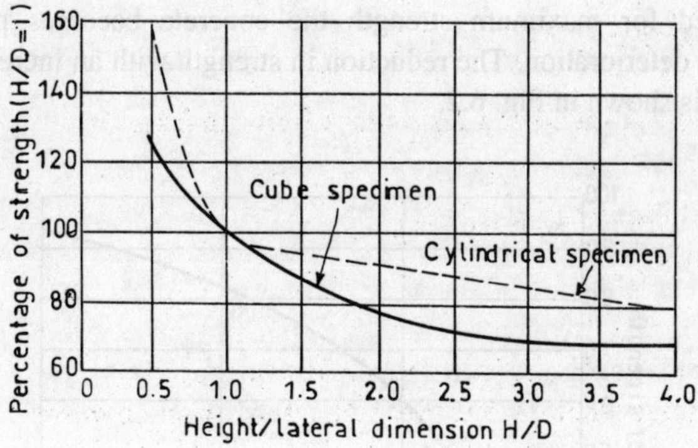

Fig. 6.1 Influence of Ratio of Height to Lateral Dimension on the Compressive Strength

Table 6.2 Relative Strength of Prisms with Different Height to Depth Ratio

Height/side ratio	0.5	1.0	2.0	3.0	4.0	5.5
Relative strength	1.5	1.0	0.8	0.72	0.68	0.6

If size of cube is decreased the compressive strength tends to increase and is given in Table 6.3.

Table 6.3 Relative Strength of Concrete from Cubes of Different Sizes

Cube size (mm)	100	150	200	300
Relative strength to 150 mm cubes	1.05	1.0	0.95	0.87

Size of Specimen Relative to Maximum Size of Aggregate: The test specimen, as regards the size and shape, is different in different countries. Generally 150 mm cubes are specified irrespective of size of aggregates. However, for aggregate less than 25 mm in size 100 mm cube is allowed.

Moisture Condition of Specimen: Strength of concrete depends upon moisture content at the time of testing. The dry cubes may have drying shrinkage and bond failure leading to smaller strength. The moisture content in concrete provides lubrication effect and reduces strength. Due to dilation of the cement gel by the absorbed water the force of cohesion between the cement particles is decreases. Probably the decrease in strength on account of this reduction is more than that of the loss of strength due to rupture of gel bond and the dry cubes give higher strength.

Also, the specimen should be tested immediately after taking it out of curing tank. To give uniformity to results as compared to testing of samples which may have dried to different degrees.

Strength of dry sample = (1.1–1.2) × strength of saturated sample

Air Voids These are formed because of the evaporation of the water used in making concrete and by entrained air. If the water used is more than the optimum water required for maximum strength the concrete becomes permeable and susceptible to deterioration. The reduction in strength with an increase in percentage air voids is shown in Fig. 6.2.

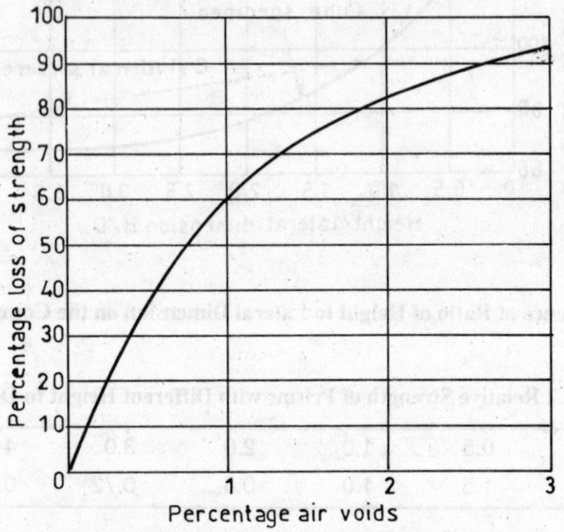

Fig. 6.2 Loss of Strength with the Percentage of Air-Voids

Rate of Loading Strength of concrete increases with that in rate of loading. The influence of rate of application of load on the compressive strength of concrete is shown in Fig. 6.3. Normally in a static loading test, rate is such that the test is completed in 2 to 3 minutes. Within this range, the effect is found to be negligible. At low rates of loading there is more time for creep to occur, so that the increase of strength with rate of loading provides evidence for the theory that failure occurs at limiting values of strain, rather than stress. The relation between observed strength and rate of loading is given by

$$S = S_1(1 + k \log R)$$

where S = strength at a given rate of loading

R = rate of loading in kg/cm^2/sec
S_1 = strength of concrete at a rate of 1 kg/cm^2/sec
k = a constant = 0.08 for 28 day test
The rate of loading of cube specimen is 140 kg/cm^2/min.

Type of Testing Machine: Compressive strength tests of concrete are made in various machines either of the lever-arm type or adoptions of lever presses. Considerable discrepancies have been found to occur between the results of compressive strength test made with different types of machines. It may be up to even 20 per cent. It may be because of errors in centering the cubes, wear of the lower compression plate, friction in the spherical seating of the upper compression plate, and inaccurate calibration of the machine itself.

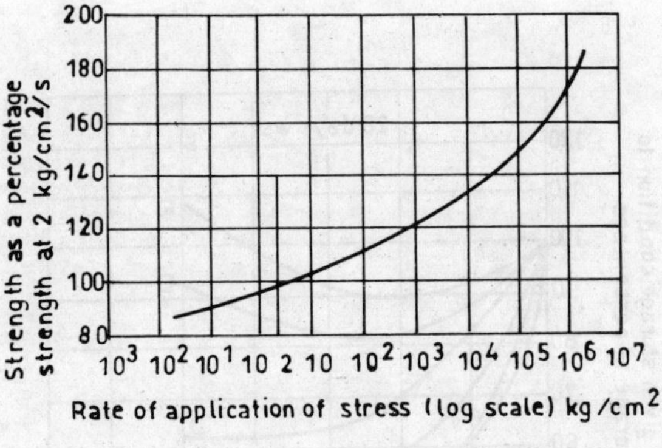

Fig. 6.3 Influence of Rate of Application of Load on Strength of Concrete

Type of Stress Situation that may Exist: Concrete is tested for uniaxial compression with the line of action of load on a cube specimen at right angles to the axis of cube about which is cast. However, in actual structure, the concrete at any point is in a complex stress condition and not in a uniaxial compression. Concrete under triaxial state offers more resistance and fails only after considerable deformations, which justifies uniaxial compression testing.

Type and Age of Cement: The effect of type of cement on the rate of strength gain is shown in Fig. 6.4. With age the strength of cement reduces since it will set

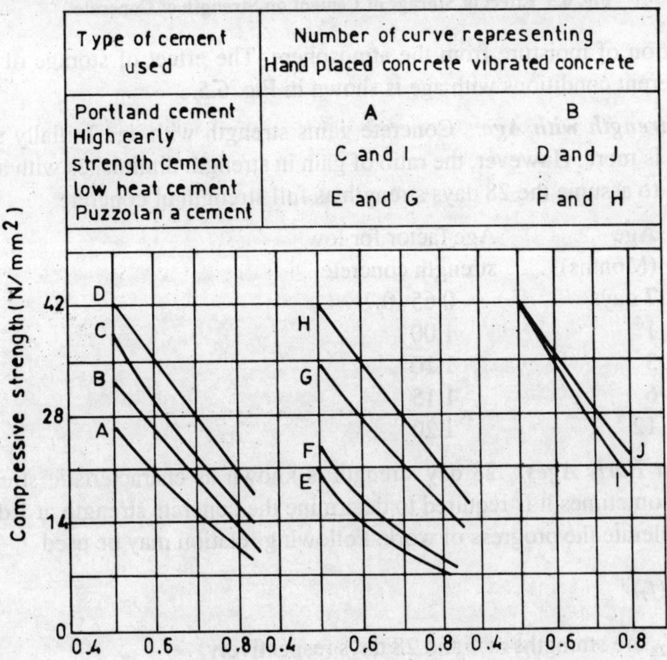

Fig. 6.4 Influence of Different Types of Cement on the Strength of Concrete

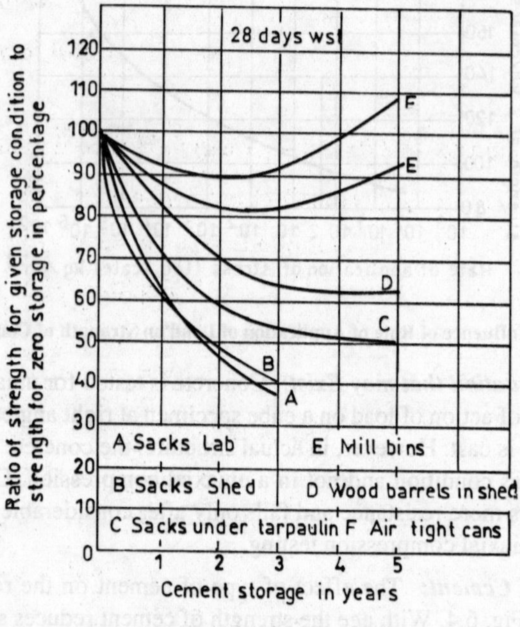

Fig. 6.5 Effect of Storage of Cement on Strength of Concrete

by absorption of moisture from the atmosphere. The effect of storage of cement under different conditions with age is shown in Fig. 6.5.

Gain in Strength with Age: Concrete gains strength with age Initially strength developed is more. However, the ratio of gain in strength diminishes with age. It is customary to assume the 28 days strength as full strength of concrete.

Age (Months)	Age factor for low strength concrete
7 days	0.65–0.7
1	1.00
3	1.10
6	1.15
12	1.20

Strength at Early Ages: 28 day strength is known as characteristic strength of concrete. Sometimes it is required to determine the concrete strength at 1 day or 7 day to accelerate the progress of work. Following relation may be used

$$f_{28} = k_2(f_7)^{k_1}$$

where f_7, f_{28} are strengths at 7 and 28 days respectively

$k_1 = 0.3\text{–}0.8$ (for different cements)

$k_2 = 3\text{–}4$ (for different curing conditions)

Cement/Aggregate Ratio: Provided other factors are kept constant, cement to aggregate ratio will greatly influence concrete strength. With an increase in cement to aggregate ratio the ultimate strength will increase to some extent as shown in Fig. 6.6.

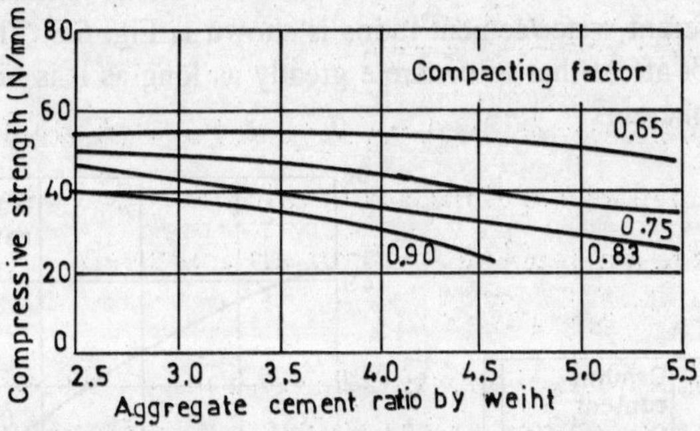

Fig. 6.6 Influence of Aggregate/Cement Ratio on the Strength of Concrete

Effect of Water/Cement Ratio: The effect of w/c ratio on the strength of concrete is great. The aim is generally to use the lowest water/cement ratio which will give a concrete sufficiently plastic to place in position. Approximate relation of strength of concrete with water/cement ratio is given by

$$S = k\left(\frac{c}{w+c+a}\right)^2$$

where S = strength of concrete
w = volume of water
c = volume of cement
a = volume of air

For concrete which is to be compacted by vibrator a lower water cement/ratio may be used.

Type and Size of Aggregate: The extent to which an aggregate will pack down and produce a minimum void content is dependent on its particle shape. The crushed stone and gravels give higher strength. A reduction in the void content of the coarse aggregate by better packing, means that the amount of mortar can be reduced and hence sand and cement. Thus the coarse aggregate to sand ratio is increased and although the overall mix may be leaner the mortar may be richer, and by virtue of reduction in water/cement ratio which may thereby be permitted, the strength of concrete may be increased.

A rounded spherical shaped aggregate when compacted contains less voids than an irregular and flaky aggregate of the same nominal size. Therefore, the former gives higher strength.

A common belief has been that the larger the maximum size of aggregate, the denser and stronger will be the concrete. The large aggregates have lower total surface area and require lower water/cement ratio resulting in higher strength. In practice it is not so. The larger aggregates give lower surface area for development of gel bonds leading to lower strength. Also larger size aggregates give heterogeneous concrete causing nonuniform distribution of load when stressed.

High strength concrete gives lower strength as compared to lean concrete if a big size aggregate is used. The influence of maximum size of aggregate on compressive strength of concrete with different cement contents is shown in Fig. 6.7 and that with different water/cement ratios is shown is Fig. 6.8. The strength of aggregate does not affect that of concrete greatly as long as it is higher than the design grade of concrete.

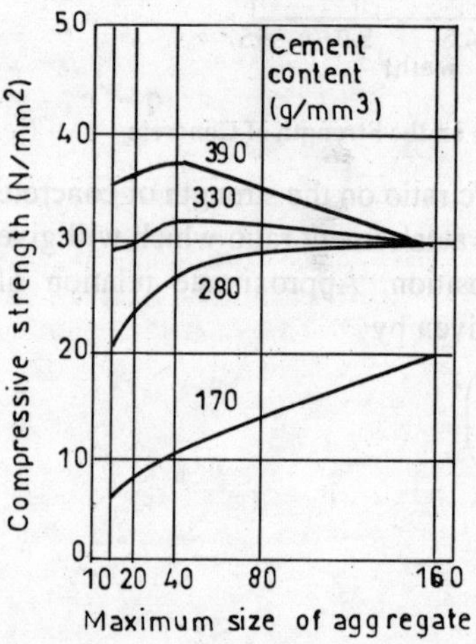

Fig. 6.7 Influence of Maximum Size of Aggregate on 28 Day Strength of Concrete of Different Richness

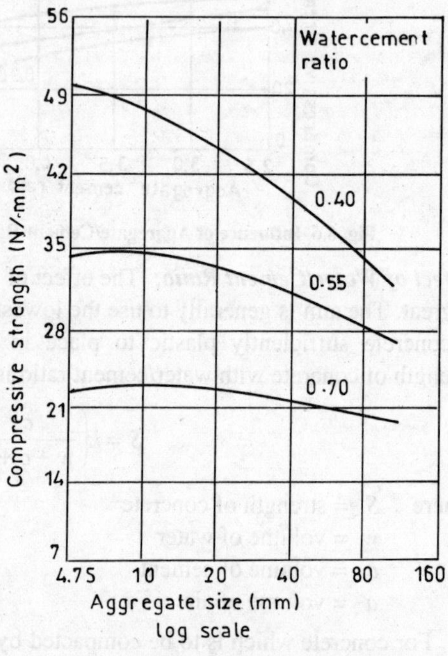

Fig. 6.8 Influence of Maximum Size of Aggregate on the Strength of Concrete

Degree of Compaction: Inadequate compacting leading to air void contents of 5 per cent and of 10 per cent result in a loss of strength of 30 per cent and 55 per cent respectively. Following equation may be used to find the variation in strength

$$S = j\left(\frac{C}{1-\rho} - n\right)$$

where S = unit compressive strength
C = volume of cement grains in a unit volume of concrete
ρ = density of concrete
j and n = constants

The effect of degree of compaction on strength is shown in Fig. 6.9.

Mixing Time: The strength of concrete increases with increase in time of mixing up to two minutes beyond which no significant improvement is observed. Fig. 6.10 shown the effect of mixing time on strength of concrete.

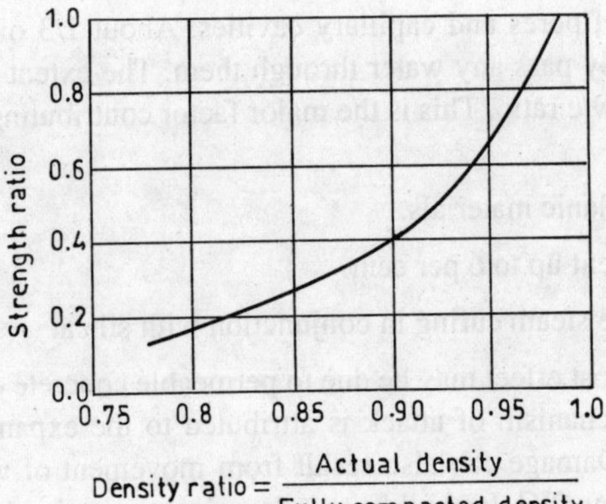

Fig. 6.9 Effect of Compaction on the Strength of Concrete

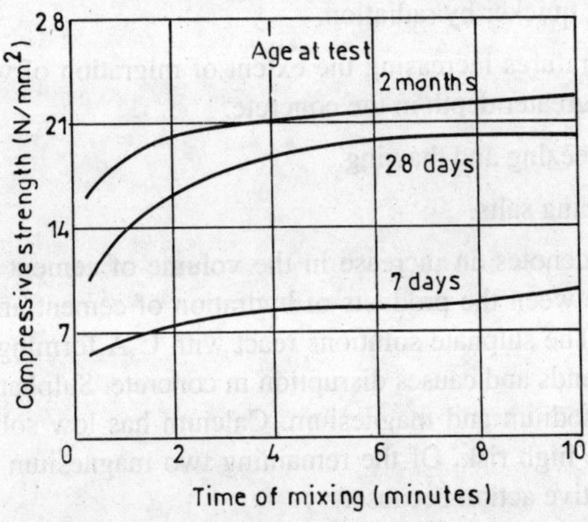

Fig. 6.10 Effect of Mixing Time of Strength of Concrete

Curing Method and Curing Temperature: The effect of method of curing on strength has been discussed in Sec 6.4. The higher the temperature the greater is the rate of hardening of concrete. 10 hours curing at temperatures of about 90°C, concrete may attain 70 per cent of its 28 day strength.

6.7 DURABILITY

Concrete should be durable. The various factors affecting the durability of concrete used in normal conditions are as follows:

Permeability Almost all forms of deterioration in concrete are due to ingress of water. The ways in which durability of concrete may be affected because of permeability are:

1. The chemicals in liquid form affect the concrete by penetrating it.
2. Frost action, rusting of steel, etc.

Concrete has gel pores and capillary cavities. About 1/3 of gel pores are so small that they hardly pass any water through them. The extent of capillary cavities depends on the w/c ratio. This is the major factor contributing to permeability. The remedies are:

1. Use of puzzolanic materials.
2. Air entrainment up to 6 per cent.
3. High pressure steam curing in conjunction with silica.

Frost Action The frost effect may be due to permeable concrete or by temperature below 0°C. The mechanism of attack is attributed to the expansion of absorbed water on freezing. Damage can also result from movement of water within concrete on cooling below 0°C. Ice builds up in large pores causing large expansion in local areas the others being dry cause disintegration. The conditions favouring frost attack are:

1. Horizontal surfaces open to sky absorbing more water in wet conditions and cooling quicker by radiation.
2. Low temperatures increasing the extent of migration of water resulting in freezing to greater depth in the concrete.
3. Repeated freezing and thawing.
4. Use of de-icing salts.

Sulphate Attack denotes an increase in the volume of cement paste due to the chemical action between the products of hydration of cement and solutions containing sulphates. The sulphate solutions react with C_3A forming a chemical enttringite which expands and causes disruption in concrete. Sulphates occur in three forms—calcium, sodium and magnesium. Calcium has low solubility so that it does not constitute high risk. Of the remaining two magnesium sulphate has the most severe disruptive action because:

(a) the reaction product is insoluble, precipitating out of solution and leaving the way clear for further attack, and

(b) magnesium sulphate reacts with the C_3S hydrate in cement.

A common source of sulphate salts is the soil around the concrete foundation, from the water used in making concrete, and by using unwashed aggregates.

The remedy is to use blast furnance slag cement, sulphate resisting cement, supersulphated cement, and to reduce the permeability.

Mineral Oils: Petroleum and its products do not directly affect hardened concrete. To overcome the deteriorating affects of mineral oils on fresh concrete permeability is reduced by surface treatment by applying at least four coats of sodium silicate.

Organic Acids: Acetic acid, lactic acid, and butyric acid severely attack concrete. Formic acid is corrosive to concrete.

Vegetables and Animal Oils and Fats: Vegetable oils contain small amount of free fatty essence and deteriorate concrete slowly. Fish and cotton seed oils are found to be the most corrosive.

Sugar: is a retarding agent and gradually corrodes concrete. The remedy is to coat the concrete surface with sodium silicate solution, tar or asphalt.

Sewage: H_2S gas evolved from septic sewage may promote formation of H_2SO_4 affecting the concrete. However, sewers running full are not affected.

Thermal Effects on Concrete: Concrete is a heterogeneous material and the ingredients have dissimilar thermal coefficients which affect durability.

Cracks: Cracks are inherent in concrete and cannot be completely prevented but can be minimized. Use of unsound materials, high w/c ratio, bad jointing techniques, freezing, thermal effects, etc. lead to cracks and make the concrete less durable.

6.8 PHYSICAL PROPERTIES

Stress-Strain Curve of Concrete

The true elastic curve for concrete in compression (Fig. 6.11) can be plotted by applying and releasing load unit the set at zero loads beeomes constant. By subtracting the set from the total deformation the elastic deformation for a given load is determined. Since mortar and concrete have no elastic limit, the modulus of elasticity for concrete must be the slope of the stress-deformation curve at zero stress and can be calculated from the formula

$$S_c = \left(1 - \frac{1}{2} q\right) E_c\, \varepsilon$$

assuming stress-deformation curve to be parabolic.

where q = ratio of ε (the unit deformation for S_c) to the ultimate unit deformation
E_c = initial modulus

The modulus of elasticity of concrete increases with the density and to some extent with age.

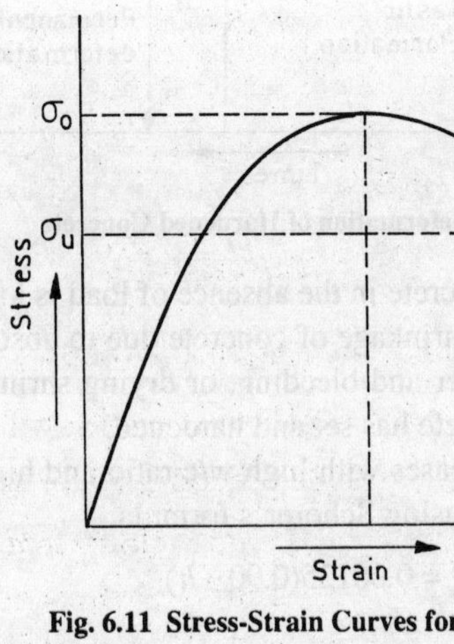

Fig. 6.11 Stress-Strain Curves for Concrete

Poisson's Ratio: Under unit compression the unit lateral expansion of concrete is about 1/6th to 1/12th of the unit strain in the direction of the applied forces for the ordinary range of working stress. The ratio increases with the richness of the mix. The value of Poisson's ratio varies between 0.1 to 0.16 for working loads on two month old, 1:2:4 concrete.

Creep: The continued deformation with time under applied load is creep. It may be defined as increase of strain in concrete with time under sustained stress. This is also known as *plastic flaw* or *time yield*. The rate of creep decrease with time and the creep strains at five years are taken as terminal values. The deformation of hardened concrete is shown in Fig. 6.12.

Creep may be due to closer of internal voids, viscous flow of the cement paste and flow of the water out of the cement gel. In reinforced concrete structures it is of advantage since it causes better distribution of stresses. For example in a R.C.C. column there is a reduction of stress in concrete and a corresponding increase of stress in steel due to creep. As another example creep relieves the high stressed portions of concrete in a continuous beam and increases the stress in the adjacent less stressed portion. Creep causes large deformations and deflections and is undesirable.

Creep increases rapidly with the stress, loading at an early age of concrete, broken ballast, soft and porous aggregate, poorly graded and improperly compacted concrete.

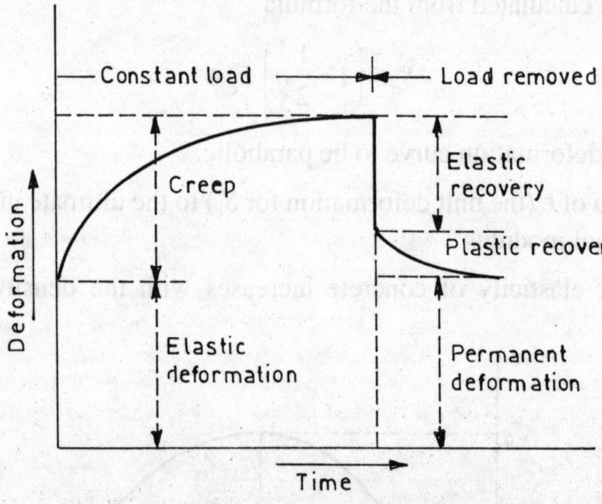

Fig. 6.12 Deformation of Hardened Concrete

Shrinkage: Contraction of concrete in the absence of load is known as shrinkage. It may be plastic shrinkage—shrinkage of concrete due to absorption of water by aggregates, evaporation of water and bleeding, or drying shrinkage—the shrinkage taking place after the concrete has set and hardened.

Shrinkage of concrete increases with high w/c ratio and high cement content. Shrinkage can be estimated by using Schorer's formula:

$$\varepsilon_s = 0.00125(0.90 - h)$$

where ε_s is shrinkage strain and *h* is relative humidity expresses as a fraction. The rate of shrinkage decreases with time.

6.9 REINFORCED CEMENT CONCRETE (RCC)

Cement concrete possesses high compressive strength but is weak in tension. To overcome this drawback concrete is reinforced with steel bars. RCC may be defined as a composite material made up of cement concrete and reinforcement in which the concrete resists compression with reinforcement resisting tension and shear.

Steel for reinforcing bars can be mild steel bars, designated as Fe-250, High yield strength deformed (HYSD) bows, designated as HYSD-415, 500, 550, or thermomechanically treated (TMT) bars, designated as TMT 415, 500, 550 (the values 250, 415, 500, 550 being in N/mm^2).

Reinforced cement concrete is the most versatile building material available and is extensively used in the construction industry ranging from small structural elements such as beams, columns to massive structures like dams and bridges. In fact, today, it is impossible to think of a structure without concrete.

EXERCISE

1. Define concrete. What are its advantages and limitations? How is concrete graded?
2. Define workability of concrete. What factors are responsible for change in workability?
3. Describe briefly the factors affecting strength of concrete.
4. What is meant by durable concrete? Describe the factors that affect it.
5. Discuss briefly the desirable properties of concrete.
6. (a) What is curing? What is its significance?
   (b) Define w/c ratio.
7. (a) What are the requirements of good concrete?
   (b) What concrete mix will you recommend in following situations?
   (a) Beam (b) Column
   (c) Slab (d) Foundation for column
8. What is reinforced cement concrete? Describe briefly about the reinforcing bars.

CHAPTER 7

Alternate Materials

7.1 INTRODUCTION

Construction is a never ending activity. Materials play an important role in the building construction. The traditional materials such as bricks timber, cement, steel, etc. are in constant demand both in developed and developing countries. The depleting resources of conventional construction materials limit their use for heavy constructions and pose a need to develop new materials. Another reason for the development of alternate materials is to take care of ecological imbalance caused because of cutting forests (to obtain wood) and loss of useful land (to manufacture bricks) for agricultural purposes. Also, the production of tradition materials involve high technology, equipments and many a times prove to be uneconomical under certain situations.

This in consequence called for exploring new alternate materials for construction activities with better productivity and equivalent quality. Many materials such as Fibre Reinforced Concrete (FRC), Ferrocement (FC), and Light Weight Concrete (LWC) have emerged in past few decades. Also, several alternate materials have come up with the use of waste materials, e.g., fly ash, red mud, rice husk ash, and blast furnace slag, as well as those available locally like bamboo, etc.

Materials play an important role in the building construction. The principal properties of the materials predetermine their applications. To save the precious materials for important projects, the materials have to be developed with greater economic efficiency and with comparable performance under specific service conditions.

7.2 FIBER REINFORCED CONCRETE

Conventional concrete is modified by random dispersal of short discrete fine fibers of asbestos, steel, sisal, glass, carbon, Poly-propylyene, nylon, etc. Asbestos cement fibres so far have proved to be commercially successful. The improvement in structural performance depends on the strength characteristics, volume, spacing, dispersion and orientation, shape and their aspect ratio (ratio of length to diameter) of fibers. The fibre content is generally 1-2% by weight.

Advantages

1. Strength of concrete increases.
2. Fibers help to reduce cracking and permit the use of thin concrete sections.
3. Mix becomes cohesive and possibilities of segregation are reduced.
4. Ductility, impact resistance, tensile and bending strength are improved.

Disadvantages

1. Fibers reduce the workability of a mix and may cause the entrainment of air.
2. Steel fibers tend to intermesh and form balls during mixing of concrete.

Applications Fiber reinforced concrete is useful in hydraulic structures, airfield pavements, highways, bridge decks, heavy duty floors, and tunnel linings. To achieve economy in low cost construction vegetable fibers are used. Some of the commonly available vegetable fibers are bast or stem fibers, leaf fibers, fruit hair fibers and wood fibers.

A variety of building elements can be made of natural fiber concrete, but its most widespread application is in the production of thinner building elements such as corrugated sheets etc.

7.3 FERROCEMENT

Ferrocement is a composite material in which the filler material, cement mortar, is reinforced with fibres, usually steel mesh dispersed throughout the composite. The fibres impart tensile strength to the mass.

In rationally designed ferrocement structures the reinforcement consist of small diameter wire meshes wherein uniform distribution of reinforcement is made possible throughout the thickness of the element. Because of the distribution of such reinforcement over the entire matrix, high resistance cracking is achieved. Toughness, fatigue resistance, impermeability, etc. are also improved.

The reinforcement may be hexagonal wire mesh (0.5-1.00 mm diameter at 5-25 mm spacing) welded wire mesh (18-19 gauge), woven mesh, expanded metal sheet and Waston mesh. The skeletal steel may be placed 300 mm apart to serve as a spacer rod to the mesh reinforcement. A rich mix of Portland cement and sand usually 1:2 to 1:1.5, is used as a mortar. The thickness is kept 10-40 mm, with a clear cover of 1.5-2 mm to reinforcement. No form work is required since the wire mesh and chicken mesh receive the mortar wherein the mortar applied with pressure is held in position by mechanical interlocking.

Steel bars are generally provided to make the formwork of the structure and is known as skeletal steel. The size of the rod varies from 4 to 10 mm and at a spacing of 300 mm apart. In highly stresses structures like boats, barges and tubular sections this spacing is reduced to 75 mm.

Portland cement, rapid hardening Portland cement, sulphate resistance Portland cements or Portland blast-furnance slag cement may be used.

The advantages of ferrocement are:

1. Easy availability of raw materials.
2. Reduction in weight consequent of this section.
3. Moulding can be done without any formwork.
4. No machinery or sophistication is required in construction.

Ferrocement acts a homogeneous material in elastic range. Stress-strain curve for ferrocement is linear in elastic range and cracks do not develop. With an increase in stress beyond this limit cracks develop. On further increasing the stress the number of cracks increase rather than the crack widths. Near yield the cracks widen and failure takes place.

The excellent crack control and impermeability characteristics of ferrocement make it suitable for liquid retaining structures, boat building, gas containers caissons, canal lining, etc. Since it is cheaper than steel and R.C.C. and can be cast in thin sections it is most suitable for low cost roofing, precast units, manhole covers, casings, etc. and is the most appropriate building material for the construction of domes, vaults, shells, grid surfaces, corrugated sheets and folded plates. It is one fo the possible substitute of timber. Sanitary furnitures, almirahs, doors can be made with ferrocement. Since ferrocement can be used as structural material, a composite construction of ferrocement and light weight concrete can be made.

7.4 SULPHUR IMPREGNATED CONCRETE

The physical properties of greatest interest of sulphur are extreme weather resistance, its great compressive strength and hardness, its capability of conversion into polymeric form having plastic properties, and its readily attainable melting point, located above ambient atmospheric temperatures. Such melting point makes it possible to work with the element in either a solid or liquid form. Sulphur's properties of hardness, stability of weather, and low melting point have suggested the possibility of making a cheap and useful concrete analogous in many ways to Portland cement concrete but having superior resistance to acids and to electrolytes. Such a sulphur concrete has been prepared by mixing approximately 84% of an aggregate with about 16% of molten sulphur. This sulphur performs best when modified by reacting it with 2–5% of a dicyclopentadiene at a temperature approximately 130–160°C. Aggregate is then added and complete mixing should be achieved. The material may be poured into any appropriate form and allowed to set. While the product is still hot and the sulphur is in a liquid form, the material may be compacted by use of vibrator and its surface may be prepared by leveling device or by heat treatment.

The high compressive strength of sulphur concrete is produced within few minutes to a few days. The sulphur concrete develops a strength of 28 MPa in about 30 minutes and achieves a maximum strength, which may be 56–63 MPa depending upon the type of aggregate used, within a maximum of 4 of 5 days. This behaviour contrasts with the slower increase in compressive strength of Portland cement concrete. It has demonstrated a high fatigue resistance compared with Portland cement concrete. It is impervious to acids, sulphates and other dissolved electrolytes over long periods of time regardless of the strength of the attacking agent; it is also impervious to water. Sulphur concrete can be made with a great variety of aggregate, such as crushed quartz, crushed lime stone, various kinds of sand and such residues as cinders. It is not attacked by bacterial action or by atmospheric oxidation. Its density is slightly greater than that of Portland cement concrete, being approximately 2400 to 2450 kg per cu.m, when lime stone aggregate is used.

Sulphur concrete blocks can be used in construction of walls for residential purposes with precise dimensions, no requirement of water, no need of mortar during construction, and no requirement for paint, plastering or finishing.

Sulphur concrete blocks may be prepared by heating 23 parts of sulphur and 2 parts of dicyclopentadiene in a concrete mixer to a temperature of 130–150°C. Seventy five parts of aggregate are then added and the heating and mixing are continued in order to produce a thoroughly reacted products. The material is then poured into oil coated steel mould of dimensions 40 cm × 20 cm × 20 cm. The moulds are designed to produce a tongue and groove joining surface and smooth exposed surface.

7.5 SURKHI (BURNT CLAY)

It is one of the artificial puzzolana obtained by burning clay soils as specified pre-determined temperatures. In doing so the water molecules are driven off and a quasi-amorphous material, reactive with lime, is obtained. However, in practice, surkhi is manufactured by grinding the brick bats in the grinding mills until an impalpable powder is obtained. This puzzolana is called surkhi in India, semen merah in Indonesia and homra in Egypt.

The best surkhi is obtained by burning clay in field or in a kiln and is classed as grade I whereas that obtained by grinding brick bats is classed as grade II. Soils containing little amount of clay are placed alternately with fuel layers in pit and are fired. The residue obtained from firing is friable and needs no pulverization. In the kiln method of production 50-100 mm clay lumps along with coal fuel are placed in shaft kiln (Fig. 7.1). Coal is fired and the clay is calcined at 700°C. The temperature is regulated by the air blower and feed input.

The clay or clay products used in the manufacture of surkhi must not contain a high percentage of silica. A good surkhi should be clean, cherry red in colour, and free from any foreign matter. It should conform generally to the following chemical requirements on an oven dry basis (at 105°C).

Silica	Not less than 40 per cent
Silica + Alumina + Iron oxide	Not less than 70 per cent
Calcium oxide	Not less than 10 per cent
Magnesium oxide	Not more than 3 per cent
Sulphuric anhydride	Not more than 3 per cent
Soda and potash	Not more than 3 per cent
Water-soluble alkali	Not more than 0.1 per cent
Water-soluble material	Not more than 1 per cent
Loss on ignition	Not more than 10 per cent

The physical requirements are same as that given in Table 9.1 except that for soundness.

Surkhi is extensively used in making mortar and concrete as an adulterant for economy. But its chief function is to impart strength and hydraulic properties to mortar. When mixed with cement to react with lime liberated during the setting and hardening of cement it makes dense, compact and impermeable concrete.

7.6 FLY ASH

Fly ash or pulverized fuel ash (PFA) is the residue from the combustion of

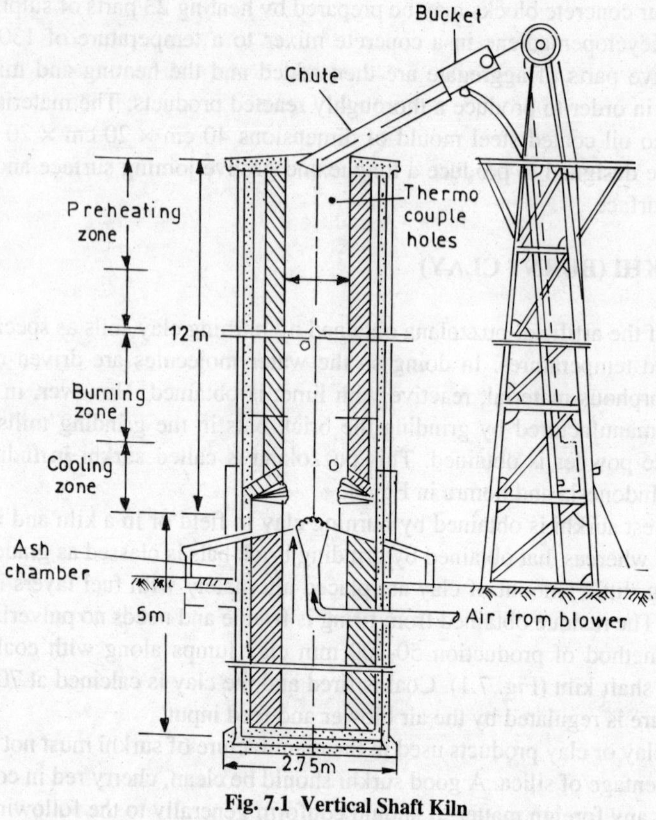

Fig. 7.1 Vertical Shaft Kiln

pulverized coal collected by mechanical or electrostatic separators from the flue gases or power plants. It constitutes about 75 per cent of the total ash produced. The properties and composition of fly ash vary widely, not only between different plants but from hour to hour in the same plant. Its composition depends on type of fuel burnt and on the variation of load on the boiler. Fly ash obtained from cyclone separators is comparatively coarse and contains a large proportion of unburnt fuel, whereas that obtained from electrostatic precipitators is relatively fine having a specific surface of about 3500 cm^2/gm or it may be as high as 5000 cm^2/g. Normally it is rather finer than Portland cement. Fly ash consists generally of spherical particles, some of which may be like glass and hollow and of irregularly shaped particles of unburnt fuel or carbon. It may vary in colour from light grey to dark grey or even brown. The principal constituents are normally:

Silicon dioxide + aluminium oxide + iron oxide	$S_1O_2+Al_2O_3+Fe_2O_3$	70%
Silicon dioxide	S_1O_2	35%
Aluminium oxide	Al_2O_3	15–30%
Carbon (in the form of unburnt fuel)		up to 30%
Alkalis	Na_2O	1.5%
Magnesium oxide (maximum)	MgO	5%
Sulphur trioxide (maximum)	SO_3	2.75%
Loss on ignition by mass (maximum)		12%

Carbon content is important consideration for use with cement; it should be as low as possible. The fineness of fly ash should be as high as possible. The silica contained in fly ash combines slowly over a very long period with the lime liberated during the hydration of the cement. For this reason silica should be present in finely divided state. Curing at a temperature of 38°C has been found to greatly accelerate its contribution to the strength of concrete. Curing at high pressure and temperature in autoclave is known to promote the reaction between the lime of cement and the silica in the fly ash. This reaction should tend to prevent the release of free lime to reduce efflorescence.

Specifications

Fly ash consists of spherical glassy particles ranging from 1 to 150 μm, most of which passes through a 45μm sieve. Physical requirements are given in Table 7.1.

Table 7.1 Physical Requirements

S.No.	*Characteristic*	*Requirements*	
		Grade I	*Grade II*
1.	Fineness-Specific surface in m^2/kg, minimum	320	250
2.	Lime reactivity-Average compressive strength N/mm^2, minimum	4.0	3.0
3.	Compressive strength for Portland-puzzolana cement	not less than 80 per cent of the corresponding plain cement mortar cubes	
4.	Drying shrinkage, maximum	0.15	0.10
5.	Soundness-Expansion of specimen, per cent, maximum	0.80	0.80

Effects of Fly ash on Cement Concrete

On Amount of Mixing Water: The use of fly ash in limited amounts as a replacement for cement or as an addition to cement requires a little more water for the same slump because of fineness of the fly ash. It is generally agreed that the use of fly ash, particularly as an admixture rather than as a replacement of cement, reduces segregation and bleeding. If the sand is coarse the addition of fly ash should have beneficial results; for fine sands, its addition may increase the water requirement for a given workability.

On Strength in Compression: Since the puzzolanic action is very slow, an addition of fly ash up to 30 per cent may result in lower strength at 7 and 28 days, but may be about equal at 3 months and increase at ages greater than 3 months provided curing is continued.

On Modulus of Elasticity: It is lower at early ages and higher at later ages.

On Curing Conditions: It is similar to Portland cement concrete.

On Shrinkage of Concrete: Coarser fly ashes and those having a high carbon content are more liable to increase drying shrinkage than the finer fly ashes and those having a low carbon content.

On Permeability: The permeability of concrete is reduced 28 days pulverised fly-ash-concrete may be three times as permeable as ordinary concrete but that after 6 months it may be less than one quarter permeable.

On Resistance of Chemical Attack: Fly ash slightly improves the resistance of concrete to sulphate attack.

On Heat of Hydration: Fly ash reduces the heat of hydration in concrete. A substitution of 30 per cent fly ash may result in a reduction of 50–60% heat of hydration.

On Air Entrainment: The presence of fly ash reduces the amount of air entraining agent.

Setting Time: A 30 per cent substitution of fly ash may result in an increase of initial setting time up to 2 hours.

Applications

The Portland cement and fly ash can be ground or mixed intimately to produce Portland Puzzolana cement. These cements have lower rate of development of strength but ultimate strength is comparable to Portland cement. It has low heat of evolution and is used in the places of mass concrete such as dams and in places of high temperature. Fly ash can also be used as part replacement of cement to increase workability. When mixed with lime and sand it forms excellent material for manufacturing bricks.

7.7 RICE HUSK ASH

The combustion of agricultural residues volatises the organic matter and a silica-rich ash is produced. Of all the agricultural wastes, rice husk yields the largest quantity of ash with about 93 per cent silica which gives its puzzolanic properties.

Burned in the ordinary way rice husks produce a crystalline silica ash but if burned under suitable conditions they give a highly reactive black non-crystalline silica residue having puzzolanic properties. Temperature and duration of combustion are of utmost importance for good quality rice husk ash. The right temperature is 700°C for 2-3 hours.

Rice husk ash when mixed with lime, gives a black cement. It can also be mixed with Portland cement and 28 day strength up to 55 MPa can be obtained. Rice husk ash cement containing not more than 20 per cent of lime are acid-resisting. To improve its reactive properties the rice husk ash should be ground in ball mills for about one hour. It can replace cement in mortars by 30 per cent.

The rice husk mixed with 20–50% hydrated lime is ground in a ball mill to produce ASHMOH, a hydrated binder suitable for masonry, foundation and general concerting. When rice husk is mixed with cement instead of lime, the hydraulic binder is termed as ASHMENT.

Rice husk ash can also be used with line sludge obtained from sugar refineries.

The dried lime sludge is mixed with an equal amount of crushed rice husk and then mixed with water and tennis ball size cakes are prepared and sun-dried. These are fired to produce powder which can be used as a hydraulic binder.

Rice husk ash when mixed with soil (20 per cent), instead of lime sludge, produces excellent binding properties. This binder when used as 30 per cent in mixture with Portland cement gives the properties of Portland puzzolana cement.

7.8 GROUND BLAST FURNANCE SLAG

Iron ore is smelt in blast furnace. By melting the iron ore at 1400–1600°C pig iron is produced and the floating impurities, containing mainly lime, silica and alumina form the blast furnace slag. By slow cooling of the slag crystalline material is produced, which is used as aggregate, Glassy pallets (> 4 mm) produced on rapid cooling form excellent light weight aggregate and granules (> 4 mm) on grinding possess hydraulic properties.

The ground blast furnace slag exhibits hydraulic action in the presence of calcium hydroxide liberated by Portland cement when hydrated. The ground slag is blended with Portland cement to produce Portland Blast Furnance cement, the proportion of the former not exceeding 65 per cent. The early strength of the cement so produced might be less but the ultimate strength is comparable. Because of low heat of hydration the ground blast furnace slag cement finds its application in mass concreting. The other advantages of addition of blast furnace slag to cement are improved workability, resistance to chemical attack and the protection provided to reinforcement making it suitable for reinforced concrete and prestressed concrete.

7.9 SAW DUST

The use of saw dust for making the light weight concrete is investigated by Dawe J.L. for making concrete masonry units. In such concrete, coarse aggregates are partially replaced with saw dust. Because of its light weight and low thermal conductivity, it offers several advantages over traditional block construction. Compressive strengths of saw dust cement and concrete are given in Table 7.2. CBRI, Roorkee has developed saw dust concrete for making door and window frames which is found to be economical.

Table 7.2 Saw dust cement and concrete 28 days compressive strength

Material	*Proportion*	*Strength (MPa)*
Saw dust cement	Cement:Saw dust	31
	1:1	8.5
	1:2	5.0
	1:3	
Saw dust concrete	Cement:Sand:Sawdust	7.1
	1:1:2.5	6.8
	1:1:3.0	4.1
	1:1:3.5	

7.10 GYPSUM

Mineral gypsum is a crystalline form of calcium sulphate which occurs naturally as a soft rock in many parts of the world. When calcined by heating to approximately 165°C gives hemi-hydrate plaster, also known as plaster of Paris. Compressive strength of gypsum plaster blocks, even when tested after soaking in water for 24 hours, is more than that required for low cost building schemes. In arid climates gypsum plaster can be used externally without excessive deterioration. In practice a small amount of repair and maintenance is required to maintain the material in good condition in the long term. Durability of gypsum plaster may be improved by addition of 5 to 10% of lime. Products made from gypsum have good fire resistance because of the inherent water of crystallisation in their structure which has to be evaporated off first before the product is weakened and damage.

Glass reinforced gypsum composite board combines the properties of fibre glass and gypsum. The basic raw material for the process are plaster of Paris (Gypsum) and glass fibre. Central Glass and Ceramic Research Institute, Calcutta has developed this board. These boards can be manufactured by spray suction technique to any desired dimensions or thickness to suit the market requirement. They are available in panels of various sizes. The thickness of the panel is usually 4 mm or above. Glass reinforced gypsum boards can be used to make door panels, false ceiling, partitions, wall panneling etc.

It is waterproof, fire proof and insect resistant. The pancels have a better dimensional stability under adverse weather conditions. This board has better screw holding capacity and can be cut easily to any size.

7.11 RED MUD

Red mud, is a waste material obtained from aluminium plants causing environmental pollution and disposal problem.
Chemical composition of red mud is as given below:

Constituents	% (by weight)
Fe_2O_3	35
Al_2O_3	19
SiO_2	8
Na_2O	5
CaO	3
Loss on ignition	10
Others	20

The specific gravity of red mud is 3.15 which approximately equal to the cement. Some of the products like aggregates, bricks and concrete can find its large scale utilisation. Red mud with carbonaceous shale forms excellent bricks as a replacement of conventional ones. These bricks are vacuum extruded, dried and fired at

1150°C having compressive strength of 250 to 260 Kg/cm^2 are free from shrinkage cracks and efflorescence. A recent development of red mud jute fibre polymer composite has been found to be suitable for shutters as a substitute of timber.

7.12 BAMBOO

Bamboo is a naturally occuring material and is a very useful building material if used properly under controlled conditions. It has high tensile strength of the order of and can be used as reinforcing material as a substitute of steel in concrete structures. The only drawback is its durability and life span. The ingress of water in the bamboo reinforced concrete elements is of concern; bamboo gets rot. To overcome this drawback, bamboo is painted with bitumen and sand sprinkled (to increase bond with concrete).

Some of the other applications of bamboo are in making, roofing items, walls, floors, and piles are as follows.

In bamboo-growing regions the climate is generally warm and humid, requiring construction materials of low thermal storage capacity and design which permit good cross. ventilation. Bamboo constructions ideally satisfy these requirements, which is why they have long been used for housing in warm humid climates. Plain bamboo walls cannot be made air-tight or waterproof, so that cross-ventilation is a natural feature, providing indoor comfort and removing moisture. The flexibility and high tensile strength makes bamboo walls highly earthquake resistant, and in case of collapse, their low weight causes less damage to people and property, reconstruction is relatively quick and easy.

Rammed earth walls generally have low earthquake resistance, but with bamboo reinforcement this problem can be overcome. Use of bamboo as a reinforcement in the rammed earth wall increases the tensile strength of the wall. The walls comprise a number of independent vertical sections, which are held together at the top by a bamboo ring beam. During an earthquake each section can respond to seismic forces individually, thus avoiding differential stresses within the whole wall, which can subsequently collapse. The roof rests on timber supports, which are structurally separated from the walls, so that any wall movement will not cause to roof collapse.

Shingles are used to cover pitched roofs (and quite often walls) on a supporting grid of bamboo or wooden laths. The appearance is typically a fishscale structure, but some types of bamboo shingles rather resemble spanish tiles. Appropriate lengths of bamboo culms or timber logs are cut and the shingles are split off vertically, whereby bamboo culms are split into quarter or half sections and wood shingles are flat tiles cut with a special knife and hammer. For fixing bamboo shingles, pre-drilled holes are needed for nailing or tying with a tough string. Quarter-cut bamboo shingles can also be made with splints which are hooked into the lathing. Timber shingles are nailed onto the battens, whereby the curvature of the shingles after drying must be taken into consideration. The minimum roof pitch for shingles is 45°.

Bamboo floors, are common in bamboo structures and to some extent in timber framed houses. The simplest method is to lay bamboo culms in parallel tied to the supporting framework. However, this gives a very uneven surface and can be uncomfortable to sit or stand on for long. More even surfaces are achieved by using bamboo board on by cutting bamboo strips, which are woven into boards. Since bamboo components cannot be joined together without leaving gaps, the floors are well ventilated, improving the indoor climate and prevent moisture accumulation.

Split-bamboo piles

Split-bamboo piles, improve the bearing capacity of soft compressible soils and reduce settlements of the buildings. The hollow bamboo culms are filled up with loosely wound coconut coir and jute thread wrapped in jute fabric, holes in the culm permit the water in the soil to trickle in, thus drying out the soil and improving its load-bearing capacity.

EXERCISE

1. Describe briefly alternate materials. Why are these needed?
2. Write short notes on:
   (a) Fibre reinforce concrete
   (b) Ferrocement
   (c) Sulphur impregnated concrete
3. What is puzzolan? Describe briefly puzzolana cement.
4. Write notes on the following:
   (a) Fly ash (b) Surbhi
   (c) Rice husk ash (d) Ground blast furnace slag
5. Briefly describe the followings:
   (a) Saw dust (b) Gypsum
   (c) Red mud
6. Describe briefly the applications of bamboo in construction industry. What precautions should be exercised while using bamboo as reinforcement?

CHAPTER 8

Composite Materials

8.1 INTRODUCTION

Composites are combinations of two materials in which one of the materials called the **reinforcing phase**, is in the form of fibers, sheets or particles and is embedded in the other material called the **matrix phase**. The reinforcing material and the matrix material can be metal, ceramic, or polymer. Typically, reinforcing materials are strong with low densities while the matrix is usually a ductile, or tough, material. If the composite is designed and fabricated correctly, it combines the strength of the reinforcement with the toughness of the matrix to achieve a combination of desirable properties not available in any single conventional material. For example, the design situation may demand, both the strength and toughness, which have inverse relation, and are not exhibited by one material. When no single conventional material is able to satisfy the competing design specifications for a given application, the solution may be a composite material.

Because of the variety of available reinforcement and matrix materials, as well as the ability to combine them in a wide range of volume fractions, composites can be produced with a broad range of elastic modulus, strength, and toughness combinations. The flexibility of tailoring properties to a specific need is one of the most important attributes of composites. Another advantage over conventional material is that composites can be designed to exhibit specific properties in specific directions (i.e., their anisotropy can be beneficial). The only drawback of the composites is that these are often more expensive than the conventional materials.

The majority of modern composites selected for its mechanical properties are designed as structural materials. Some of the other classes of composites are also there and are used for their electrical, thermal or magnetic properties. Superconducting wires are the example of latter. However, the scope of the book permits discussion regarding composites designed for structural applications only.

Application of composite materials in civil infrastructure system is not only technologically sound but also economically justifiable. The economic impact to evolve from the use of composite infrastructure is:

1. Reduction in installation and construction cost
2. Reduction in maintenance cost
3. Reduction in insurance liability
4. Market expansion for the raw
5. Creation of the fabricators for composite sections materials suppliers
6. Development and sale of the service equipment for inspection of composite structures

8.2 COMPOSITION

Most fiber-reinforced composites consist of fibers, matrix and interfaces and are collectively responsible for the unusual characteristics of composites.

Fiber

The first man made polymer reinforcing fibers were nylon and polyester. The most frequently used fibers are of glass, boron, carbon, ceramic, metal etc.

For any class of fiber reinforced composites the ones with the highest specific strength and modulus values generally have all their fibers aligned in one direction, (i.e., unidirectional fiber reinforced composites). If the loading direction is known (and always the same) then the composite can be designed and fabricated so that the strong and stiff (fiber) direction coincides with the loading direction. In this case the ''weakness'' in the perpendicular directions is not a problem. If, however, the loading direction is not known, or varies with time, then a nearly isotropic composite is required. The fibers must be arranged so that a portion is oriented in several directions within the material. This type of fiber architecture yields properties that are between those of the ''strong'' and ''weak'' directions in aligned fiber composites.

In most unidirectional fiber reinforced composites the fibers do not run continuously from one end of the component to the other. If the fiber length is significantly less than the component dimensions, then the material is known as a discontinuous fiber reinforced composite. When a discontinuous fiber with a high elastic modulus is embedded in a low modulus material, and the resulting composite is loaded in the fiber direction, the fibers carry a higher load than does the matrix,. This is the principle of fiber strengthening. There are, however, several conditions that must be met in order to achieve the maximum extent of strengthening of the composite from the fibers. One requirement is that the fiber length exceed some critical (minimum) value.

The minimum or critical fiber length, l_c, required for effective strengthening is a function of several variables including the diameter of the fiber d, the ultimate tensile strength of the fibers, σ_{fu}, and the matrix shear yield strength, τ_{my}. The second factor sets the limit on the load carried by the fibers, and the third factor describes the ability of the matrix to transmit the load from one fiber to the next. The relationship among these variables is found using a force balance. The force carried by the fibers is equal to the normal stress in the fibers multiplied by their cross-sectional area. This force is transferred to the fibers via a shear stress acting on the fiber surface. Mathematically, the force balance is:

$$\pi d(l_c/2)\tau_{my} = (\pi d^2/4)\sigma_{fu}$$

or equivalently,

$$l_c/d = \sigma_{fu}/(2\tau_{my})$$

The quantity (l_c/d) is known as the **critical aspect ratio**. Typically, the critical aspect ratio ranges from 20 to 150 for most fiber and matrix materials. Since a typical fiber diameter is between 10 and 30 mm, critical fiber lengths are on the order of 0.2–4.5 mm. In most cases, the fiber length l, is much larger than l_c.

The most common geometrical shape for the reinforcing phase in a high-performance structural composite is a fiber. The reason for this is that the strength of a brittle material is inversely related to the square root of its maximum flaw size and the chance of having a large flaw in a given length of fiber decreases as the cross sectional area decreases. When this observation is combined with the need for significant surface area for load transfer from the matrix to the reinforcing phase, the advantages of long slender fibres becomes apparent. In fact, the strongest materials on earth are fibers. A small diameter is also important in providing a fiber with much needed flexibility; the fiber bending stiffness is proportional to the fourth power of diameter. As the diameter of a fiber decreases, its elastic modulus, strength, and flexibility all increases.

Ideal fiber material for strengthing and stiffening a matrix requires the following attributes: low density, high tensile strength, high modulus of elasticity, high flexibility, and the ability to form a strong interface with the matrix.

Matrix

Like fibers, matrix materials can be polymers, ceramics, or metals. Carbon is also used as a matrix material with carbon fibers in a class of composites known as carbon-carbon composites. The primary purpose of the matrix materials is to provide lateral support to the fibers and transfer loads. They also are a source of toughness in the composite, since the majority of fiber materials are brittle. Cracks that have propagated through a brittle fiber are stopped when their tips encounter tough matrix materials. An exception to ductile matrix materials is ceramic matrix materials, which are inherently brittle. Composites using ceramic matrices, such as reinforced concrete, are used in compressive load applications, or the brittle behavior is countered by carefully tailoring the interface properties.

Interfaces

Interfaces play and important role in determining the properties of composites, consider 1 cm^3 of a unidirectional composite made from 25 μm diameter, continuous fibers embedded in a matrix. If the fibers are arranged in a square array and are on average 50 μm apart, the volume fraction of fibers is approximately 20 per cent. The total fiber-matrix interface area is approximately 314 cm^2 compared to the 6 cm^2 external surface area. The typical fiber volume fraction in composites is two to three times the above amount, and the interfacial area increases proportionally. The large interfacial area significantly affects the properties of composites, in particular the crucial properties of toughness and ductility.

The fiber-matrix interface bond strength can be characterized by the shear stress required to cause sliding between the fiber and the matrix. The interface bonding can be mechanical, chemical, or both.

8.3 TYPES

Unidirectional Composites

When all the fibers are oriented in a single direction, the composite is called unidirectional. Their properties in the fiber direction are vastly superior to those in the transverse plane. These are not suitable for applications involving multidirectional loading.

Laminate Composites

These consist of unidirectional composites stacked in an arrangement so that fibers are oriented along 0°, 90° and ± 45° directions. The thin sheets, called the laminae, can be stacked together in a regular arrangement to make up the laminate composite. If the fibers are stacked in the 0° and 90° directions only, the strength of the composite is high along these directions, but such composites have poor shear resistance. To obtain good shear resistance, the fibers must also be stacked along the ±45° orientations. Such laminates are strong in all directions within the plane containing the fibers but are weak in the direction normal to the fiber planes.

Aggregate Composite

When the reinforcing phase has roughly equal dimensions, in all directions the material is known as aggregate composite.

An example of such a composite is concrete, which is a mixture of fine aggregate in the form of sand and coarse aggregate in the form of gravel in a matrix of cement (calcium aluminosilicate). By mixing large and small particles the volume fraction of reinforcement can be increased because the small particles can fit into the cavities among the large particles. Typical total aggregate (fine plus coarse) volume fractions in concrete are 0.60 to 0.75. The aggregate strengthens the composite in two ways. First, the inherent compressive strength of the aggregate is simply higher than that of the matrix. Second, the high volume fraction of irregularly shaped aggregate particles provides strength in compression through mechanical interlocking. As an added bonus, the aggregate is less expensive than the cement, so high aggregate volume fractions reduce the composite cost.

Concrete is one of the most common structural materials. Its use, however, is somewhat limited, since its tensile strength is much lower than its compressive strength. The composite's tensile strength can be increased by "reinforcing" the aggregate composite with steel rods. These reinforcing rods are placed in the composites in the locations that will bear the largest tensile loads. The interfacial strength between the "fibers" and the matrix is increased by having ridges on the steel rods, which form a mechanical bond with the surrounding matrix. The addition of the steel rods transforms the composite from nearly isotropic to highly anisotropic.

Particulate Composite

It is a particular type of aggregate composite. The examples include SiC or Al_2O_3 particles embedded in an aluminum alloy matrix and rubber particles in epoxy. In these composites the volume fraction of the reinforcing phase is much lower than that in cement. Typically, reinforcing phase volume fractions are about 0.15. Since the aspect ratio of the reinforcing particles is approximately one, particulate composites are essentially isotropic.

On the basis of matrix following are the types of composites.

1. Metal-matrix composites
2. Polymer-matrix composites
3. Ceramic-matrix composites
4. Carbon-carbon composites

8.4 PROPERTIES

Testing of materials is both expensive and time-consuming. It also causes delays in the time required for bringing a new material into the marketplace. Therefore, materials scientists and engineers have always spent a good part of their efforts directed toward predicting material properties. This section presents some elementary approaches for estimating properties of composite materials. For simplicity, we limit our discussion to properties of unidirectional fiber reinforced composites.

Important factors in any model for predicting composite properties are the volume fraction of fibers (V_f), the volume fraction of the matrix (V_m), and the fiber diameter.

Elastic Modulus

If each component behaves in a linear elastic manner, the composite will also behave in a linear elastic manner, the Young's modulus, E_c, having a value intermediate between those of the separate composites. Consider a composite that contains a volume fraction f of a material with Young's modulus E_f in a matrix with Young's modulus E_m. Bounds on the value of E_c can be calculated for two extreme cases.

With continuous fibres aligned in the direction of stress (Fig. 8.1 (a)) and considering unit area of cross-section of the composite, the mean stress σ_c is given by

$$\sigma_c = f\sigma_f + (1 - f)\,\sigma_m \tag{1}$$

where σ_f and σ_m are the stresses in the fibre and in the matrix, respectively. We can assume that the strains in the fibres and in the matrix will be equal so that if ε is the strain

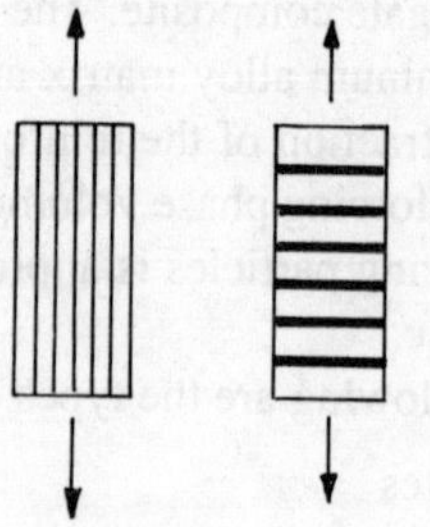

Fig. 8.1 (a) Aligned fibre composite, (b) Lamellar composite.

$$E_c\varepsilon = fE_f\varepsilon + (1-f)E_m\varepsilon$$

or
$$E_c = fE_f + (1-f)E_m$$

If all the reinforcing material is in flat sheets perpendicular to the direction of stress (Fig. 8.1(b)) then the stress σ in the two materials will be equal but the strains ε_f and ε_m will differ. The total strain ε is given by

$$\varepsilon = f\varepsilon_f + (1-f)\varepsilon_m$$

or
$$\frac{\sigma}{E_c} = f\frac{\sigma}{E_f} + (1-f)\frac{\sigma}{E_m}$$

or
$$E_c = \frac{E_f E_m}{fE_m + (1-f)E_f}$$

Strength

Stress-strain for a composite material is shown in Fig. 8.2. Fibres are assumed to be brittle, the matrix ductile, and the failure strain of the fibre greater than the yield strain of the matrix.

In a more general case of a brittle fibre in a ductile matrix which shows some strain hardening (Fig. 8.3(a)), let σ_{fu} be the breaking stress of the fibres at a strain σ_f, σ_m be the stress in the matrix at the strain at which the fibres fail and σ_{mu} be the tensile strength of the matrix. When the composite is strained to ε_f, the stress will be given by the law of mixtures as

$$\sigma_c = f\sigma_{fu} + (1-f)\sigma_m \tag{2}$$

and is represented by line AB on Fig. 8.3 (b). When the fibres have all broken and the load is taken entirely by the matrix, then at the point of matrix failure the stress calculated on the total cross-sectional area of the composite will be

$$\sigma_c = (1-f)\sigma_{mu}$$

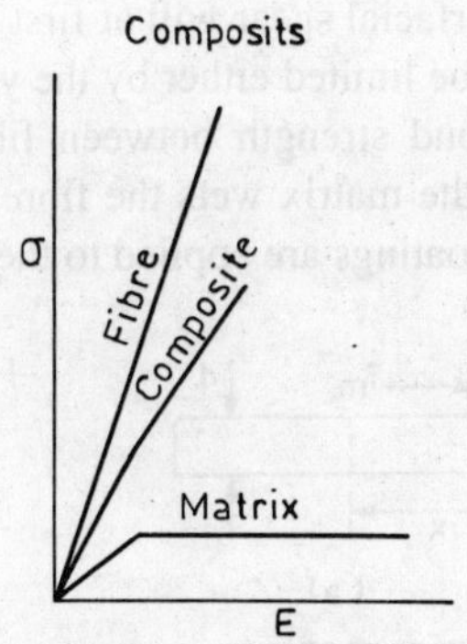

Fig. 8.2 Stress-strain curves of fibre, matrix and 50% volume fraction composite

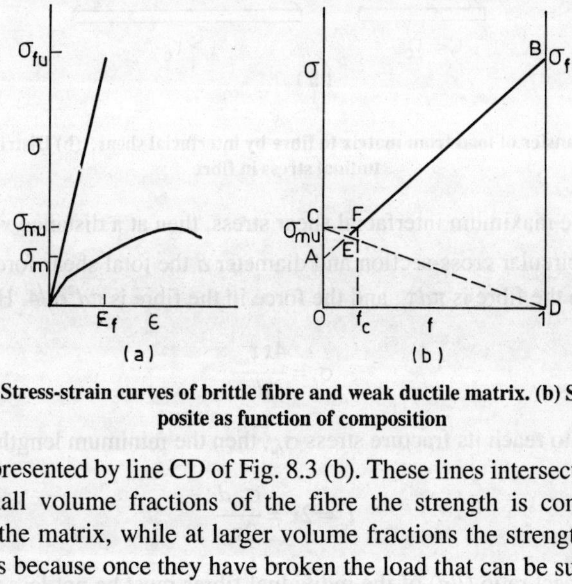

Fig. 8.3 (a) Stress-strain curves of brittle fibre and weak ductile matrix. (b) Strength of composite as function of composition

which is represented by line CD of Fig. 8.3 (b). These lines intersect at E and show that for small volume fractions of the fibre the strength is controlled by the strength of the matrix, while at larger volume fractions the strength is controlled by the fibres because once they have broken the load that can be supported by the matrix is less. The strength of the composite thus exceeds the strength of the matrix only if the volume fraction of fibres exceeds a critical value f_c given by point *F*, i.e.,

$$\sigma_c = f_c \sigma_{fu} + (1 - f_c)\, \sigma_m = \sigma_{mu}$$

$$f_c = \frac{\sigma_{mu} - \sigma_m}{\sigma_{fu} - \sigma_m}$$

There is also obviously a maximum volume fraction that is possible if all interstices between fibres are to be filled and if all the fibres are to be separated by matrix material. In the case of fibres with cylindrical cross-sections and all of the same diameter, the maximum value of *f* is just over 0.9.

If the fibres are discontinuous, the stress will not be uniform along the length of each fibre. The tensile stress builds up from the fibre ends, the force being transmitted to the fibre by the interfacial shear from the matrix in the manner shown in Fig. 8.4(a). The interfacial shear will at first increase as the strain in the composite increases, but will be limited either by the yield strength in shear of the matrix material or by the bond strength between fibre and matrix. This bond strength will be high only if the matrix wets the fibre during manufacture. If the combination is non-wetting, coatings are applied to the fibres to promote wetting.

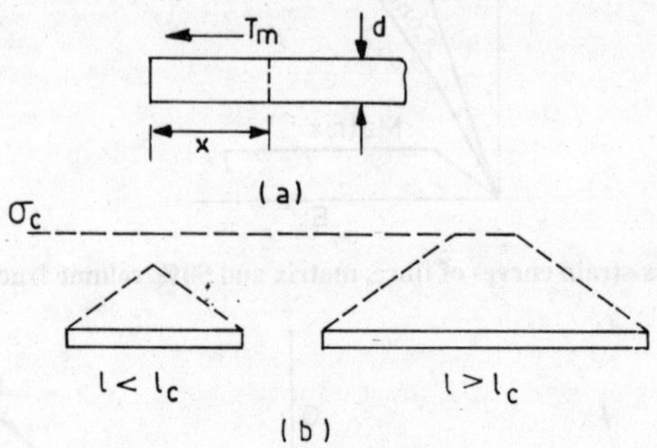

Fig. 8.4 (a) Transfer of load from matrix to fibre by interfacial shear. (b) Distribution of longitudinal stress in fibre

If τ_m is the maximum interfacial shear stress, then at a distance x from the end of a fibre of circular cross-section and diameter d the total shear force that can be transmitted to the fibre is $\pi d\tau_m$ and the force in the fibre is $\pi d^2\sigma_f/4$. Hence

$$\sigma_f = \frac{4x\tau_m}{d}$$

If the fibre is to reach its fracture stress σ_{fu}, then the minimum length of fibre is

$$l_c = 2x = \frac{\sigma_{fu} d}{2\tau_m}$$

Hence the aspect ratio (l/d) of the individual fibres must be not less than the critical value $\sigma_{fu}/2\tau_m$ if they are to develop their full strength. Otherwise the matrix will fail without breaking the fibres.

The average stress in a fibre of length l ($> l_c$) when the stress in mid-length is equal to the breaking stress is given by

$$\sigma_f = \sigma_{fu}\left(1 - \frac{l_c}{2l}\right)$$

and σ_f can be substituted for σ_{fu} in Eqn. 2 to give the strength of the composite. We see that discontinuous fibres produce less strengthening than continuous ones.

For fibres with a brittle matrix, the fracture strain of which is less than the failure strain of the fibres, the matrix will crack and transmit all the load to the fibres. These can stretch, with failure of the interfacial bond until they also reach their failure strain and may break at some distance from the plane of the matrix crack, so exhibiting pull-out. Where fibres run at right-angles to the direction of stretching, there may be debonding which initiates matrix cracking.

In compression, the strength of the composite depends on the ability of the matrix to constrain the fibres from buckling. Failure is usually preceded by delamination of individual fibres in unidirectional composites or sheets of fibres in bi-directional composites.

Specific Stiffness and Specific Strength

Composite materials are frequently of advantage where mass or weight saving is a prime requirement and their usefulness can be assessed in terms of their specific stiffnesses and specific strengths, that is, the ratios of Young's modulus and tensile strength respectively to the density.

Toughness

The fracture of a fibre composite may involve plastic deformation of the matrix, fracture of the matrix, plastic deformation of the fibres, pull-out of the fibres, and fracture of fibres, all of which contribute to the energy required. Fracture in composites usually begins with cracking of the most brittle phase. The metal-matrix and polymer-matrix composites this implies cracking in the brittle fibers; in ceramic-matrix composites this usually means cracking in the matrix. The manner in which this initial fracture progresses determines the toughness of the composite. To illustrate this point, consider a system of brittle fibers in a ductile matrix. When fracture occurs in an isolated fiber at any point along its length, the stresses carried by the fiber in the vicinity of the crack must be transferred to the surrounding matrix and the other fibers. If the surrounding matrix and fibers are able to withstand the stresses, the fracture will stabilize at that location.

In a composite of brittle fibres in a brittle matrix in which both matrix and fibres break along the same plane, the toughness may be assumed to be given by the law of mixtures

$$G_c = fG_{cf} + (1-f)G_{cm}$$

where G_{cf} and G_{cm} are the toughness of fibre and matrix, respectively.

Composites of brittle fibres in a tough matrix in which the fibres break ahead of an advancing crack will exhibit heavy deformation of the remaining matrix which bridges the crack. These bridged regions will neck down to fracture in a ductile manner. Most of the work done will be associated with this plastic deformation. Hence, a smaller volume fraction of fibre would give a larger G_c and higher strength in the presence of a notch, even though it would give a lower un-notched strength.

Ductile fibres in brittle matrices will also give high toughness associated with the necking of the fibres after the matrix fractures. Here increased volume fracture of fibres would increase both toughness and unnotched strength.

In composites with a weak fibre-matrix interface or with low matrix shear strength, the transverse tensile and shear stresses in the vicinity of an advancing crack tip may cause longitudinal cracks to form along the plane of the fibre. This debonding blunts the crack and temporarily stops its further advance. After travelling a small distance, this secondary crack may initiate further cracks perpendicular to the tensile stress at each of its ends, the whole process repeating when these cracks meet other fibres. This process will increase the toughness G_c.

Thermal Expansion

The coefficient of thermal expansion is an important design parameter for all structural material subjected to even modest variations in temperature during service. As a first estimate, the expansion coefficient of the composite, α_c, can be obtained by the rule of mixtures:

$$\alpha_c = V_m\alpha_m + V_f\alpha_f \tag{3}$$

where α_m and α_f are the expansion coefficients of the matrix and fiber materials. In most common composites $\alpha_m \gg \alpha_f$, so in a uniaxial composite, the fibers constrain the expansion of the matrix in the longitudinal direction. This causes the composite to expand less in the longitudinal direction and more in the transverse direction than predicted by Equation 3. More accurate expressions for estimating the expansion coefficients in the longitudinal and transverse directions have been derived, but are outside the scope of this introductory chapter.

8.5 USES AND APPLICATIONS

Many composites occur naturally. For example, wood which essentially consists of cellulose fibers in a matrix of lignin, and bones, which are composites of soft protein collagen and hard minerals called apatites. However, the majority of advanced composite materials with unique properties are man-made.

Straw-reinforced mud bricks for construction of huts, and Laminated bows made from wood, animal tendons and silk are examples of composites that were in use as early as several centuries BC.

The most widely used composite materials in construction industry are concrete, reinforced cement concrete. Some of the other special concretes such as ferrocement and fiber-reinforced concrete also are composites.

Composite materials are known to have more environmental resistance that traditional civil engineering materials such as steel, concrete, masonry, and plaster. Degradation in strength and stiffness for steel structures due to the corrosion problem requires frequent inspection, maintenance, and repair. Similarly, stress cracking due to the warm/cold weathering limits the service life of concrete structures. Use of composites in structures minimizes property and life loss induced by earthquake.

Currently, composite materials are being used to retrofit and/or reinforce existing infrastructures. For example, composite materials have been used to provide an external reinforcement to concrete columns to increase their earthquake resistance. Similarly, flat composite laminates have been bonded to the exterior surface of reinforced concrete deck to increase its bending stiffness. Furthermore, throughout the world, several pedestrian bridges have been built successfully using low modulus pultruded composite sections. In our view, composite materials are suitable for construction of infrastructures such as highway bridges, power transmission towers, office/residential buildings, retaining walls, etc.

Based on the fiber type in the composite, the material can be selected for specific uses. Glass fibers are used to strengthen polymeric materials in composites. Glass fiber reinforced plastics and glass fibre reinforced cement concrete are the excellent composite materials. Alumina-nickel composite is used where high temperature as stipulated. Boron fibers are used for stiffening aluminium composites. Carbon fibers are used with epoxy resins or metals such as aluminium as matrix for compressor blades in gas turbines, high strength to weight ratio is the requirement. Carbon fibers in very small quantity in concrete forms *Smart Concrete*, a composite of the era. Ceramic fibers such as fibers of Al_2O_3 and SiC, are becoming very useful. A variety of metals e.g., steel can be used to draw high strength wires which serve as metal fibers.

Based on the type of matrix the application of composites in various fields is as follows.

Metal-matrix composite

The applications include piston heads for automotive engines, connecting rods in engine crankshafts, gas turbine blades, radar domes, and electronic packages.

Polymer-matrix composite

Aerospace industry is the one where the materials with high specific strength and stiffness are required and composites are the ideal materials. These are also used for I-section beams, various automative parts and sports goods such as tennis rackets, golf sticks, fishing poles, boats etc.

Ceramic-matrix composite

The most common example is reinforced cement concrete. Carbon-fiber glass is also gaining popularity.

Carbon-carbon composite

Typical applications include brake components, heat shields, and rocket nozzles.

EXERCISE

1. What are composite materials? Why are these required?
2. Describe the composition of composite material. Give the functions of the ingredients of composites.
3. (a) Describe briefly the importance of interfaces in composites.
   (b) What are aggregate composite materials. Give examples.
4. What are the various types of composites? Describe them briefly.
5. Describe briefly the properties of composite materials.
6. Discuss briefly the importance, uses and applications of composite materials in construction industry.

CHAPTER 9
Smart Materials

9.1 INTRODUCTION

The term smart materials, sometimes also called intelligent materials or active materials or adaptive materials, describes a group of material systems with unique properties. The technological field of smart material is not transparent or clearly structured. It has evolved over the past couple of decades with increasing pace during 1990's to become what it is today.

Generally these materials respond with a change in shape upon application of externally applied driving forces. In addition some material systems do not exhibit a shape change, but rather have other significant properties. The examples are electro- and magneto-theological fluids. These fluids can change viscosity over many orders of magnitude upon application of an external magnetic or electrical field.

Consequently, the term ''smart materials'' is not very well defined and frequently used to describe different systems and systems' behaviors. Although there have been approaches to quantify and classify different levels of smartness or intelligence in systems, from a practical standpoint it is most important to understand that one of the classifications is established and used as a standard in the academic, scientific, or industrial community.

To add to the confusion the terms ''smart devices'', ''smart systems'' or smart structure'' are often careless used. Here one should note that in general the system complexity increases from the unit ''material'' to ''device'' to ''systems'' to ''structures''.

9.2 CLASSIFICATION

To get a better understanding of the active materials field it is appropriate to introduce and approach to classify different smart materials. Ideally the classification should be collectively exhaustive and mutually exclusive. The most common way of structuring is by looking at the input and the output of a material system as illustrated below.

The input or stimulus can be for example a change in temperature or in magnetic field. The material then intrinsically responds with an output, which in turn can be for example a change in length of the material, change in viscosity or change in electrical conductivity.

Active materials can be divided into two groups. One group comprises the ''classical'' active materials as viewed by the academic community and is characterized by the type of response these materials generate. Upon application of a

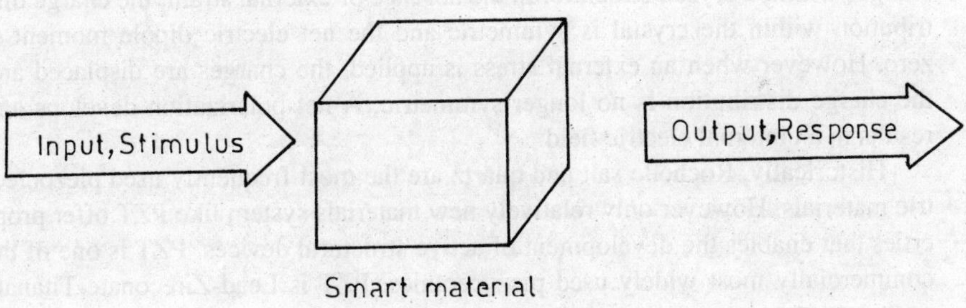

Fig. 9.1 Approach for classification

stimulus the materials respond with a change in shape and/or in length of the material.

Thus input is always transformed into strain, which can be used to introduce motion or dynamics into a system. These materials are the most widely used group for design of smart structures, where active materials are integrated into a mechanical host structure (for example a building or a helicopter rotor blade) with the goal to change the geometrical dimensions of the structures.

The second group consists of materials that respond to stimuli with a change in a key material property, for example electrical conductivity or viscosity. While they are equally important from a scientific point of view, they are less frequently integrated into mechanical structures but rather used to design complex modules, for example clutches, fasteners, valves or various switches. Frequently these materials are used as sensors.

Although materials in this group do not produce strain upon application of an external stimulus they are sometimes also referred to as actuator systems. Examples include the electro- and magnetorheological fluids, which respond with an increase in viscosity upon application of an external electrical or magnetic field.

The applied driving forces for smart materials can be broadly classified as electrical fields-common materials include piezoelectric ceramics and piezoelectric polymers, thermal fields-materials are mainly shape memory alloys (SMAs), and magnetic fields-common materials include magnetostrictive materials and magnetic shape memory alloys.

Piezoelectric Effect

Piezoelectricity is defined as a change in electric polarization with change in applied stress, which is usually referred to as the direct piezoelectric effect.

The converse piezoelectric effect is in analogy the change in strain for a free crystal (or stress for a clamped system) with change in applied field. Thus the converse piezoelectric effect us typically used when a material system is to be used an actuator: Input is a voltage gradient, output a strain. For low fields, there is a linear relationship between strain and electrical field. Reversing the field also reverses the direction of the strain.

The microscopic origin of the piezoelectric effect is the displacement of ionic charges within a crystal structure. In the absence of external strain, the charge distribution within the crystal is symmetric and the net electric dipole moment is zero. However when an external stress is applied, the charges are displaced and the charge distribution is no longer symmetric. A net polarization develops and results in an internal electric field.

Historically, Rochelle salt and quartz are the most frequently used piezoelectric materials. However only relatively new materials system like PZT offer properties that enables the development of active structural devices. PZT is one of the commercially most widely used piezoceranics. PZT is Lead-Zireconate-Titanate or Pb_x (Ti, Zr)$_{1-x}$ O_3.

Thermal Fields

The most common group of materials that respond to a temperature change with a shape change or elongation are Shape Memory Alloys (SMAs).

These alloys, for commercial use mostly Titanium-Nickel alloys, undergo a phase transformation upon temperature change. This austenitic to martensitic transformation can directly lead to a volume and change shape of the sample. Furthermore, if properly designed, the material (e.g. in wire or spring form) can be trained to transform fully reversibly in an exactly reproducible way, thus representing a micro-motor that can be driven by temperature increase and decrease.

Although the transformation is diffusionless and extremely fast the system frequency is mostly limited by the time heat can be extracted from the material. Therefore wires with a low thermal mass are preferably used. Furthermore they have the advantage of small cross-section and thus relatively high electrical conductivity which allows resistive heating with electrical current.

Magnetic Fields

In magnetostrictive materials as this material class is called there is a strong magneto-mechanical coupling. When excited by a longitudinal magnetic field, a magnetostrictive cylinder would experience an elongation.

The origins of the effect are on the atomic and molecular level. Inside the rod, the applied magnetic filed causes the internal magnetic domains to rotate. This movement at the crystal structure provides proportional, positive and repeatable expansions within microseconds.

The most common commercially available giant-magnetostrictive material is called Terfenol-D (Tb).3Dy0.7Fe2). The name is a mixture of the alloying elements Tb and Fe and the letters ''NOL'' which stand for the Naval Ordnance Laboratory where the material was discovered.

Ferromagnetic Shape Memory Alloys (FMSA) on the other hand are a recently discovered class of actuator material. Strain is generated by applying magnetic fields to the crystal structure.

The origin is a characteristic martensitic transformation which is similar to that found in traditional *Shape Memory alloys* (hence the name). However as mentioned before FSMA's are NOT driven by a thermal field.

In ferromagnetic materials, the existence of local magnetic moments is inextricably related to the electronic configurations. The wavefunctions of the electrons in a multi-atom system are related to the interatomic distance in self-consistent manner. They differ significantly from atomic orbitals for short interatomic distances, such as are found in solids.

As the name suggests FSMAs are ferromagnetic alloys which also support the shape memory effect. This means that they undergo the characteristic martensitic transformation upon cooling, and show all features of conventional shape memory alloys. However, the actuation mechanism in these materials is radically different to ordinary shape memory alloys. Rather than being heat driven, they are magnetically driven. This difference allows for increases frequency response (fast actuation). So far only alloys in Ni-Mn-Ga ternary have been proven to show satisfactory performance.

9.3 APPLICATIONS

The broad but strongly interdisciplinary field of materials seeks to apply multifunctional capabilities to existing and new structures. By definition, smart structures and materials are those which can sense external stimuli, via internal sensing and/or actuation, and respond with active control to that stimuli in real or near-real time. Current activities in the field range from the design, fabrication, and test of fully integrated structural systems to enabling research in individual discipline areas i.e., materials, sensing and actuation techniques, control algorithms and architectures, etc.

The typical approach to achieving smart structures synthesizes composite materials and structures from known constituents. The active elements are either embedded in or attached to conventional structural materials. Typical smart structure sensors include fiber optics and piezoelectric ceramics and polymers. Embedded sensors can be used in discrete or distributed locations to provide built-in structural quality assessment capabilities, bot during composite processing ans system operation. In terms of system performance, it is important that the right feature be measured (e.g., strain associated with curvature of deflection) and its signal interpreted with respect to the desired performance, it is important that the right feature be measured (e.g., strain associated with curvature or deflection) and its signal interpreted with respect to the desired performance outcome (e.g., a change in lift). Typical smart structure actuators are shape memory alloys (SMAs), piezoelectric and electrostrictive ceramics, magnetostrictive materials, and electro- and magneto-rheological fluids and elastomers. When embedded with a sensor/signal processing network and an appropriate control system, actuators allow structural performance to be changed (e.g., to compensate for damage) or adapted to meet various operational performance criteria (e.g., to change wing lift). A typical example of application of smart materials is of smart concrete.

Smart concrete was developed Dr. Deborah D.L. Chung from State University of New York at Buffalo. Smart concrete is reinforced by carbon fiber as much as 0.2% to 0.5% of volume to increase its sense ability to strain or stress while still has good mechanical properties.

By adding small amount of short carbon fiber into concrete with a conventional concrete mixer, the electrical resistance of concrete increases in response to strain or stress. As the concrete is deformed or stressed, the contact between the fiber and cement matrix is affected, thereby affecting the volume electrical resistivity of concrete. Strain is detected through measurement of the electrical resistance. So, the smart concrete has the ability to sense tiny structural flaws before they become significant, which could be used in monitoring the internal condition of structures and following an earthquake.

In addition, the presence of the carbon fibers also control the cracking so that the cracks do not propagate catastrophically, as in the case of conventional concrete.

Even by adding carbon fibers, the extra cost of material will increase about 30% this expense is still significant cheaper than attaching or embedding sensors into structures.

Smart concrete is stronger than conventional concrete by the use of carbon fibers. It takes greater force for smart concrete to bend, and it absorbs more energy before fracture.

Monitoring can be a real-time and continuous effort.

Some fo the application of smart concrete are:

The use of smart concrete for the purpose of weighting vehicle on the highway. The highway made by this concrete could be able to determine where each vehicle was, and what its weight and speed were. Vehicles could be weighed while traveling normally on the highway.

The use of smart concrete for real-time vibrations sensing of bridges or other highway structures. It could also be used in buildings to dampen vibrations or reduce earthquake damage.

EXERCISE

1. What are smart materials? How are these different from conventional materials?
2. What is the principle behind the working of smart materials?
3. Briefly describe the classification of smart materials.
4. Write a not on applications of smart materials.
5. What is smart concrete? Why is it used?

UNIT–III

CHAPTER 1

Surveying

1.1 INTRODUCTION

Surveying is the art of making measurements of the relative positions of points on, above, or beneath the surface of earth. When these measurements are plotted to the scale, natural or artificial features may be exhibited in their correct horizontal or vertical relationship. The observations are made for the measurement of horizontal and vertical distances, angles and directions. The points may be located for the purpose of compiling a map or they may be set to define lines and grades for engineering works. To achieve all this it is required to precisely define the coordinates of various points of interest. These may be obtained by making measurements and generating data using the conventional instruments (chain, tape, compass, level, vernier or optical theodolete), electronic instruments (Total Station), or remote sensing and geographical Information system. This chapter deals with the fundamentals of surveying and introduction to Total Station, remote sensing and GIS are presented in the following chapters.

Measurement of Horizontal Distances

Measurement of horizontal distance is probably the most basic operation performed in surveying and perhaps the most difficult as well. The horizontal distance between two points is the distance between the plumb lines through the points. It is important to emphasize that in plane surveying the distances measured should be horizontal. When distances are measured on slopes, sufficient data should be collected so as to compute horizontal projections.

History reveals that the measurement of horizontal distance has taken a variety of forms with marked variations in the accuracies achieved. Various methods such as rope stretching, bamboo, pacing, chaining, optical (tacheometry) and electronic (electronic distance measurement) exist, varying from crude to the highly sophisticated. The cost of making a measurement increases with the desired precision of work. Therefore it is important to know the methods available and their accuracies so as to obtain the required precision with economy.

Rope stretching and bamboo measurements are very crude methods and are obsolete. Pacing can be recommended if an error of five percent is permissible and if the ground is flat. Tacheometry may be employed when the ground is rough, undulating and not suitable for chaining. Electromagnetic distance measurement (E.D.M) method yields results with high precision but is expensive.

SI units are used for the linear measurements. Accordingly, the distances are expressed, as far as possible, in meters.

Measurement of Vertical Distances

The relative position of a point in terms of the vertical distance, above or below another point is designated by its elevation. The *elevation* of a point may thus be defined as its vertical distance above or below a reference surface (datum) having zero elevation. Therefore, elevation of a point may be considered as its vertical coordinate. It is treated as positive if the point is above and as negative if the point is below the datum.

Usually sea level is considered to be the standard datum, but sometimes an arbitrary assumed surface is taken as the reference. The value for mean sea level (m.s.l.) as datum is obtained by averaging the elevations of high and low tides, at several points, for a long period of time say about 19 years. However, any surface may be assumed as datum when relative elevation over a limited area is to be established. This is done by assigning an elevation to some permanent control point and determining the elevation of the desired points in the area with respect to the assigned value of the control point. The assigned value can be of any magnitude. For convenience a sufficiently large value should be used so that no point in the system has a negative value. The operation of determining the difference of elevation of points with respect to each other on the surface of the earth is called *levelling*.

The most common instrument for the measurement of vertical distances in a level. A level may be a mechanical one, e.g. surveyor (Dumpy) level, tilting level, or an electronic level.

Measurement of Angles

Transit or theodolite is the instrument used to measure horizontal and vertical angles. It was invented by Roemer, a Danish astronomer in 1690. The instrument was used to observe the passage (transit) of stars across any portion of the celestial meridian. About a century later it was modified to suit the surveying requirements and even horizontal angles could be measured with it. In 1830, further additions were made to this basic instrument to make possible the measurement of several things apart from the measurement of vertical and horizontal angles. Because of its variety of uses, such as the measurement of horizontal and vertical angles, prolonging a straight line, measurement of bearings, and measurement of horizontal and vertical distances and to determine the direction of true north it is sometimes referred to as *universal instrument*.

Depending upon the facilities provided for reading of observations the theodolites may be classified as simple vernier theodolite, micrometer theodolite, optical (glass arc) theodolite and electronic theodolite. Though the first two are obsolete and all the modern theodolites are of the optical or electronic digital type.

In surveying any of the following system can be used for angular measurements.

Sexagesimal System: One complete circle is divided into 360°. Degrees are further subdivided into minutes and seconds. This is the system used in India. 1°= 60′ and 1′= 60″.

Centesimal System: One complete circle is divided into 400 grades. The grade is further subdivided into centesimal minutes and seconds.

1 grade (1^g) = 100 centesimal minutes (100^c)

1 centasimal minute (1^c) = 100 centasimal seconds (100^{cc})

this system is used in Europe.

Hour System: This system is used in navigation and astronomy. One complete circle is divided into 24 hours (24^h). The hour is further subdivided into minutes and seconds.

1 hour (1^h) = 60 minutes (60^m)

1 minute (1^m) = 60 seconds (60^s)

Measurement of Directions

The direction of a survey line in measured with the help of an instrument known as compass. The types of compass in general use are surveyor compass and Prismatic compass. However, a theodolite may also be used for measuring directions.

1.2 HORIZONTAL AND VERTICAL CONTROLS

Every survey, from mapping a continent to a small plot of land, depends upon a carefully measured framework which is thereafter treated as free from error. Subsequently, the details are filled in the framework by less elaborate methods. The fixation of a framework for a survey is known as *horizontal control.* It can, in general, be established either by traversing or triangulation or by a combination of both. Some of the other methods to provide horizontal control are trilateration and electromagnetic distance measurement (E.D.M.) traversing.

The word traverse literally means ‘passing across’. In surveying it means ‘determining the lengths and directions of consecutive lines’. The linear measurements are made with a chain or tape and the relative directions of the lines are measured with a chain, compass or theodolite. The integrated measurement of distance and direction provides the essential two dimensional data for determining land area and for providing horizontal control, i.e., the relative location of points in the horizontal plane.

In triangulation the entire area to be surveyed is covered with a framework of triangles. If the length and direction of one side, and all three angles of a triangle are measured precisely, the lengths and directions of the remaining two sides of the triangle can be computed. The length of the first line, which is measured precisely is known as *base line*. The other two computed sides are used as new base lines for two other triangles interconnected with the first triangle. By extending this process, i.e., the measurement of the further interconnected triangles and using the computed sides, a chain or network of triangles can be spread over the entire area. The apex of the triangles so located with a relatively greater accuracy provide horizontal control of the survey. Thus triangulation may be defined as a system of multiplying ground controls on the earth's surface. As a check, the length of one of the sides of the last triangle is also measured and compared with the computed one.

Horizontal control usually consists of a combination of triangulation and traverse. For most of the surveys of small extent, e.g., plane surveys, where direct linear measurements are impossible, triangulation is most suited. It is also suitable when long sight are taken. In triangulation, a number of lines of sight are required at each station. When the sights are long, the stations are elevated by building towers. In case the distances are short, the expense of towers offsets any saving and traverse becomes economical. Therefore, the character of the physical conditions encountered rather than the accuracy desired, governs the choice of the method of horizontal control. Triangulation is most suited for hilly areas, whereas traversing is suitable for flat areas.

The vertical control in surveying is established with levelling. The results of levelling are referred to a standard datum, usually mean sea level. The vertical heights of the points observed above the datum are called elevations. When the two dimensional data obtained from horizontal control is supplemented with the results of a level network, a three dimensional representation of the area surveyed is obtained.

1.3 IMPORTANCE OF STUDYING SURVEY

Surveying is one of the oldest arts practised by man. Surveying is of special importance and interest to the civil engineer. Surveys are required prior to and during the planning and construction of buildings, dams, highways, railways, bridges, canals, tunnels, drainage works, water supply and sewerage systems, etc. They may also be required for planning and construction of factories, assembly lines, jigs, fabrications, missile ranges, launch sites and mine shafts. Surveying is the starting point for any project or constructional scheme under consideration. Details of the proposed work are plotted from the field notes. The reliability of the estimation of quantities and the effectiveness of the design depends upon the precision and thoroughness exercised during the survey.

The introduction and practice of surveying is indispensable to all branches of engineering. The training that a student receives, irrespective of his branch of engineering, in the art of observing, recording, and computing data, as well as in the study of errors, their causes and effects, directly contribute to his success in other professional courses. He develops *inter alia* such qualities as self-reliance, initiative and the ability to get along with the others. This also helps an engineer get acquainted with the reasonable limits to accuracy can best be obtained by making measurements with the surveying equipment employed in practice as these measurements provide a true concept of the theory of errors. An engineer must also know when to work to thousands, hundredths or tenths of a metre and what precision in field data is necessary to justify carrying out computations to the desired number of decimal place. With experience, he learns how the funds, equipment, time, and personnel available will govern the procedure and results. Taking field notes under all sorts of field conditions trains a person to become an excellent engineer capable of exercising independent judgments.

Today, the art of surveying has become an important profession. An introduction to the principles and practices of surveying is, therefore desirable as an integral part of engineering education and training, irrespective of the branch of specialization. A knowledge of surveying trains the ability of engineers to visualize, think logically and pursue the engineering approach. It promotes a feeling of confidence, a habit of working in groups, neatness and care in documentation, and begin interpersonal relations by way of simultaneous and tactful handling of clients.

Objects of Surveying

The prime object of survey is to prepare a plan or map. Both plans and maps are the graphical representations of the features on a horizontal plane. The former is a large scale representation whereas the latter is a small scale one. Some of the other objects of surveying are establishing boundaries of the land, measurement of area and volume, and selection of a suitable site for an engineering project.

1.4 PRINCIPLES OF SURVEYING

There are two basic principles of surveying. These find their inherent application in all the stages of a project i.e., from initial planning till its completion.

1. To work from whole to part.
2. To locate a point by at least two measurements.

To Work from Whole to Part

It is the main principle of surveying and a method violating the principle of working from whole to part should not be adopted until and unless there is no alternative. The main idea of working from whole to part is to localize the errors and prevent their accumulation. On the contrary if we work from part to whole the errors accumulate and expand to a greater magnitude in the process of expansion of survey and consequently the survey becomes uncontrollable at the end.

Location of a Point by Two Measurements

Two control points (any two important features) are selected in the area and the distance between them is measured accurately. The line joining the control points is plotted to the scale on drawing sheet. Now the desired point can be plotted by making two suitable measurements from the given control points.

Let *A* and *B* be the two control points, whose positions are already known on the plan. The position of *C* can be plotted by any of the following methods.

1. By measuring distance *BC* and angle α as shown in Fig. 1.1(a).
2. By dropping a perpendicular from *C* on the line *AB* and measuring either *AD* and *CD* or *BD* and *CD* as shown in Fig. 1.1(b).
3. By measuring the distances *AC* and *BC* as shown in Fig. 1.1(c).

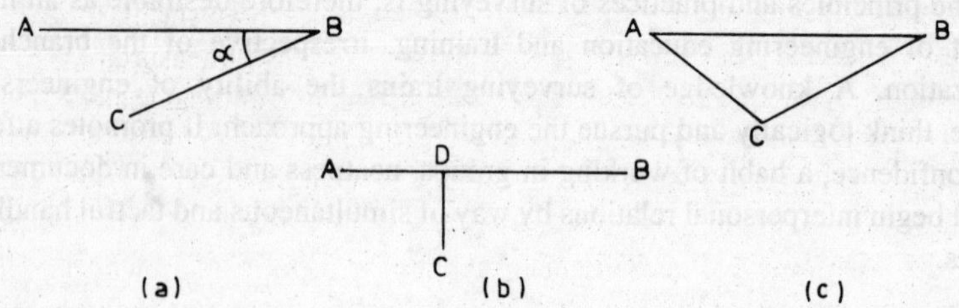

Fig. 1.1

1.5 SCALE

Scale is defined as the fixed proportion which every distance between location of the points on the map bears to the corresponding distances between their positions on the earth's surface. Primary considerations in choosing the scale for a particular project are those to which the map will be put and the extent of territory to be represented. Frequently the choice is restricted to some scale small enough so that the whole map will fall within a rectangle of given size, the dimensions of which are determined by the size of blue print frame, or by the size of sheet most convenient for handling. The following rules may be observed in deciding the scale:

1. Choose a scale large enough so that in plotting or in scaling distances from the finished map it will not be necessary to read the scale closer than 1/100.
2. Choose as small a scale as is consistent with a clear delineation of the smallest detail to be plotted, due regard being paid to rule 1.

When a scale is represented by a fraction whose numerator is invariably unity, it is called a representative fraction. In forming the representative fraction both the numerator and denominator must be reduced to same denomination. For a scale of 1 cm = 1 km, the representative fraction is 1/100,000.

The representative fractions and scales recommended for various types of maps are as follows:

Type	Representative Fraction (R.F.)	Scale
Geographical map	1 : 160,00,000	1 cm = 160 km
Topographical map	1 : 250,000	1 cm = 2.5 km
Location map	1 : 500 to 1 : 2500	1 cm = 5 m to 25 m
Forest map	1 : 25000	1 cm = 0.25 km
Cadastral map	1 : 1000 to 1 : 5000	1 cm = 10 m to 50 m
Town planning	1 : 5000 to 1 : 10000	1 cm = 50 m to 100 m
Buildings	1 : 1000	1 cm = 10 m
Mines	1 : 1000 to 1 : 25000	1 cm = 10 m to 25 m
Preliminary survey of rails and roads	1 : 1000 to 1 : 6000	1 cm = 10 m to 60 m

The scales given above are called engineer's scale. For example 1 cm = 160 km implies that 160 km distance on ground is represented by 1 cm on map. However, it is necessary to draw a scale on map, because as the map shrinks or expands the scale line also shrinks or expands with it and thus the measurements made from the map are not affected. Such scales are known as graphical scales.

1.6 ERRORS IN SURVEYING

It is understood that every measurement contains errors of unknown magnitude due to several reasons and hence no measurement in surveying in exact. A surveyor, should therefore, understand thoroughly the nature of the source and behaviour of the errors which may effect the results. A knowledge of the errors and procedures necessary to maintain a required precision aid the surveyor to develop a good judgment in his work.

The difference between a measured value and the true value is the true error. The true error cannot be determined as the true value is never known. However, using theory of probability, it is possible to determine a value called most probable value which is close to the true value. The most probable error and the standard error can also be determined.

The surveyor must know how various types of errors occur and how these can be controlled. The surveyor must have the skill and the judgment to keep the error within the limits and to make reasonably accurate measurements.

Sources of Error

The sources of error in surveying may be classified as natural, instrumental, and personal.

Natural Errors: These result from the temperature, refraction, obstacles to measurements, magnetic declination etc. For example, the length of a steel tape varies with changes in temperature. Such sources of error are beyond the control of the surveyor, but by taking precautionary measures and adopting suitable methods to fit into the conditions the errors can be contained within permissible limits.

Instrumental Errors: These result from the imperfect construction and adjustment of the instrument. The incorrect graduation of a steel tape and the improper adjustment of the plate levels of a transit are few examples. The effect of most of the instrumental errors can be brought within the desired limits of precision by applying proper corrections and selecting suitable field methods.

Personal Errors: These arise from the limitations of the human senses such as sight, touch, and hearing. For example, improper bisecting of the object by fixing the line of sight of a transit while measuring angles is a personal error.

Types of Errors

Errors in a measurement may be positive or negative. The former occurs if the measurement is too large and the latter if too small. Errors are classified as systematic errors and accidental errors.

Systematic Errors: These are the errors which occur from well-understood causes and can be reduced by adopting suitable methods. For example, the error due to sag of a tape supported at ends can be calculated and subtracted from each measurement. However, the tape can be supported throughout its length at short intervals and the sag error may be reduced to a negligible quantity. It always has the same magnitude and sign so long as the conditions remain same and such as error is called constant systematic error. Whereas, if the conditions change the magnitude of the error changes and this is known as variable systematic error. A systematic error follows a definite mathematical or physical law and, therefore, a correction can always be determined and applied. It is also known as *cumulative error.*

Accidental Errors: These are the errors due to a combination of causes and are beyond the control of surveyor. It can be plus or minus. Calibration of a chain is an example of an *accidental error.*

Mistakes: Mistakes are caused by the misunderstanding of the problem, carelessness or poor judgment. These can be corrected only if discovered. The best way is to compare several measurements of the same quantity and do away with the odd measurement which does not follow any law. On the other hand, error is defined as the difference of the measured and the true value of the quantity. The distinction arises from the fact that mistakes can be avoided by being careful, whereas errors result from sources which can be minimized but not avoided.

1.7 PRECISION IN SURVEYING

Surveying, although simple in concept, requires great skill and practice for doing the work accurately and economically. It requires basic knowledge of various disciplines such as mathematics, physics, geodesy and astronomy. The need for accurate surveying is increasing rapidly with the development in technology. The construction of modern buildings, highways, railways, high dams, long bridges, and tunnels requires accurate surveying. As the cost of the land and property is increasing rapidly, high accuracy is required in marking the boundaries and the land subdivision.

Both precision and accuracy are used to describe physical measurements. Precision is referred to as the degree of fineness and care with which any physical measurement is made, whereas accuracy is the degree of perfection obtained. It follows that a measurement may be accurate without being precise and vice versa.

1.8 EARTH SURFACE

The basic aim of the survey is to acquire data to portray the earth surface or its portion as precisely as possible.

The earth is not a true sphere and is slightly flattened at the poles. Its polar axis is somewhat smaller in length (about 43.45 km) than that of its equatorial axis. Any section of the earth parallel to the equator is a circle and any of its section parallel though the poles is an ellipse. Such a figure may be generated by revolving about its minor axis and is called an oblate spheroid. Precisely the equatorial section is also slightly elliptical and therefore such a figure should be called an ellipsoid. Precise observations indicate that the southern hemisphere is a trifle larger than the northern. Therefore, all the polar sections are oval and can be called ovaloid.

In fact, no geometrical solid represents the true shape of the earth. The earth is also recognized by a new name, geoid. However, for all measurement purposes in surveying, the irregularities of the earth's surface, as discussed above, may be assumed to be absent and the resultant surface be considered a spheroid.

Based on the accuracy desired the survey is classified as plane or geodetic.

Plane Survey: Survey in which the mean surface of earth is regarded as plane surface and not curved as it really is, is known as plane surveying. Following assumptions are made.

(i) A level line is considered a straight line and thus the plumb line at a point is parallel to the plumb line at any other point.

(ii) The angle between two such lines that intersect is a plane angle and not a spherical angle.

(iii) The meridians through any two points are parallel.

When we deal with only a small portion of earth's surface the above assumptions can be justified. The error introduced for a length of an arc of 18.5 km is only 0.0152 m greater than the subtended chord and the difference between the sum of the angles of spherical triangle and that of plane triangle is only one second at the earth's mean surface for an area of 195.5 km^2. Therefore, for the limits of the provisions stated above, the survey may be regarded as a plane survey.

Plane surveys are done for engineering projects on large scale such as factories, bridges, dams, location and construction of canals, highways, railways, etc., and also for establishing boundaries.

Geodetic Survey: Survey in which the shape (curvature) of the earth's surface is taken into account and a higher degree of precision is exercised in linear and angular measurements is termed as geodetic surveying. Such surveys extend over large areas.

A line connecting two points is regarded as an arc. The distance between two points is corrected for the curvature and is then plotted on plan. The angles between the intersecting lines are spherical angles. All this necessitates elaborate field work and considerable mathematical computations.

The geodetic surveying deals in fixing widely spaced control points, which may afterwards be used as necessary control points for fixing minor control points for plane survey. This is carried out by the Department of National Survey of India.

1.9 ELECTRONIC SURVEYING MEASUREMENTS

The application of electronics in surveying equipments in recent years has enabled the surveyor to measure distances and angles much more easily and to higher precision than is possible using mechanical or optical methods. With the advent of electromagnetic distance measurement (EDM) equipment, electronic theodolites—for digital display of angles, and EDM mounted electronic theodolites or with their improved and modified form as total station more efficient method of producing maps and plans have come up.

Electromagnetic Distance Measurement (EDM)

To use an EDM system, the instrument is set over one end of the line to be measured and some form of reflector is set over the other end such that the line of sight between the instrument and the reflector is unobstructed. An electromagnetic wave is transmitted from the instrument towards the reflector where part of it is reflected back to the instrument. By comparing the transmitted and received waves, the instrument is able to compute and display the required distance. An EDM system is shown in Fig. 1.2.

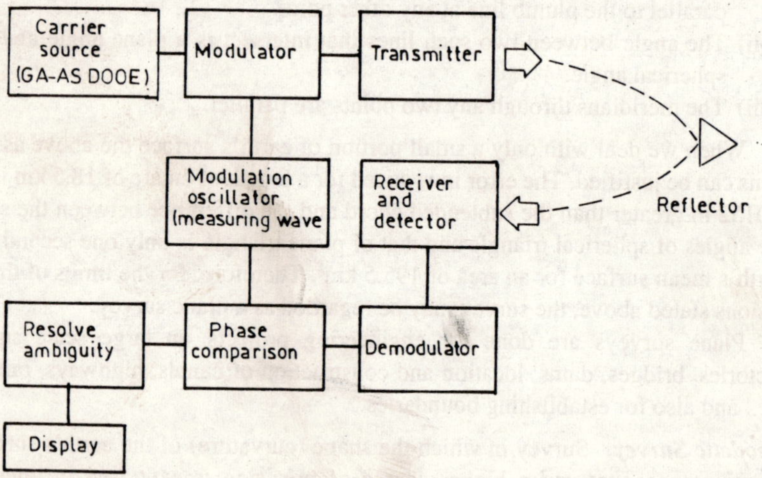

Fig. 1.2 EDM system

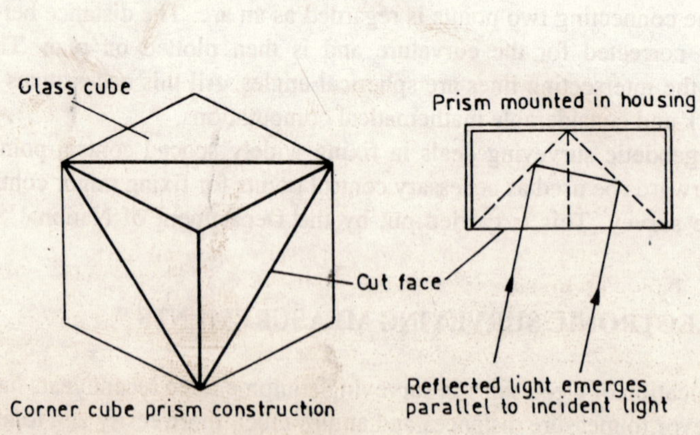

Fig. 1.3 Retroreflector—returns the beam, beam along a path exactly parallel to the incident path over a range of angle of incidence of about 20°

Since the infra-red carrier waves used in most EDM systems have wavelengths close to visible light, a simple mirror could be used for reflection. However, because of the narrow spread of the transmitted beam an accurate alignment of mirror would be required. To overcome this a *corner cube prism* (or retroreflector), shown in Fig. 1.3 is used. These can be mounted over a pole or tribrach. The height of the prism is normally set to equal the height of the instrument.

Electronic Theodolite

An electronic theodolite has arrangements to automatically resolve angular values and display them in a digital form. Display is achieved with the help of light emitting diodes (LEDs) or liquid crystal diodes (LCDs). In electronic theodolites, absolute angle measurement is provided by a dynamic system with opto-electronic scanning. The electronic theodolites are provided with control panels with key boards having multi-function keys. The angle display can be had with a single key-stroke. The electronic theodolites have opened the way to electronic data acquisition and data processing. An area where the use of electronic measurements is increasing is deformation monitoring.

The introduction of EDM into surveying has replaced or changed many traditional methods of surveying. Generally, the use of EDM in engineering surveying results in a saving in time and, in most cases, an improvement in the accuracy of distance measurement when compared with taping over distances greater than a tape length and when compared with optical methods.

The use of EDM for large-scale route and site surveys (scales of 1:1000 and larger) is now standard practice among all but the smallest survey organisations. Such surveys are normally carried out using total stations with some form of data recorder which can store and then transmit field data directly to a computer for processing and plotting, Chapter 2. Compared with conventional tape and tacheometric methods, electronic detail surveying is quicker, has a better precision and is more reliable, especially when using electronic data transfer.

EXERCISE

1. Define Surveying. What are the principles of surveying?
2. What are the various measurements made in surveying and what instruments are required to make these measurements?
3. Describe briefly the horizontal and vertical controls in surveying.
4. What is the importance and object of surveying? Why its study is indispensable to all branches of engineering?
5. Define scale and representative fraction. Why is it necessary to draw a scale on map?
6. What are various types of errors in surveying. Distinguish accuracy and precision.

7. (a) Describe briefly the earth-surface.
   (b) Differentiate between the following:
      (i) Plane and geodetic surveys
      (ii) Map and plan
8. Write a note on importance of electronic measurements in surveying.

CHAPTER 2

Total Station

2.1 INTRODUCTION

The electronic distance measurement (EDM) instrument (geodimeter) was first introduced in late 1950's. The early instruments, which were capable of very precise measurement over long distances, were large, heavy, complicated, and expensive. Rapid advances in related technologies have provided lighter, smaller simpler, and less expensive instruments. Current EDMs use infrared (light) waves or microwaves (radiowaves). The microwave systems require a transmitter/receiver at both ends of the measured line, whereas the infrared systems require a transmitter at one end and a reflecting prism at the other end. Microwave systems have a usual upper measuring range limit of 50 km. Infrared EDMs come in long range (10–20 km), medium range (3–10 km) and short range (0.5–3 km). These can be directly mounted in a tribarch (a three legged stand).

When EDM instrument is directly mounted on the standards or on the telescope of a theodolite, the EDM can provide both the horizontal and vertical position of one point relative to other. The slope distance provided by the EDM can be reduced to its horizontal and vertical equivalents by using the slope angle provided by the theodolite.

The electronic digitized theodolite, introduced in 1960's, set the stage for modern field data collection and processing. When the electronic theodolite is used with a built in EDM or an interfaced EDM, the surveyor has a very powerful instrument. With an onboard microprocessor, that automatically monitor the instruments operating status, and a data collector that records and processes field data, the instrument is called Total station—also known as Electronic Tacheometer instrument (ETI). The microprocessor not only controls both the angle and distance measuring systems but is also used as a small computer that can calculate slope corrections, vertical components, rectangular coordinates and, in some cases, can also store observations directly using an internal memory.

Plate 1 shows the Nikon DTM–300 Total Station. Its technical specifications are summarised in Table 2.1

2.2 FEATURES OF TOTAL STATION

The total stations currently available have different technical specifications, but have a similar format. The features common to the majority of Total Stations are described below.

Table 2.1 Specifications of Total Station Nikon DTM–300

Telescope	
Tube length	150mm/5.9 in.
Effective diameter of objective	36mm/1.41 in.
	(EDM aperture 40mm/1.57 in.)
Magnification	26× (standard)
Field of view	1°30′
Resolving power	3.5"
Minimum focusing distance	1.0m/3.3′
Angle measurement	
Reading system	Incremental encoder
Unit of reading	Degree/Gon/6400MIL
Least count (selectable)	
(360°)	5" or 10"
(400G)	5ccG or 10ccG
(6400 MIL)	0.02MIL or 0.05MIL
Accuracy	5"/16ccG/0.02MIL
(Standard deviation based on DIN 18723)	
Tilt sensor (Automatic Vertical	Liquid type
Compensator) Working range	±3′
Distance measurement	
Range (with Nikon prism)	
(Normal conditions: ordinary haze, visibility 20 km/12.5 ,miles)	
with mini prism	
with single prism	380m/1,300 ft.
with triple prism	800m/2,600 ft.
(Good conditions: no haze, visibility 40km/25 miles)	1,100m/3,700 ft.
	450m/1,500 ft.
with mini prism	1,000m/3,300 ft.
with single prism	1,200m/3,900 ft.
with triple prism	±(5+3ppm × D),
Accuracy	−10°C to +40°C/+14°F to + 164°F
	±(5+5ppm × D),
	−20°C to +50°C/−4°F to + 122°F
Maximum measurement display	1,230m/4,000 ft.
Least count	
Standard MSR mode	1mm/0.005 ft.
Tracking mode	10mm/0.05 ft.
Measuring intervals	
Standard MSR mode	4 sec.
Tracking mode	1.2 sec.
Measuring mode	Continuous/Single/Average (2-99)
Ambient temperature range	−20°C to +50°C (−4°F to + 122°F)
Atmospheric correction range	
Temperature range	−40°C to + 60°C (−40°F to + 140°F)
Pressure range	400 to 999mmHg (1mmHg step)
	15.8 to 39.3 in.Hg (0.1 in.Hg step)
	533 to 1,332hPa(1hPa step)

(Contd...)

Display	Dot-matrix LCD 16 characters × 4 lines
Level vial	
Sensitivity of plate level vial	30"/2mm
Sensitivity of circular level vial	10'/2mm
Leveling base	
Tribrach	Detachable
Optical plummet	
Image	Erect
Magnification	3×
Field of view	5°
Focusing range	0.5m/1.6 ft. to ∞
Power sources	
Type	Ni-Cd DC 7.2V
Continuous operating time	5.0 hrs. or 4,700 measurements (for distance/angle measurement) 19 hrs. (for angle measurement only)
Quick battery charger (Q-7U/Q-7E)	
Input voltage	115V for Q-7U, 220/240 V for Q-7E
Recharging time	1.5 hrs.
Dimensions (W × D × H)	
Main unit (with carrying handle)	164 × 177 × 335mm (6.5 × 7.0 × 13.2 in.)
Weight	
Main unit w/battery	5.3 kg/11.7 lbs.
Battery BC-4	0.3kg/0.7 lbs.
Carrying case	2.5kg/5.5 lbs.

Angle Measurement

This is carried out using an electronic theodolite. Typically a Total Station can record angles with a resolution of between 1" and 20" and all instruments incorporate some form of compensator—single axis or dual axis one—for automatic levelling of the instrument.

Distance Measurement

Generally all Total Stations use coaxial optics in which the EDM transmitter and receiver are combined with the theodolite telescope. This makes the instrument much more compact and easier to use on site. Normally a Total Station will measure a slope distance and the microprocessor uses the vertical angle recorded by the theodolite along the line of sight to calculate the horizontal distance. In addition, the height difference is also calculated and displayed. All instruments use some form of signal attenuation to protect the receiver.

Three modes are usually available for distance measurement.

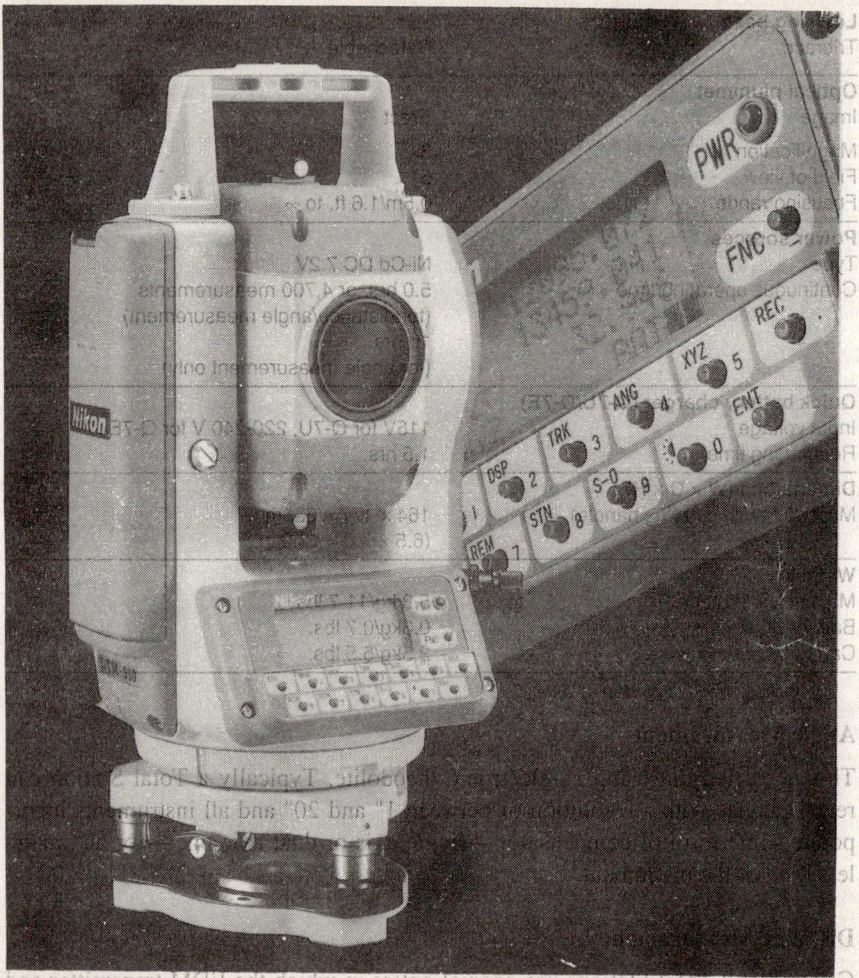

Plate 1 Nikon DTM–300 Total Station

Standard or Coarse mode: It has a resolution of 1mm and a measurement time of 1–2 seconds.

Precise or fine mode: It has a resolution of 1mm and a measurement time of 3-4 seconds. This is more accurate than the standard mode since the instrument repeats the measurement and refines the arithmetic mean value.

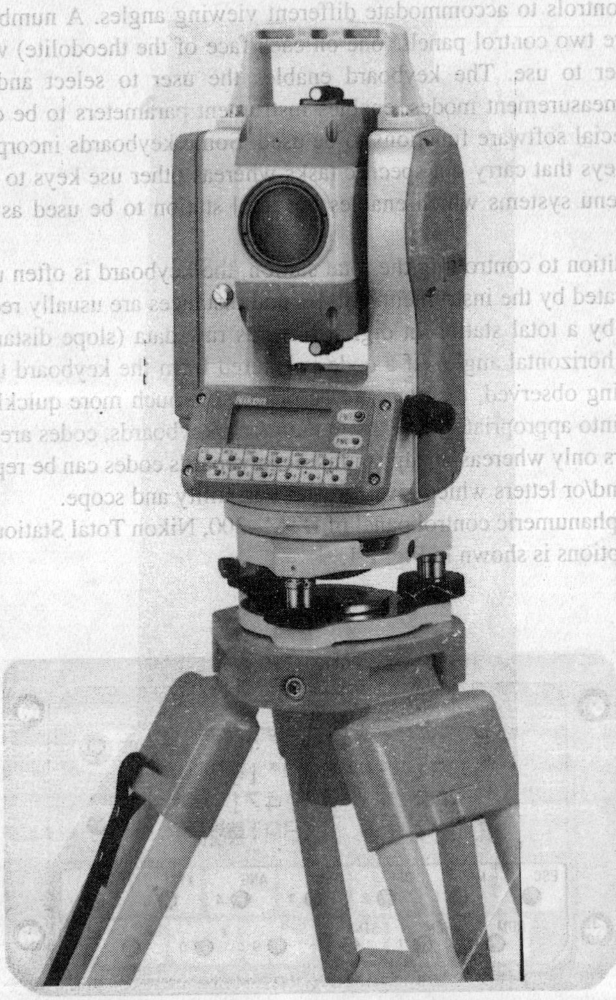

Plate 1 Nikon DTM–300 Total Station

Tracking or fast mode: The distance measurement is repeated automatically at intervals of less than one second. Normally, this mode has a resolution of 10 mm and is used extensively when setting out.

The range of a total station is typically 1–3 km to a single prism, assuming good visibility up to a range of 500 m. Most instruments allow for the input of temperature and pressure which enables the distance readings to be automatically corrected for atmospheric effects.

Control Panel

A total station is activated through its control panel which consists of a keyboard and multiple line liquid crystal display (often abbreviated to LCD). The LCD of a total station is moisture proof, it can be illuminated and some incorporate contrast controls to accommodate different viewing angles. A number of instruments have two control panels (one on each face of the theodolite) which makes them easier to use. The keyboard enables the user to select and implement different measurement modes, enables instrument parameters to be changed and allows special software functions to be used. Some keyboards incorporate multi-function keys that carry out specific tasks whereas other use keys to activate and display menu systems which enables the total station to be used as a computer might be.

In addition to controlling the total station, the keyboard is often used to code data generated by the instrument. Angles and distances are usually recorded electronically by a total station in digital form as raw data (slope distance, vertical angle and horizontal angle). If a code is entered from the keyboard to define the feature being observed, the data can be processed much more quickly by downloading it into appropriate software. On numeric keyboards, codes are represented by numbers only whereas on alphanumeric keyboards codes can be represented by numbers and/or letters which gives greater versatility and scope.

The alphanumeric control panel of DTM–300, Nikon Total Station, alongwith key descriptions is shown in Fig. 2.1.

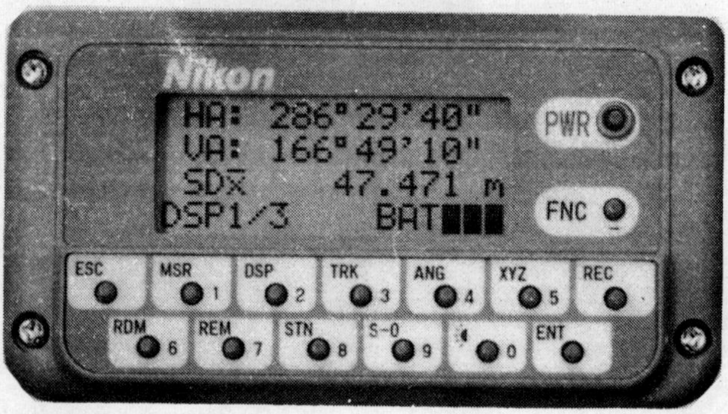

Fig. 2.1 Alphanumeric control panel

Key Descriptions

MSR : Distance measurement in normal measurement mode
DSP : Selects display item by scrolling display
TRK : Distance measurement in high-speed measurement mode
ANG : Horizontal angle zero-set, user-defined angle input
XYZ : Conordinate measurement
RDM : Remote distance measurement
REM : Remote elevation measurement
STN : Station setup (known point, 2-point resection or 3-point resection)
S-O : Stake-out by inputting angle distance or coordinates

Illumination: Display illumination

REC : Data record to internal memory
ENT : Data entry or sending observation data to communication port in normal observation mode.
FNC : Input of temperature and barometric pressure, prism constant and height of target, simple COGO calculation, settings, view/edit stored data, communication (internal memory), vertical collimation correction.

Power Supply

Rechargeable nickel-cadmium (NiCad) batteries are now standard for surveying instruments and these are connected directly to the total station without using cables. For angle and distance measurements, between two and ten hours use can be obtained from a battery, depending on the instrument. Most Total Stations are capable of giving a battery power indication and some have an auto power save power save feature which switches the instrument off or into some standby mode after it has not been used for a specified time.

It is a good practice, no matter what assurances a manufacturer may give about the life of a battery, to have a fully charged spare with the instrument at all times.

2.3 USEFUL ACCESSORIES

Tracklight

Tracklight is a visible light which enables a pole-mounted prism to be set directly of the line of sight without the need for hand signal from the total station. The device consists of a flashing three colour light in the prism is to the left of centre of the line of sight, a green light is seen; and if the prism is to the right, a red light is seen. When the prism is on line, a flashing white light is seen, the frequency of which doubles when it strikes the prism giving confirmation that the prism is in the correct position.

Geotronics

To *Geotronics Unicom* is communication system which allows speech to be transmitted from a Geodimeter instrument to a prism. This consists of a small microphone on the control panel which is activated by pressing a key and a receiver with small loudspeaker mounted on the prism pole. The usual method of communication on site, however, is by use of short range VHF hand-held radios.

2.4 SUMMARY OF TOTAL STATION CHARACTERISTICS

A. Input Parameters

1. Angle unit: degree or gon.
2. Distance unit: feet or meter
3. Pressure unit: Hg or mm Hg
4. Temperature unit: °F or °C

5. Prism constant (offset) usually – 0.3m
6. Automatic point number incrementation
7. Height of Instrument
8. Face 1 or Face 2 selection
9. Height of reflector
10. Point numbers and code numbers of occupied and sighted stations
11. Date and time of settings—for Total Stations with on-board clocks

B. Capabilities

1. Monitor: batter status, signal attenuation, horizontal and vertical axes status, collimation factors
2. Compute coordinates—northing, easting, elevation
3. Traverse closure and adjustment
4. Topographic reductions
5. Remote object elevation
6. Distances between remote points
7. Inversing
8. Resection
9. Layout
10. Horizontal and vertical collimation corrections
11. Vertical circle indexing
12. Records search and review
13. Programmable features
14. Transfer of data to the computer
15. Transfer of computer files to the data collector for layout purposes

2.5 ON-BOARD SOFTWARE

In addition to controlling the angle and distance functions of a Total Station, the microprocessor is also programmed to perform coordinate and other calculations. Some of these are described below.

Slope Corrections and Reduced Levels

From raw data (slope distance, vertical angle and horizontal distance), a Total Station will calculate and display horizontal distance and vertical distance. If the reduced level of the instrument station, the height of the instrument and the height of the prism are entered, the reduced level of the prism station can also be calculated and displayed. These are the most basic functions of a total station.

Horizontal Circle Orientation

The horizontal circle of a total station can be set to read a known bearing by entering the easting (*E*) and northing (*N*) coordinates of the station occupied followed by the *E* and *N* of a reference station. The reference station is then sighted and the orientation program is activated to calculate the bearing from the station occupied to the reference station and to set the horizontal circle to display this bearing (Fig. 2.2). The instrument would now be ready for further coordinate measurements or fro setting out.

Coordinate Measurement

Having orientated the horizontal circle of a total station, the coordinates of other points can be determined fairly easily. A new point is sighted, and distance and circle readings taken: when using the coordinate measurement program the instrument will now display the coordinates of the new point (Fig. 2.3). This can be extended to three dimensions if the reduced level of the instrument station and appropriate instrument and prism heights are also entered into the total station which will then display the reduced level of the new point.

Traverse Measurements

Traversing consists of a series of distance and angle measurements taken between successive points that enable the coordinates of those points to be calculated. Shown in Fig. 2.4, this type of coordinate determination is performed by a total station as a series of horizontal circle orientations and coordinate measurements taken at each traverse station.

For traverse measurement the Total Station is set up at station 1 and is oriented on to reference point. Angles A and distance D_1 are measured and coordinates of station 2 obtained. The Total Station is moved to station 2, and is oriented onto station 1 and the angle A_2 and Distance D_2 are measured to obtain coordinates of station 3. The instrument is moved to other stations and procedure repeated.

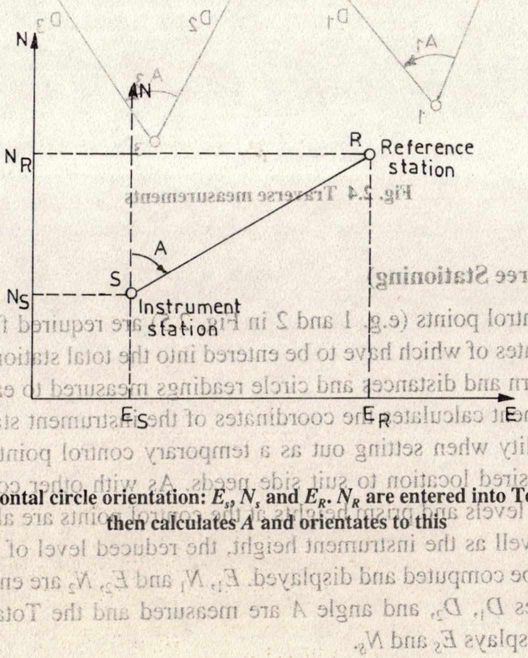

Fig. 2.2 Horizontal circle orientation: E_s, N_s and E_R. N_R are entered into Total Station which then calculates A and orientates to this

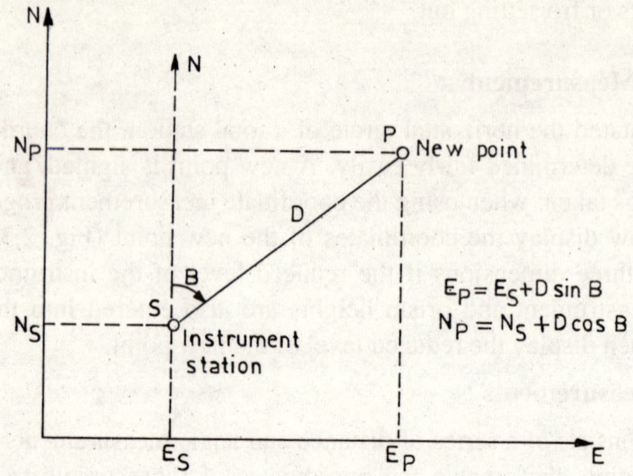

Fig. 2.3 Coordinate measurement Total Station determines *D* and *B* and calculates E_P, N_P from instrument station

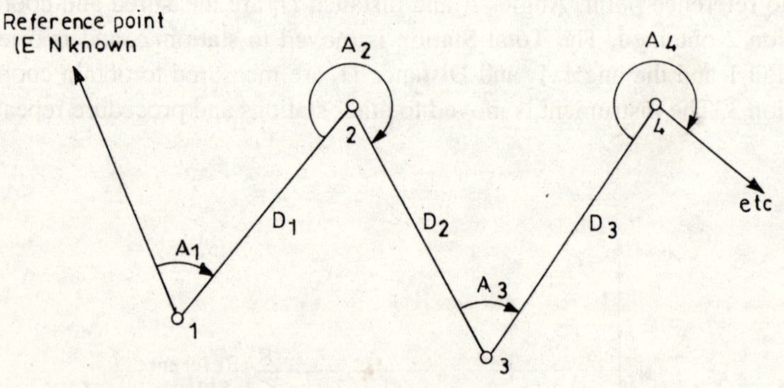

Fig. 2.4 Traverse measurements

Resection (or Free Stationing)

At level two control points (e.g. 1 and 2 in Fig. 2.5) are required for free stationing, the coordinates of which have to be entered into the total station. These points are sighted in turn and distances and circle readings measured to each. Using this data, the instrument calculates the coordinates of the instrument station. This can be a useful facility when setting out as a temporary control point can be established in any desired location to suit side needs. As with other coordinate functions, if reduced levels and prism heights at the control points are also fed into the total station as well as the instrument height, the reduced level of the instrument station can also be computed and displayed. E_1, N_1 and E_2, N_2 are entered into total station. Distances D_1, D_2, and angle A are measured and the Total Station then calculates and displays E_S and N_S.

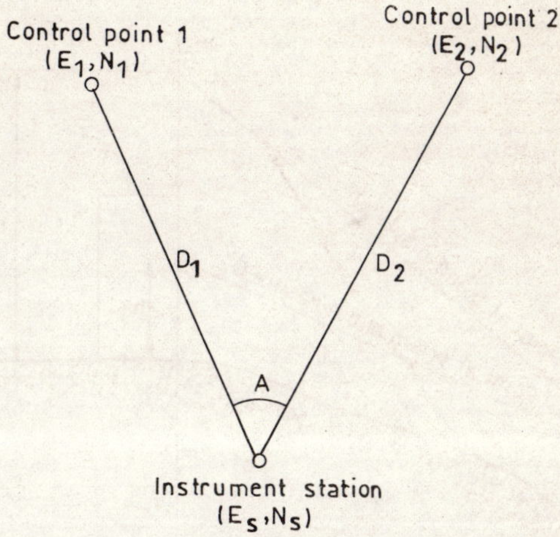

Fig. 2.5 Resection with Total Station

Remote Elevation Measurement (REM)

This function is used to determine heights at inaccessible points where it si not possible to locate a prism. Since measurements are taken along an extended plumb line through the prism, the prism must be positioned vertically above or below the point(s) to be surveyed. The prism height h_1 is entered into the instrument and the horizontal distance D to the prism determined (Fig. 2.6). In REM mode, the total station will now display the height from the ground at the prism to any point along the vertical through the prism. In Fig. 2.6, the top of a structure S could be sighted directly above the prism and its height h recorded. REM can also be used to set profile boards at their correct heights.

Setting Out Functions

Total stations can be used for setting out with given horizontal angle and distance values or with given coordinate.

When the horizontal angle and distance to be set out are known, these are entered into the instrument which has already had its horizontal circle oriented to a reference station. As soon as the appropriate key(s) are pressed to activate the setting out mode on the total station, it displays the difference (dHA) between the entered and measured horizontal angle values. In order to set the required direction for setting out, the telescope is rotated until a difference of zero is displayed (that is, dHA = 0). Following this, a pole-mounted prism is located on the line of sight as near to the required distance as possible: devices such as the Tracklight or Unicom are useful for lining in the prism but VHF radios with hand signals between instrument and prism are often used. Once aligned, the prism is sighted and the distance to it measured by the total station. The difference between the measured and entered distances is displayed and by moving the prism this difference is reduced to zero to locate the point.

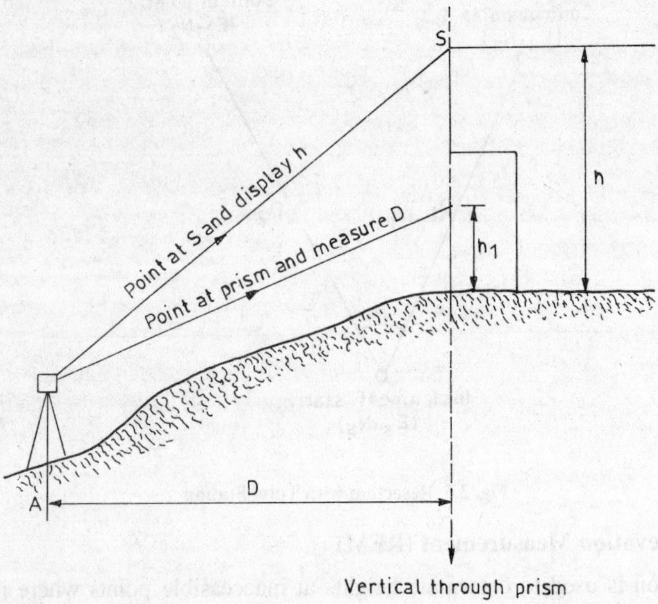

Fig. 2.6 Remote elevation measurement with Total Station

When the coordinates of the point to be set out are known, those of the instrument and reference stations should also be known. Prior to measurement, the station coordinates are entered into the instrument as well as those of reference station and the horizontal circle is oriented to the coordinate grid such that it displays bearings directly. Next, the coordinate values of the point to be set out are entered into the total station. When the setting out mode is selected the instrument displays the difference between calculated and measured bearings. There is no need for the observer to calculate any bearings as the total station does this automatically. As with the horizontal angle and distance mode, the telescope is rotated until this difference is zero such that it is pointing in the required direction for setting out. With the pole-mounted prism located on this line of sight, the horizontal distance to it is measured and the difference between this and the value calculated by the total station is displayed. This is reduced to zero by moving the prism.

2.6 CONSTRUCTION LAYOUT

Total stations are particularly well suited for collecting data in topographic survey. The collected data can be readily downloaded to a computer and processed into point coordinates—northing, easting, elevation (X, Y, Z)—alongwith point attribute data.

The significant increases in efficiency made possible with Total Station topographic surveys can be transferred to layout surveys when the original point coordinates exist in computer memory or floppy disk—together with the coordinates of all the key design points. To illustrate, consider the example of a road construction project.

First, the topographic detail is collected by using Total Stations set up at various control points; the detail is then transferred to the computer, adjusted, if necessary, and converted into *Y*, *X*, and *Z* coordinates. Various coordinate geometry and road design programs can then be used to design the proposed road. When the proposed horizontal, cross section, and profile alignments have been established, the proposed coordinates (*Y*, *X*, and *Z*) for all key horizontal and vertical (elevation) features can be computed and stored in computer files. The points coordinated will include top-of-curb and center-line positions at regular stations, as well as all changes in direction or slope. Traffic islands, and the like will also be included, as well all curved or irregular road components.

The computer files will now include coordinates of all control stations, all topographic detail, and finally all design component points. The layout will be accomplished by setting up at a control point, sighting another control point—with the correct azimuth on the horizontal circle—and then turning the computed angle to locate the desired layout point; the computed distance is, at the same time, measured out with the Total Station. When the Total Station is set to tracking mode, the surveyors will set the prism on target by rapid trial and error measurements.

Modern Total Stations offer an **even more efficient technique**. Instead of having the layout data only on a print-out, the coordinates for all layout points can be uploaded into the Total Station microprocessor. The surveyor can then, in the field, identify the occupied control point and the reference backsight point(s), thus orienting the Total Station. The desired layout point number is then entered with the required layout angle and distance being inversed from the stored coordinates and displayed. The layout can proceed by setting the correct azimuth and then, by trial and error, the prism is moved to the layout distance (the Total Stations is set to tracking mode for all but the final measurements); with some Total Stations the prism is simply tracked with the remaining left/right (±) distance being displayed alongside the remaining near/far (±) distance. When the correct location has been reached, both displays show 0.000 m (0.00 ft).

2.7 LIMITATION OF TOTAL STATION

Even though a total station can perform many of the calculations often done manually on site, this does not mean the surveyor or engineer should lose this ability. For this reason, it is strongly emphasized to carryout calculations throughout. A knowledge of these may avoid situations where too great a reliance is place on the digital readout obtainable from a Total Station and may avoid obvious mistakes when generating data using setting-out functions. In other words, the ability to process and manipulate coordinates by hand gives a surveyor or engineer the 'feel' for a correct orientation and distance, especially when setting out.

It is accepted that the total station is a very sophisticated instrument but this in itself can create problems. As an example, traversing is a recognised surveying procedure for obtaining the coordinates of control points. Using onboard software, a Total Station can produce a set of coordinates much more quickly than by traversing since observations can be taken from a single instrument position.

Although the methods used by Total Stations for obtaining coordinates have their applications, they are very dangerous if used in the wrong circumstances, especially when fixing the positions of control points. There are many cases where the time taken to arrive at the end result when using a Total Station is reduced by such a large factor compared with an established field procedure that the temptations are often too great and mistakes occur. While every opportunity should be made to make full use of a Total Station, any field procedure involving a Total Station that does not include an independent check on fieldwork must be treated as incomplete.

EXERCISE

1. What is Total Station? Describe briefly.
2. Describe briefly the features of Total Station.
3. Write notes on the following:
   (a) Track light
   (b) Geotronics
4. What are the typical characteristics of Total Station?
5. Describe briefly the following operations with regards to Total Station.
   (a) Horizontal circle orientation
   (b) Coordinate measurements
   (c) Resection
6. Describe briefly how a layout is made with the help of Total Station.
7. What are the advantages and limitation of Total Station.

CHAPTER 3

Remote Sensing For Data Extraction

3.1 INTRODUCTION

Ability is a capacity to resolve. To know is not what we see but what we understand. Also, this capacity depends on how much we can see and analyze. Man has been exploiting the natural resources to the maximum for his well being while maintaining without destroying nature. However, with the rapid increase in population it has become inevitable that the available resources should be managed in a comprehensive way.

Remote sensing is advanced and widely used technique for the acquisition of data without coming in physical contact with the object. The human eye is a typical example of a remote sensor but with limited capacities. The reflected energy from the objects carries information about the object. It is also important that different objects on the earth behave/interact differently in the same region of the EM spectrum enabling us to detect/delineate the objects under observation. Our Indian Remote Sensing Satellites (IRS) are extensively used for various applications, for example wasteland identification and management, watershed prioritization, crop acreage field estimates, the growth and direction of urban region, soil potentiality, ground water potentially, snow-melt, runoff, etc. Also, temporal or seasonal data collection at regional level, manually is tedious, time consuming, error prone and expensive job. Remote sensing overcomes all the above problems and makes us available with a very precise data very easily.

3.2 DEFINITION

''Remote sensing is the science and art of acquiring information about an object, area or phenomenon emitting energies in different wave lengths without making any physical contact with the object, area or phenomenon under investigation.''

The term remote sensing is commonly restricted to methods that employ electromagnetic energy (such as light, heat and radio waves) as the means of detecting and measuring target characteristics. It has two facets; the technology of acquiring data through a device which is located at a distance from the object and analysis of the data for interpreting the physical attributes of the object, both these aspects being intimately linked with each other. Aircraft and Satellites are the common platforms for remote sensing observations. The sensors flying on aerial or space platforms use different wavelengths of electromagnetic waves to perceive the data of different objects. The huge amount of data received is processed and analyzed with the use of the computers or simply by visual interpretation.

3.3 BASIC PRINCIPLE

The basic principle involved in remote sensing methods is that in different wavelength ranges of the electromagnetic spectrum, each type of object reflects or emits a certain intensity of radiation which is dependent upon the physical or compositional attributes of the object. Thus using information from one or more wavelength ranges, it may be possible to differentiate between different types of objects (e.g. dry soil, vegetation etc.) and map their distribution on the ground.

The radiation's are recorded by sensors such as scanners or radiometers mounted on suitable platforms. This remotely sensed radiance data are digitally processed for rectification and enhancement and combined with ground truth and other reference data.

3.4 ELECTROMAGNETIC (EM) RADIATION

Electromagnetic energy is used as the basic medium to extract information about earth's surface features. In remote sensing the energy transmission from the source to the feature and then to the sensors is by the means of radiation. EM radiation occurs as continuum of wavelength and frequencies in a spectrum ranging for 10^{-7} to 10^{3} μm. However, human eye is sensitive to only visible portion of spectrum (0.4–0.7 μm) here colour ''blue'' is ascribed to 0.4–0.5 μm, ''green'' to 0.5–0.6 μm, and red to 0.6–0.7 μm. Ultra violet region adjoins the blue portion and the red portion of the spectrum adjoins near infrared region. Infrared region is divided into near infrared (0.7–1.3 μm), middle-infrared (1.3–3.0 μm) and thermal infrared beyond 3.0 μm.

The range for different type of electromagnetic waves is shown in Fig. 3.1. To portray all the wavelengths horizontal scale is the log one. Figure 3.2 shows the expanded spectrum part used in remote sensing.

3.5 ENERGY INTERACTION WITH ATMOSPHERE

Solar energy is the main source of electromagnetic energy. The radiations reflected and emitted pass through the atmosphere before being detected by remote sensors. In this process they interact with atmospheric constituents like gases, suspended materials like dust particles, and are partly scattered, absorbed and transmitted. The degree of attenuation depends on path length and wavelength. The effect of scatter is inversely proportional to the fourth power of the wavelength; short wavelength are scattered more. Although the speed of EM radiation remains unaltered by the atmosphere, the several other characteristics like radiation of propagation and intensity of radiation may get affected.

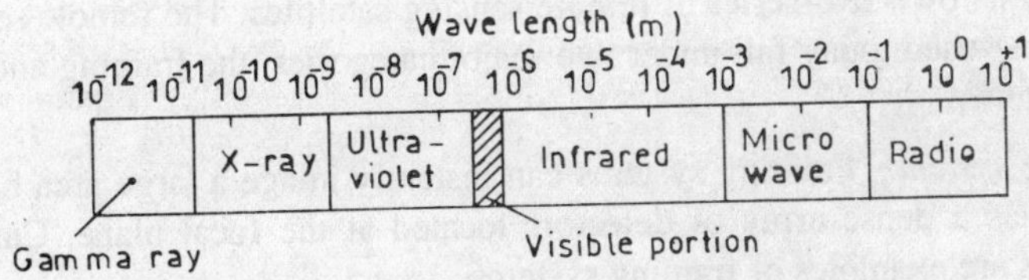

Fig. 3.1

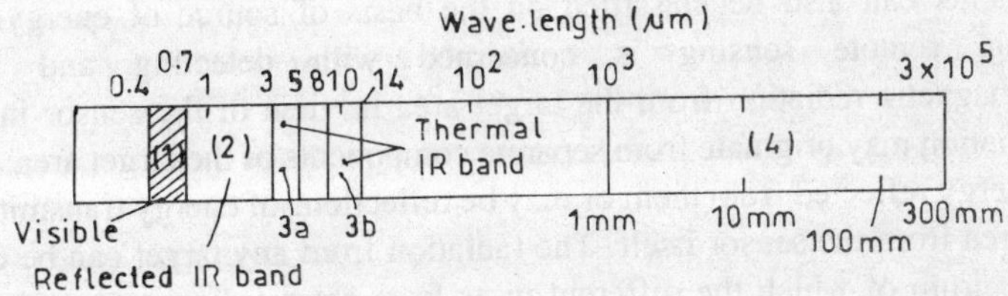

Fig. 3.2

3.6 ENERGY INTERACTION WITH EARTH SURFACE FEATURES

Electromagnetic energy which encounters matter, whether solid, liquid, or gas is called incident radiation. The interaction of EM energy with matter can change intensity, direction, wavelength, polarization and phase of the incident radiation. These interactions are recorded on remote sensing images, from which one may interpret the characteristics of matter.

The electromagnetic energy incident on the earth is reflected, absorbed and transmitted by the earth's materials to various degrees. Following the law of conservation of energy, the energy balance can be written as

$$Ei_{(0)} = Er_{(0)} + Ea_{(0)} + Et_{(0)}$$

where $Ei_{(0)}$ is the spectral incident energy, $Er_{(0)}$ is the energy component reflected. $Ea_{(0)}$ is that absorbed and $Et_{(0)}$ is that transmitted. These components differ for different objects at different wavelengths.

Since the proportions of energy reflected, absorbed, and transmitted will vary for different earth features, it is these differences that make us to distinguish different features on an image. Also, since most of the remote sensing systems operate in the wavelength regions in which reflected energy predominates, the reflectance properties are of great importance.

3.7 REMOTE SENSING SYSTEMS

The technology of remote sensing has developed rapidly during the recent years. With concern for the environment growing everyday, remote sensing is fast becoming the most effective tool for monitoring pollution. Aircraft were originally used for remote sensing but have become obsolete nowadays and specialized remote sensing satellites are becoming increasingly common. Even India has launched its own IRS-series of remote sensing satellites. The remote sensing systems being used today fall under two major categories, the framing and the scanning systems.

Framing system: Framing systems can instantly image a large area because the system has a dense array of detectors located at the focal plane. Cameras and Videcons are examples of framing systems.

Scanning system: A scanning system employs a single detector, with a narrow field of view, which sweeps across the terrain to produce an image. Remote sensing systems can also be classified on the basis of source of energy. Broadly speaking, remote sensing is concerned with detecting and recording electromagnetic radiation from the target area in view of the sensor instrument. The radiation may originate from separate components of the target area; it may be solar energy reflected from them or may be reflections of energy transmitted to the target area from the sensor itself. The radiation from any target can be expressed as a spectrum of which the different parts from short to longer wavelengths are ultraviolet, visible, infrared and radio sectors. Each kind of sensor records radiation from a different sector of the electromagnetic spectrum.

Active system: A system which utilizes man made sources of energy for data collection is called active system, e.g., taking photographs in the dark places with the help of a flash bulb.

Passive System: A system that uses an existing source of energy e.g., sun rays is called a passive system, such as taking photographs on a clear bright day.

As we have discussed, photographic cameras are passive systems that record images in the visible part of the spectrum and the immediately adjacent parts, i.e., ultraviolet and infrared. In this system, the focussed image is usually recorded by a photographic emulsion on flexible film base. Television cameras are also passive systems that operate in the same wave bands as photographic cameras. Unlike the aerial photographs, images of the T.V. camera are formed as patterns of electrical charges on an image plate which is scanned by an electron beam and converted to electrical signals. These can be relayed over long distances by radio links or stored as videotape recordings that can be later displayed.

Line scanners are passive systems that employ a rotating system to scan successive strips of ground along the track of the aircraft or satellite. Scanners can record in a wide range of spectral wave bands from the ultraviolet to the microwave. Unlike the camera, a scanner system builds up the picture from the spectral information as it is collected directly from the target area.

So far, we have discussed the passive systems, i.e., those receiving earth emission. It is possible however, to devise an active sensing system in which waves are propagated near the sensor and are bounced on the earth's surface to be recorded on their return. This is the essence of the RADAR (Radio Detection and Ranging) system. Images of landscape derived from airborne side looking radar (SLR) resemble aerial photographs with low angle solar illumination; in that, the shadow effects are produced. But the geometry of the radar pictures is entirely different from those of aerial photographs. This is because, radar is essentially a ranging device which consequently produce lateral distortion of elevated objects.

3.8 IDEALIZED REMOTE SENSING SYSTEM

An ideal remote sensing system is illustrated in Fig. 3.3, wherein, electromagnetic energy of all wavelengths and of known uniform intensity is produced by an ideal source; the energy propagates from the source without loss to a homogeneous target; and the energy of various wavelengths selectively interacts with the target, resulting in a return signal of reflected and emitted energy.

The returned signal propagates without loss to a sensor that responds linearly to energy to all wavelengths of any intensity. In real time, an intensity versus wavelengths response is recorded, processed into an interpretable format, and recognized as being unique to the particular target in its particular chemical state. The information obtained without the particular target is made readily available in a useful form to the users.

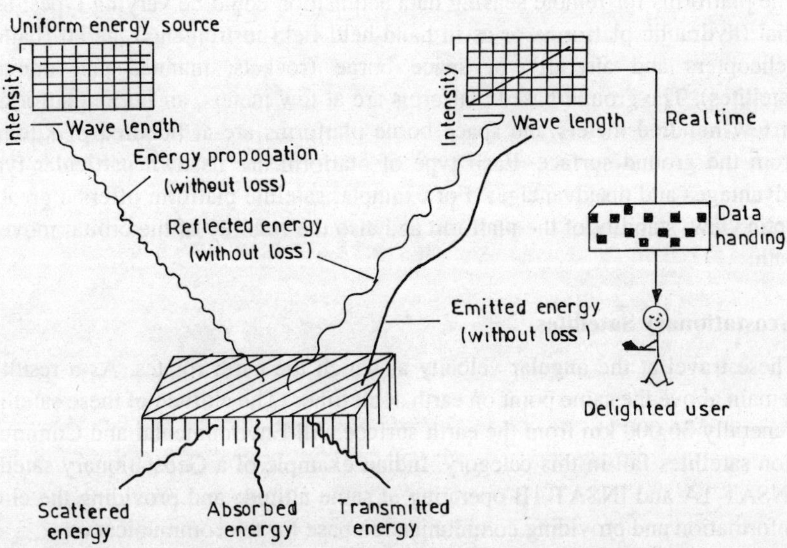

Fig. 3.3 Idealized remote sensing system

3.9 DEVELOPMENT OF REMOTE SENSING TECHNOLOGY

In recent years, the study of land, oceans, atmosphere and their interaction has made rapid advance due to continuous regional and global observations through air borne and space borne remote sensing systems. The space based remote sensing consists of acquiring information about earth surface, through instruments positioned on earth observation satellites. Beginning with the US Television (TIROS) in 1960, remote-sensing technology has reached a state where application of satellite remote sensing became operational in many countries. The

National Oceanic and Atmospheric Administration (NOAA), USA's Nimbus series of satellite served as test bed for the current NOAA satellite used in meteorological applications and the LANDSAT (American) series of satellite subsequently have proved their ability to survey land resources from space. With the advent of high spatial resolution SPOT (French Satellite) data, many new areas of application like urban sprawl mapping have emerged.

India, with the experience gained from its experimental remote sensing mission Bhaskara I and II, has now established satellite based operational remote sensing system in the country with the launch of Indian Remote Sensing Satellites, IRS-IA in 1988, IRS-IB in 1991, IRS-IC in 1995 and IRS-ID in 1977. These have been developed mainly for land applications. Under Technology Demonstration Mission IRS-P series has been developed. IRS-P2 and IRS-P3 were launched in 1994 and 1996 respectively. IRS-P4/OCEANSAT-I was launched in 1999 for oceanographic applications. Future missions are launching of IRS-P6/RESOURCESAT-I, IRS-P5/CARTOSAT-I, and IRS-2 series.

3.10 PLATFORMS AND SENSORS

The platforms for remote sensing data acquisition could be varying types: terrestrial (hydraulic platforms or as in hand-held field instruments), aerial (balloons, helicopters and aircraft), or space borne (rockets, manned and unmanned satellites). The ground based platforms are at few meters, air borne platforms are at few hundred meters and space borne platforms are at hundred of kilometers from the ground surface. Each type of platform has its own particular type of advantages and disadvantages. For example, satellite platform offers a great synoptic view, stability of the platform and also the stability of the orbital movement path.

Geostationary Satellites

These travel at the angular velocity at which the earth rotates. As a result they remain above the same point on earth at all times. The altitude of these satellites is generally 36,000 km from the earth surface. All Environmental and Communication satellites fall in this category. Indian example of a Geostationary satellite is INSAT 1A and INSAT 1B operating at same altitude and providing the climatic information and providing communication base for telecommunications.

Sunsynchronous Satellites

Satellites orbiting in a path North Pole to South Pole at about 9° of the N and S poles. These circles around once every 103 minutes resulting in 14 orbits everyday. The ground track speed is about 6.46 km/sec. As the sensor image only 185 km broad area (swath), with each revolution the satellite shifts slightly towards west just overlapping the previous orbit. With above speed, the satellite repeats same orbit every 18 days (Temporal resolution). The spacing between the successive orbits causes the satellites to keep precise pace with the sun's westward progress as the earth rotated. As a result, the satellite always crossed the equator at precisely the same local sun time. This Sunsynchronous orbit. All remote sensing satellites for ex. LANDSAT, IRS, SPOT fall within this category.

Sensor is an instrument which records the incident energy from the earth's surface in a particular wavelength remote sensing techniques use data from different sensors which are boarded on appropriate platforms situated at different altitudes.

3.11 ADVANTAGES OF REMOTE SENSING TECHNIQUES

The major advantages of the remote sensing technique over other methods of ground investigations are:

Synoptic overview: Remote sensing permits the study of various spatial features in relation to each other and the delineation of regional features/trends phenomena.

Feasibility aspect: Some areas may not be accessible to ground survey, the only feasible way to obtain information about such area is from remote sensing platforms.

Time saving: The technique saves time and manpower, as information about a large area is delineated quickly.

Multi disciplinary application: The remote sensing data can be used by researchers/workers in different discipline, like geology, forestry, land use etc. and therefore the overall benefit to cost ratio is better.

3.12 APPLICATIONS OF REMOTE SENSING

Remote sensing provides continuous and accurate monitoring of the earth's natural and other resources. Also, it helps to establish the impact of human activities on land, water, and air. The data obtained provide information necessary for making sound decisions and in formulating policy regarding land use and resource development. Remote sensing, as a technique for surveying, has opened up new frontiers and has provided means for investigating those regions and phenomena which were earlier out of our bounds due to the limitations of ground surveying. Some specific applications of remote sensing are as below.

Natural Hazards

Natural hazards though cannot be prevented but can be minimized by analyzing the remote by collected data regarding geological formation of the area and identifying the risk prone area.

Archaeology

Archaeological patterns of prehistoric land use may be recognized in remote sensing images. Many towns and roads of the old era are now covered with agricultural fields. These can be detected with the remote sensors by observing the data of change in moisture content of soil layers and other characteristics.

Geological Applications

1. Lithologic units
2. Geomorphological units
3. Geological structures
4. Tectonic and nontectonic land forms
5. Mineral exploration

Environmental Applications

1. Environmental impact assessment and carrying capacity based studies.
2. Spread and dilution of smoke plumes from industries and power plants.
3. Discharge of thermal plumes in rivers, lakes, etc.
4. Landuse/Lancover estimation and their changes due to some activity.
5. Evolution of waste disposal sites.
6. Deforestation and erosion in river catchment.
7. Land degradation and land suitability analysis.
8. Water quality based analysis on regional scale.
9. Hazard zonation.

Engineering Projects

1. Dams and Reservoirs.
2. Seismic status of the area.
3. Landslide zone identification.
4. Route location canal and pipeline alignment.
5. Town planning studies.

3.13 LIMITATIONS OF REMOTE SENSING

The practice of remote sensing is young enough that the basic facts and methods are not known completely. Scientists are still investigating to define many of the fundamental methods and concepts central to remote sensing. Following are some of the limitations of remote sensing.

1. Remote sensing depends upon observed spectral differences in the energy reflected or emitted from features of interest.

   The landscape characteristics change geographically and are temporally dynamic. During one season the contrast between cropland and pasture areas may be very sharp, and it may be possible to separate them clearly on aerial photographs. However, in some other season they may appear to be quite similar, and the separation on aerial photographs may be much more difficult and much less reliable.
2. The atmospheric interfence.
3. The remote sensing system records neither full range of brightness in the scene, nor all the details, and it includes geometric errors in the representation of the landscape. This results in the imperfect image of the landscape.
4. The interpreter has to be very knowledgeable.

EXERCISE

1. Define remote sensing. Describe briefly its applications in surveying.
2. What is the principle of remote sensing?
3. Discuss briefly the following:
   (a) Energy interaction with atmosphere
   (b) Energy interaction with earth surface features
4. Describe briefly the various remote sensing systems.
5. What is an ideal remote sensing system?
6. Write a note on applications of remote sensing.
7. What are the advantages and limitations of remote sensing?

CHAPTER 4

Geographic Information System

4.1 INTRODUCTION

The technological development and progress of the geographical world is taking place at a very fast pace. This necessitates to have a system capable to analyse and manipulate the spatially referenced data or remotely sensed information, and give the output for its information in a very short time.

Geographic information systems (GISs) are specialized data bases that preserve locational identities of the information they record. GIS may be defined as a computer based information system which attempted to capture, store, manipulate, analyze and display spatially referenced and associated attribute data for solving complex research, planning and management problems. The word *geographic* implies that location of the data items are known, or can be calculated in terms of the geographic coordinates (the latitude and longitude). The word *information* implies that the data in a GIS are organized to yield useful knowledge — often as coloured maps and images, statistical graphics, tables and various on-screen activities. The word *system* implies that a GIS is made up from several interrelated and linked components with different functions.

GIS is capable of acquiring spatially indexed data from a variety of sources, changing the data into useful formats, storing the data, retrieving and manipulating the data for analysis, and then generating the output required by the user. Their great strength is based on the ability to handle large multilayered, heterogeneous data base and to query about the existence, location and properties of a wide range of spatial objects in an interactive way. A digital computer provides the basis for storage, manipulation and display of large amounts of data that have been encoded in digital form. GIS consists of a package of computer programs with a user interface that provides access to particular function. The main advantage of GIS is a rapid analysis and display of data, with a flexibility not possible using manual methods.

GIS does not hold maps or pictures — it holds data base. That is how it is different from computer mapping, which can produce only good graphic output.

4.2 DATA FOR GIS

A GIS consists of information depicting geographic distributions of significance to those who use it. Although such data are widely available as lists and tabulations, the purpose of a GIS requires that data must be coded by geographic location, so that each item of information can be associated with a specific geographic location. A GIS retains locational identity by means of coordinate system. The fineness of the coordinate system relative to actual ground distances determines the resolution of GIS, or of the variable coded at a particular level of detail. Obviously it would be desirable to encode locations at very fine levels of details, to provide high precision in specification of location.

Data for a GIS can be derived from many alternative sources. Data can be gathered directly in the field by an original survey tailored specifically for the purposes of GIS. It is very time consuming and costly method and rarely resorted to. Generally, data is derived from a source that is already available, such as topographic maps, aerial photographs or other images, or maps prepared by Department of Survey of India.

4.3 ALTERNATIVE DATA STRUCTURES

Data for a GIS must be represented in a form that preserves locational identities of each unit of information, so that it is possible to retrieve data by location and therefore to depict and analyse geographic patterns. Because data are frequently derived from a "conventional" (nondigital) map or image, it is necessary to convert them into digital form suitable for use by a GIS. This process, known as "geocoding", records the pattern on each map in a form that can be accepted and manipulated by digital computers. Geocoding is often one of the most expensive and time-consuming aspects of GIS management because it requires considerable manual labour for coding and entering data, and for checking and editing errors.

Two alternative GIS data structures offer contrasting advantages and disadvantages. *Raster* (or "cellular") data structures consist of cell-like units, analogous to the pixels of an multi scanner system scene. The region of interest is subdivided into a network of such cells of uniform size and shape; each unit is then encoded with a single category or value (Fig. 4.1). Raster structures offer ease of data storage and manipulation, and therefore permit use of relatively simple computer programs.

Also, raster structures lend themselves to use with remotely sensed data because digital remotely sensed data are already presented in a raster format. Disadvantages are primarily related to losses in accuracy and detail due to the coding of each cell with a single category, even though several may be present (Fig. 4.2).

The alternative format is the *vector* (or "polygon") format, which records the boundaries, or outlines of parcels on the source document by listing the coordinates of the boundaries or coordinates of the vertices of polygons (Fig. 4.3). Vector format tends to provide more efficient use of computer storage than does the

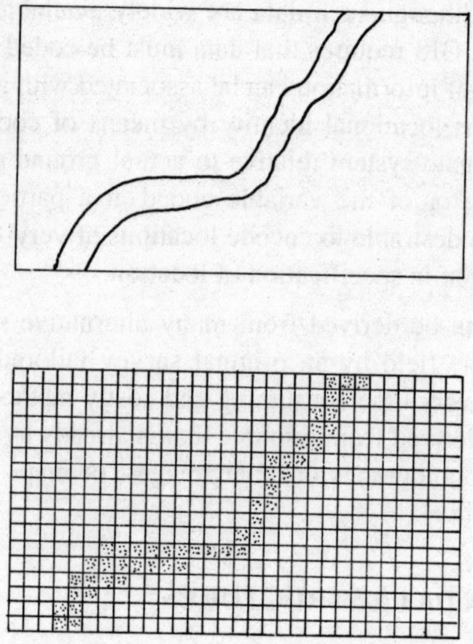

Fig. 4.1 Raster data structure

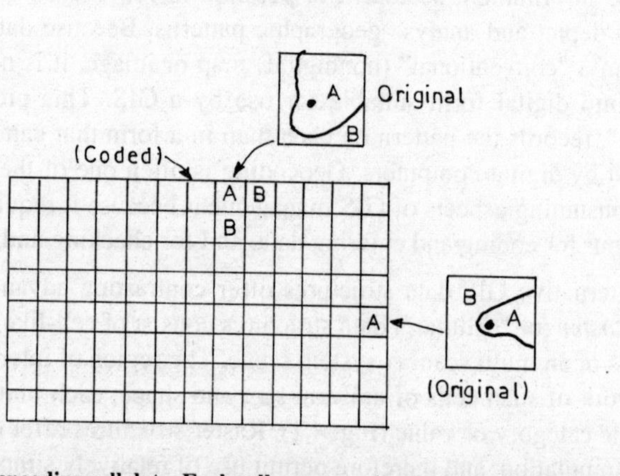

Fig. 4.2 Errors in encoding cell data. Each cell can record only a single identifier, so that cells cannot record the fine detail of the original pattern. Here cells illustrate coding by means of selecting the single class identified by the dot at the center of each cell. The two examples show how cells with the same labels can correspond to quite different ground conditions.

raster format, as well as finer detail and more accurate representation of shapes and sizes. Its main disadvantages are (in some instances) higher costs of encoding data and the greater complexity required for the computer programs that must manipulate data. Furthermore, it may be difficult to overlay, or superimpose, separate data distributions because of the small silvers, or strips, that are formed where two polygon files may not match exactly because of minor errors in digitization (Fig. 4.4).

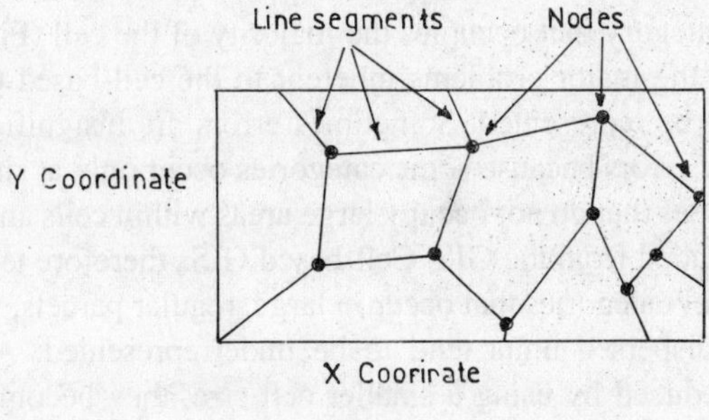

Fig. 4.3 Polygon ("vector") data structure

Usually a GIS must be designed on either raster or vector formats because of differences in equipment, computer programs, and expertise required for the two different approaches. The choice between the two depends upon the facilities available, the kinds of data to be examined, and the purposes of establishing the GIS. It is possible to convert from vector to raster format by applying relatively straightforward computer algorithms, although raster to vector conversion is at present much more difficult.

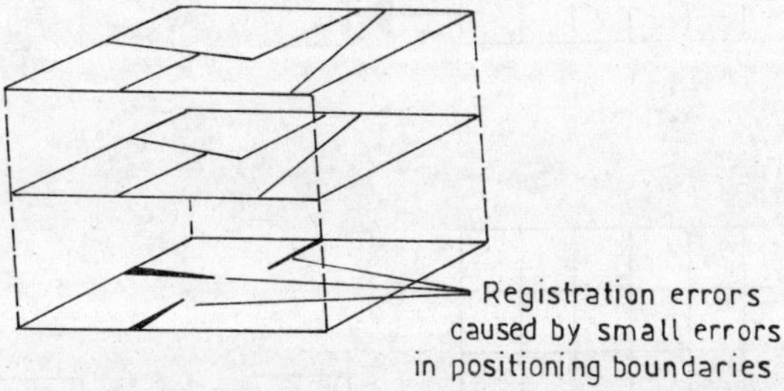

Fig. 4.4 Overlays of polygon files may not register exactly

Cell based, or raster, data bases establish a uniform network of units as the locational framework for the data base.

Cellular Data Structures

Each cell in the GIS must be encoded with a single value, or a single category that represents conditions as that location (Fig. 4.5). For continuously varying information, such as topographic elevation, the single value could be the elevation at the center of the cell, or possibly the average value of elevation within the cell. However, once a choice is made, the same procedure is applied uniformly to all cells within the map. A cell can also encode nominal data, such as a land cover class, or a geologic or pedologic category. Because a given cell may include several categories, it is necessary to define a rule for consistent selection of which category represents the cell in the data base. The "predominant category" rule selects the single category that occupies the majority of the cell (Fig. 4.5(a)). It is easy to understand the major problems inherent to the cell-based GIS. Variation within cells cannot be represented. Sometimes errors are insignificant, but often there are systematic errors because some categories occur only at small parcels, or as long, narrow shapes that do not occupy large areas within cells and therefore are systematically excluded from the GIS. Cell-based GISs therefore tend to exaggerate the importance of categories that occur in large, regular parcels, whereas small, linear, or widely dispersed units tend to be underrepresented. Although these problems can be reduced by using a smaller cell size, they become insignificant only when a cell size is very small in relation to the pattern to be studied, and the costs of using such a small cell size are often prohibitive.

An alternative procedure selects the category by means of a dot in the center of each cell; the category that falls beneath the dot is entered as the category for the cell (Fig. 4.5(b)). A third procedure uses a dot randomly positioned within each cell. (Of course, for a given data set, only a single method will be used.) Relative to the predominant category rule, these procedures improve accuracy of the data base as a whole, but of course the error at each cell can be quite large.

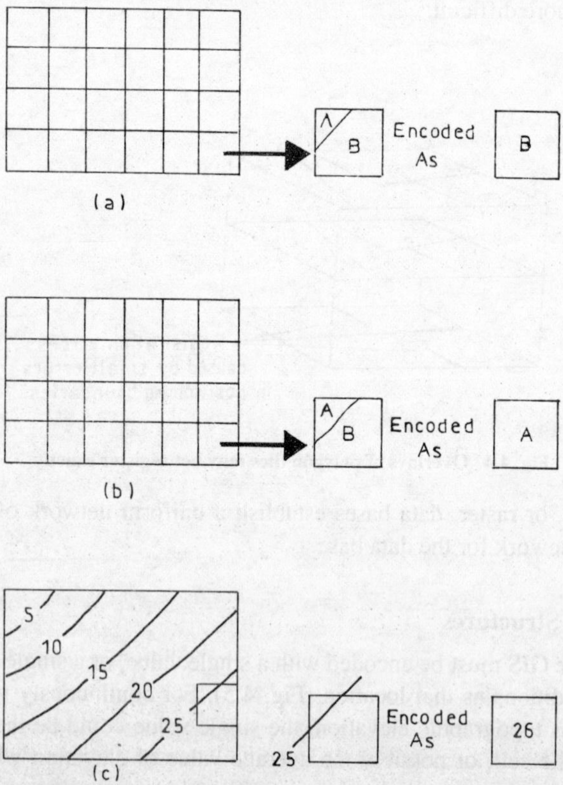

Fig. 4.5 Encoding data for cell-based GIS: (a) predominant category rule, (b) fixed dot, (c) encoding continuous data

The error can be inventory error or classification error as shown in Fig. 4.6.

Inventory error is error for the data base as a whole in reporting total areas occupied by specific categories. Inventory error is important for users of the GIS who are interested in a regional overview, who wish to learn about the character of specific regions rather than small sites within the region. Use of the dot randomly positioned within each cell reduces inventory error; for large areas inventory errors may be quite small. In contrast, *classification error* is the error in reporting the contents of each cell. Classification error is important for users who require detailed information concerning specific sites.

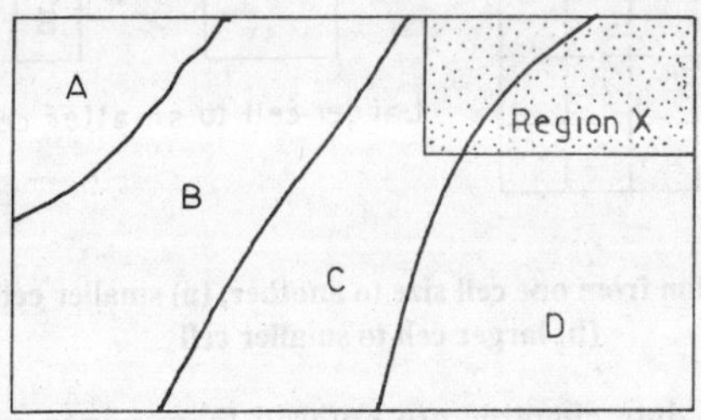

INVENTORY:
"REPORT TOTAL AREA OF CATEGORY A"
(LOCATION IS NOT IMPORTANT)

CLASSIFICATION:
"REPORT CATEGORIES PRESENT WITHIN REGION X"
(LOCATION IS IMPORTANT)

Fig. 4.6 Inventory error and classification error

The classification error is usually larger than inventory error. Clearly, classification error increases with use of the random dot method, even though the randomized selection may reduce inventory error. Both kinds of errors can be reduced if cell size is small relative to the pattern to be studied, as illustrated in Fig. 4.2. However, even modest decreases in cell size greatly increase the numbers of cells; as a result, costs and practicality soon limit efforts to reduce cell size. Therefore, selection of cell size is made with great care, because once a GIS is committed to a specific cell size, costs and inconvenience of changes preclude conversion to a smaller cell size.

It should be noted that it is possible to interpolate data from one cell size to another (Fig. 4.7). Data encoded at a 1-km grid size can be represented, for example, at a 4-km grid, as long as the analyst consistently applies a rule for deciding how to encode data for the larger cell (Fig. 4.7(a)). However, interpolation from larger to smaller cells is a different matter. It is operationally possible to create 1-km cells for data encoded at 4-km cells, but the change cannot improve the level of detail in the data; it simply creates a false impression of finer detail. There can be no method for assigning values for the smaller cells, except by replicating information of the larger cells. The visual detail has been improved, but the logical detail remains the same (Fig. 4.7(b)).

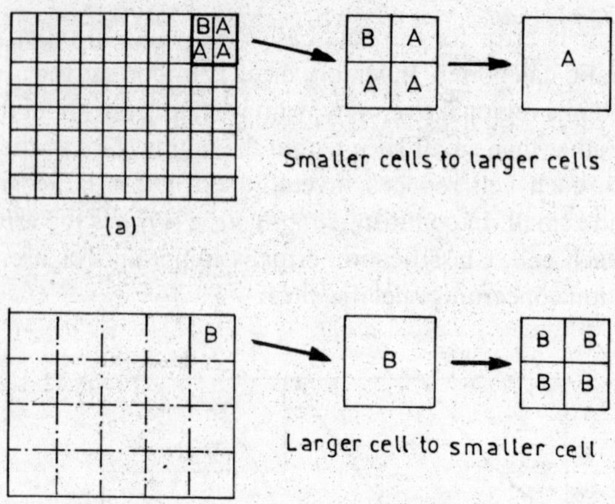

Fig. 4.7 **Interpolation from one cell size to another, (a) smaller cell to larger cell, (b) larger cell to smaller cell**

Data for cell-based data systems are entered from paper maps, aerial images, or similar documents depicting the pattern to be encoded. Often a template showing the grid system is superimposed and registered to the map. The spacing of the grid of course must be appropriate for the scale of the map.

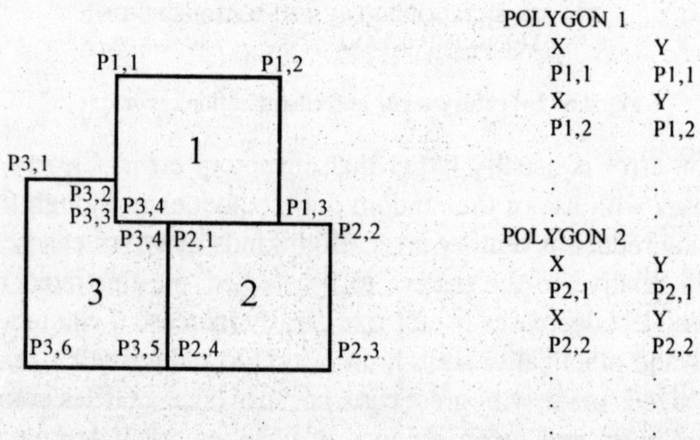

Fig. 4.8 **Location list file. Each polygon is treated as a separate unit; many points must be encoded several times**

Vector Data Structures

Four of the most widely used vector data formats illustrate different strategies for digital encoding of the boundaries that form a geographic pattern. All depend upon use of a geographic reference system that permits definition of locations at a fine level of detail relative to the sizes of polygons. The reference system may be arbitrarily defined, or may be tied to an established geographic reference system, such as latitude and longitude, state plane coordinates, etc.

The location list data structure (LLS) (Fig. 4.8) simply encodes the coordinates of the polygons that record the geographic pattern. Each polygon is encoded individually, so interior boundaries are encoded twice. Thus, although LLS is simple and convenient to use, it makes inefficient use of computer storage and creates boundary slivers that must be removed.

The point dictionary structure (PDS) (Fig. 4.9) records each vertex of each polygon only once. One file is used to record a list of coordinates of all vertices, with a label for each. A second file then identifies each polygon in location list form using labels of points that define its outline. PDS usually provides a great improvement in efficiency over LLS at only a small cost in program complexity.

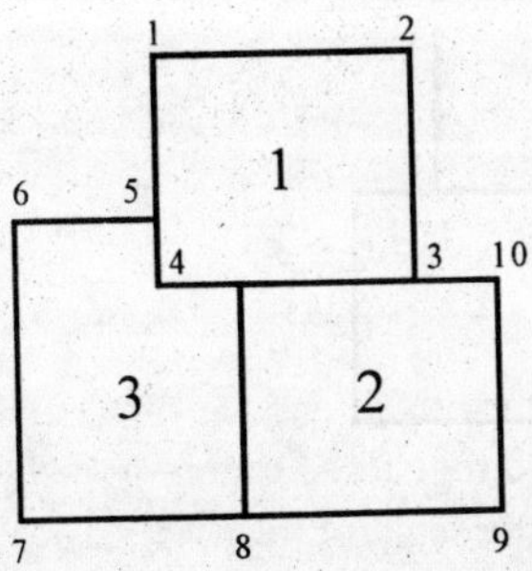

POINTS

1 x_1 y_1
2 x_2 y_2
3 x_3 y_3
. .
. .
n x_n y_n

POLYGON NAMED 3
6,7,8,11,4,5,6

Fig. 4.9 Point dictionary file. Locations of vertices are encoded independently of polygons, the "dictionary" list then defines each polygon by the points that form its perimeter.

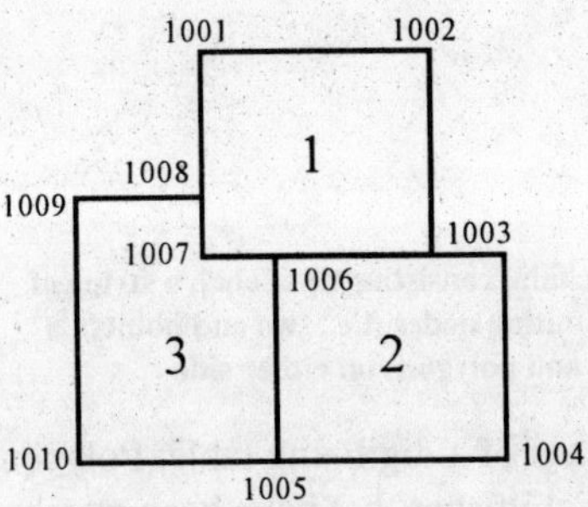

NODES		POLYGONS			
FROM	TO	LEFT	RIGHT	COORDINATES	
1006	1005	3	2	xy	xy
1007	1008	3	1	xy	xy
1006	1003	1	2	xy	xy
1006	1006	3	1	xy	xy

Fig. 4.10 Dual independent map coding (DIME). Lines are encoded as separate entities, with polygons identified as areas on either side of each line segment.

One of the most widely used formats for vector data is the dual independent map encoding structure (DIME) (Fig. 4.10). The basic unit for the DIME format is not the polygon itself, but rather the line segments that form boundaries between polygons. The DIME format encodes a set of coordinates for each line segment, with end points in each stream forming vertices of polygons. Each segment is labeled with the identifies of polygons on either side of the line segment. The complete pattern organize line segments into a network of polygons.

Finally, the chain file structure (CFS) is organized around "chains" of line segments that connect key nodes within the network of boundaries (Fig. 4.11). A chain then is simply a sequence of locations that defines the boundary between two polygons. Each chain of points has an identifying label, a sequence of points and line segments that define its shape, and labels that define the identities of polygons on either side of the chain. This forms an important improvement over DIME format, at less coding is required during data entry.

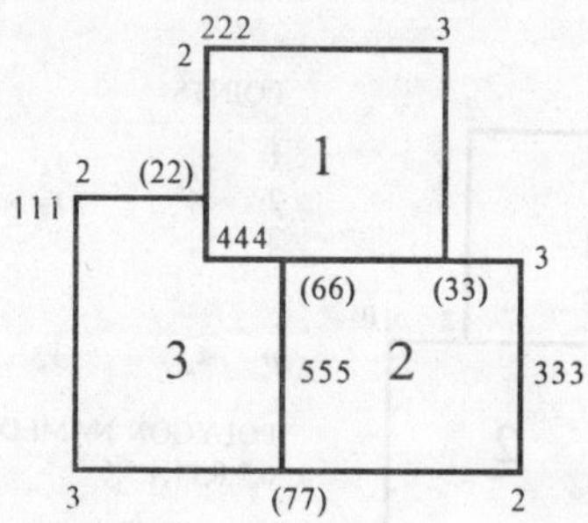

CHAIN NAME	LENGTH	FROM	TO	LEFT	RIGHT
111	3	22	77	0	3
	x_1y_1		x_3y_3		
222	3	22	33	1	0
	x_1y_1		x_2y_2		
333	3	33	77	2	0
	x_1y_1		x_3y_3		

Fig. 4.11 Chain code file. Every line is defined by a chain, consisting of a label, a string of coordinates that form its inner path, two limiting nodes (i.e., two end points), a length (number of coordinate pairs), and polygons on either side

Data for vector systems are encoded by means of a digitizing table. Polygons are digitized from paper, copies of maps, or aerial images that have been attached to the digitizing table. The analyst moves the cursor over the document, tracing the outlines of polygons and labeling each line segment with the identities of the polygons on either side of the line. Usually the digitizer is connected to a computer, so that the analyst can enter information at a keyboard, and can store the digitized information on a computer file. After the document has been digitized, the analyst must carefully edit the computer file that records the data because some line segments may have been omitted, misidentified, or incorrectly recorded.

The analyst must identify, then correct, all such errors before the data can be considered completely digitized. Equipment has been manufactured for automated digitization of polygons, although it is expensive and data still require extensive editing.

4.4 REPRESENTATION OF FEATURES

GIS can represent spatial data as the data, which have physical dimension on the earth. Figures 4.12 to 4.14 show various components of geographic data, which can be reduced in the form of point, arcs or polygons for effective processing in computer. The data may be available in tabular, geographical, digital or remotely sensed in the map form. Due to non-compatible formats of these data, it is considered as a major problem to integrate in GIS.

Point Data

Point data consist of observations that occur only at points, or occupy very small areas in relation to the scale of data base. Features such as wells, for example, illustrate data that occupy a single point even at the largest levels of detail. In contrast, features such as buildings sometimes occupy significant areas, even though they may be represented as points in the data base. A vector data base can depict point data as points and can position them accurately. In contrast, a cellular data base can depict point data only at the level of detail of a single cell, so for example a cell can show the presence of a well within the cell, but cannot show the placement of the well within the cell. Some point observations may represent quantitative measurements, such as the amount of rainfall observed at a specific point, or the depth of a specific geological unit below the surface (Fig. 4.12)

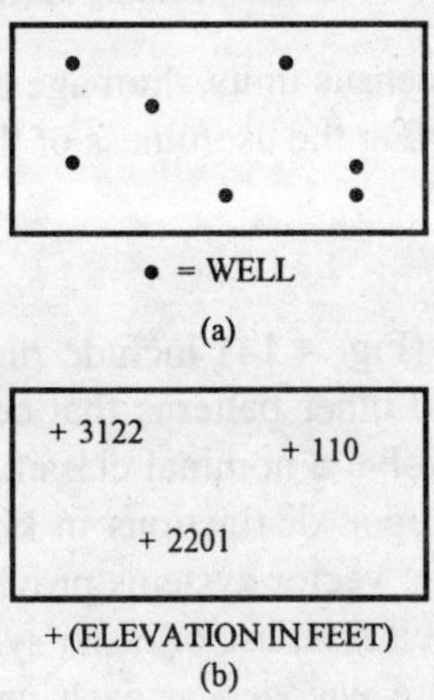

Fig. 4.12 Point data: (a) nominal, (b) quantitative

Line Data

Line data are of course exemplified by features such as highways, rivers, pipelines, or power lines. Vector systems can show line data in fine detail, whereas cellular systems can show linear data only at the resolution of the cell. Thus a raster data base can depict a linear feature only as a chain of cells encoded with a code that shows the presence or absence of a specific feature. Both raster and vector systems can depict different classes of linear features. For example a data base can encode different classes of highways or streams that permit the analyst to examine not only the entire highway or stream network but also to select only specified classes of streams or highways for analysis (Fig. 4.13).

The significance of certain forms of linear data may not be immediately appreciated. Boundaries, including political and administrative boundaries, do not necessarily form physical features on the landscape, yet often they are extremely important in determining the usefulness of a GIS.

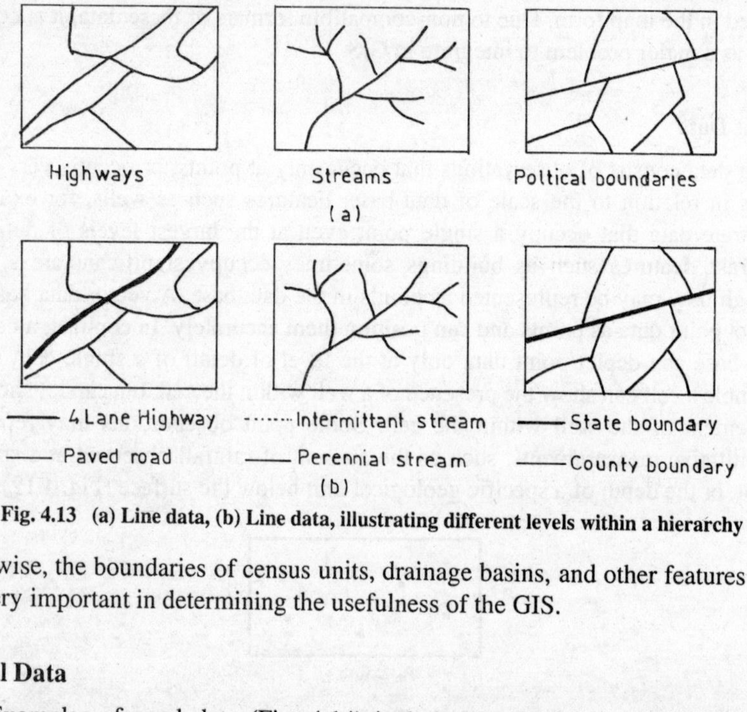

Fig. 4.13 (a) Line data, (b) Line data, illustrating different levels within a hierarchy

Likewise, the boundaries of census units, drainage basins, and other features may be very important in determining the usefulness of the GIS.

Areal Data

Examples of areal data (Fig. 4.14) include distributions such as soil, land cover, vegetation classes, and other patterns that occupy area at the scale of the GIS. Areal patterns often involve a nominal classification, such as land cover, in which the classes are based upon distinctions in kind, rather than differences in quantity. As mentioned above, vector systems provide detailed delineation of the boundaries between nominal classes; use of raster systems inevitably involves loss of information as the data are encoded at each cell. Quantitative data that are grouped by areal units, such as census units, are depicted in finer detail by polygon structures. However, continuous data, such as topographic elevation, where a continuous surface is represented by a network of equally spaced observations, are probably most directly represented by the cell-based data structure.

4.5 COMPONENT SUBSYSTEM OF GIS

The software for a GIS consists of Data input and verification, Data storage and database management, Data output, and Data transformation.

The captured data can be transformed from existing maps, detailed observations, satellites and aerial photography into digital or computer compatible format. Using digitizer/scanner etc. these data can be stored in the computer.

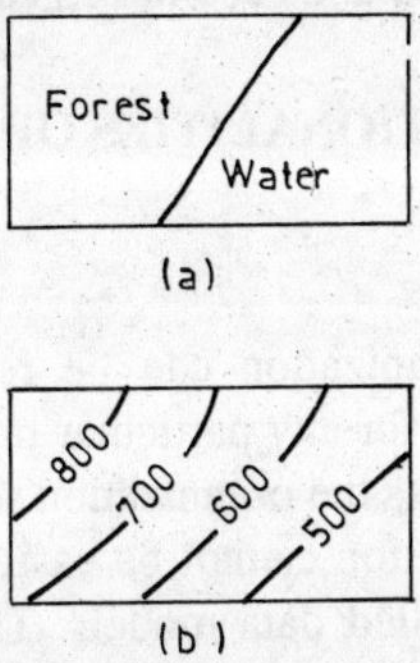

Fig. 4.14 Areal data: (a) nominal, (b) continuous

Once the data is transformed into the computer, this data has to be stored to create a permanent database for further data analysis and manipulation. The computer program that is employed to organize the database is commonly known as database management system (DBMS). The analyzed and manipulated result of the data has to be displayed or presented to the user in a user-specified format for decision making purpose. Either the data may be presented as maps, tables and figures or the screen, or recorded on magnetic media in digital format or as hard copy output drawn on printer or plotter.

A GIS may be considered to have five major component sub system (Fig. 4.15) and are as follows.

Input : Which deals with creating an image based GIS from multi-geodatasets.

Management : The purpose is efficient storage, retrieval and database management.

Processing : Data manipulation, feature enhancement and classification, etc.

Display : Display and product generation.

Output : Provides thematic maps, images, etc. for application.

4.6 PURPOSE OF GIS

The main purpose of GIS is:

1. To support making decisions based on spatial data e.g. an engineering geologist may evaluate slope stability conditions though GIS for deciding best new route.
2. To support general research.
3. To collect, manipulate and use spatial data in database management.
4. To produce standardized and customized cartographic production.

4.7 CAPABILITIES\FUNCTIONALITIES OF GIS

Organization

The importance of data organization can be revealed by a person who has collected a large mass of data for any particular purpose. Data can be arranged in many different ways, and unless the organization scheme is suitable for the application at hand, useful information cannot be easily extracted. Schemes for organizing data are sometimes called data models. The principle characteristics for

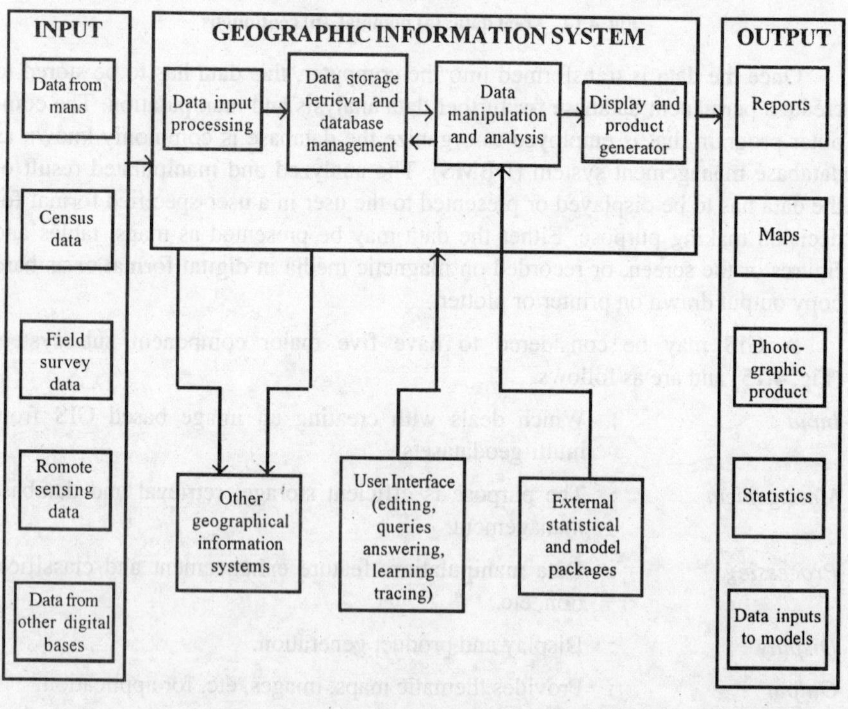

Fig. 4.15 Component subsystem of GIS

organizing GIS data is spatial location. Data models organize observations both by spatial and non-spatial attributes. Thus data organization has a fundamental importance.

Visualization

Visualization is achieved in GIS with colour and by specialized methods using perspective, Shadowing and other means. The graphical capabilities of computers are exploited by GIS for visualization. Visual display is normally carried out using the video monitor, but other output devices such as colour printers are used for hardcopy displays.

Spatial Query

Spatial query is a complementary activity to data visualization. Visualization reveals spatial pattern amongst collections of organized data items. However, visualization is not so harmful for answering question about special instances in the data, such as the value of particular data items. A GIS has the capacity to satisfy following queries:

About location: What exist at a particular location. The location of the particular region can be described in many ways using place name, post or pin code, or geographic references, such as latitude and longitude.

Condition: This query is converse to the first and requires spatial analysis to answer. Instead of identifying what exists at a certain location, one seek to find a location where certain conditions are satisfied (e.g. wasteland of at least 1500 sq. meters in size of forested area).

Pattern: This query is more sophisticated and important as one might want to know how many anomalies within an area over a time.

Trends: This query might involve both location and conditions and seeks to find different within an area over a time.

Modeling: This query is posed to determine what happens if some addition or changes are done in the existing network e.g. if a new road is added to a network or if toxic substances seeps into the ground water to local water supply. For answering these queries both geographic and other information and possibly even scientific laws.

This queries required efficient search of data items and capability for deriving their geometric and topological attributes.

Combination

The ability to merge spatial data sets from quite different sources and display and manipulate combination can often lead to an understanding and interpretation of spatial phenomena that are simply not apparent when individual spatial data types are considered in isolation. For example, by superimposing a digitized geological map on to a satellite image, it may become clear that a particular lithology has a distinctive texture or spectral response on the image. The process of combining layers of spatial data is sometimes called data, integration and can be carried out either by visualizing composite displays or various kinds, or with integration models that effectively create a new map from two or more existing maps.

The one of the really powerful features of GIS is the ability to link several maps together to form more complex algorithms. Several maps and tables of attribute data can be combined in a single processing step. The process of combining maps together is often called map or Cartographic modeling.

Prediction

Prediction is one of the purpose of GIS. For example, a Number of data layers indicative of gold deposits might be combine together to predict the favourability for gold as a new map. Such a map may then be used as a basis for making exploration decisions or land use decisions, such as "Is these region suitable as national park?" Prediction is some times research exercise to explore the out come of making a particular set of assumptions, often with the purpose of examining the performance of a model. For example one might be interested in the number and area of sites predicted as unsuitable with a slope and soil saturation, simply to evaluate model performance and sensitivity.

Basic Notations

GIS are the information systems, which offer the functionality and tools to collect, store, retrieve, analyze and display geographical information.

Features events and activities with spatial components are modeled as points, lines polygons, nets or links to form the geographical database under a geo-relational schemes. The geometrical part is defined by a series of coordinates and is connected through the feature code with the attribute tables where the non-spatial information such as the properties, symbolism etc. is stored.

Because of the complexity of the attribute tables that may be required to model the geographic information they are formed and operated under a Relational Data Base Management System (RDBMS). Thus complicated relation can be created and the same geographical data can be seen from various views to fulfill various requirements. The schematic of RDBMS is shown as below Fig. 4.16.

The geographical features are modeled as points when only their location is of interest, and their shapes or dimensions. Wells, surveying points, rain guages, etc. are example of using points. Lines are used when location and the dimensional shape is required or exist. Borders, networks, etc. are some example where lines are used. Polygons are used squares, buildings, etc. are the cases where polygons are used.

For each X, Y, pairs which defines the location of a point, and identification number (ID) exist for that and connects the line with the related record of the attribute table, the point attribute table (PAT). For each line which is defined by a chain of x, y pair an ID exists for it and connects the line with the related record of the Arc Attribute table.

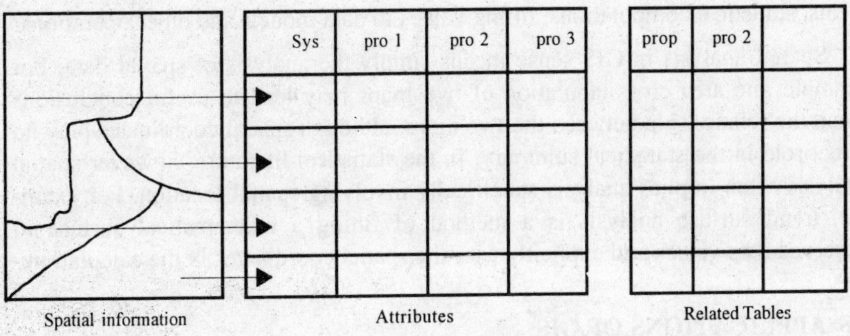

Fig. 4.16 Schematic of RDBMS in GIS

Geometric information is acquired by the means of manual digitizing, scanning, analytical coordinate data built is ready to be related to the corresponding attribute records. During this topological built a series of frequent used fields are generated with their various update versions with information which is related with geometry and the topological; relation between entities. Therefore, the generated polygon give the area, perimeter, etc. and the poly lines give the length, the representative number of the section from node to node and the numbers of the polygons (left and right) which are drawn separately by this arc.

The polygon and line topology are defining the net features in order to have complete geographical information about the aerial features as well as about their borders. Hence, the information about neighbouring polygon can be retrieved and identities of their borders can be defined and be available in any query. The link features have been proved to be beneficial for defining line features and connecting with tables of points information.

The above discussion has made it clear that for designing a geographical data base to be used for a special application, a serious of description have to be decided and definition to be given in order to fulfill the specific requirements of the application.

The definition and decisions concern:

1. The necessity operations and function of the application.
2. The structuring of the spatial information of the geographical data.
3. The structuring of the attribute information of the geographical data.

These three requirements lead to conceptual data base design and the physical data base design.

Analysis

Analytical possibilities in GIS basically and usually refer to polygon overlay with logical operation, buffering in vector maps, interpolations, zoning and simplified network analysis. It is a process of inferring meaning from data. It is often carried out visually in GIS. Analysis in GIS can also be carried out by measurements, statistical computations, fitting values to data models and other operations.

Spatial analysis in GIS sense means simply the analysis of spatial data. For example, the area cross-tabulation of two maps may lead to useful conclusions about the relationship between the two maps, although spatial coordinates play no direct role in the statistical summary. In the statistical literature, however spatial analysis often implies analysis specifically involving spatial location. For example, trend surface analysis is a method of fitting a mathematical surface to observed data values and explicitly uses the spatial coordinates in the calculation.

4.8 APPLICATIONS OF GIS

Some of the applications of GIS are as follows:

1. In business
2. In rural development
3. In risk assessment
4. In infrastructure development
5. In urban land use and transport planning
6. In Indian railways
7. A national GIS (as a map)
8. Application of digital terrain modeling for automated mapping facility mapping (AM/FM) programs.
9. In environmental protection and management
10. In small hydroplanning and resource management
11. In malarial vector distribution.

4.9 GIS BASED SOFTWARES

ARC/INFO

ARC/INFO from ESRI (Environmental Systems Research Institute), is an example of a system employs the hybrid GIS design. In one of the most popular GIS software systems, ARC/INFO was first released in 1982 and has been extended and substantially redeveloped on several occasions. Although best known as vector based system, it also have extensive complementary raster processing capabilities. The overlay, buffering, aggregation, classification descriptive statistics, and graphic and cartographic display tools have all been widely used to assist in spatial analysis. These remain extremely valuable tools in the spatial analysis tool base.

ArcView

ArcView is a powerful easy to use tool that brings GIS to the desktop. ArcView gives the power to visualize, explore, query and analyses data spatially. ArcView can be used by anyone who wants to work spatially. A key feature of ArcView is that it is easy to load tabular data. Such as dBase files and data from data base servers into an ArcView so that you can display query, summarize and organize this data geographically.

4.10 REMOTE SENSING AND GIS

GISs have evolved side by side with remote sensing systems, so that there are many close relationships between the two.

1. Remote sensing systems contribute data to GISs. Remotely sensed data can provide timely information at low cost and in a form that is compatible with requirements of GIS.
2. Both GISs and digital remote sensing systems use similar equipments and similar computer programs for analysis and display. Therefore, expertise in one field tends to form the foundation for work in the other.
3. The nonremote sensing data from a GIS can be used to assist in the analysis of remotely sensed images.

Thus, remote sensing and GISs, have natural, mutually supporting relationships to each other.

EXERCISE

1. Define geographic information system. What are the various sources from which data can be derived to be used for GIS system?
2. Describe briefly the rater and vector data structures. What are the relative advantages and disadvantages of the two data structures?
3. Describe briefly the following with regards to representation of geographical features:
   (a) Point data
   (b) Line data
   (c) A real data
4. What are the component subsystems of GIS? Describe them briefly.
5. What are the function abilities of GIS?
6. Briefly describe the applications of GIS. Discuss, how closely GIS is related to remote sensing.

PART– B

Engineering Mechanics

CHAPTER 1
Fundamental Concepts

Why do bodies move as they do? Why do bodies remain at rest as they do? These questions had engaged the attention of philosophers, mathematicians and scientists for many centuries. However, only by the year 1686, Sir Issac Newton, the principal architect of mechanics, consolidated the philosophy and experimental findings developed around these questions and put forth them in the form of three laws of motion. Newton's first law of motion provides us the basis for developing a practical definition of force and the second law provides us the basis for developing a practical definition of the magnitude of force.

Newton's First Law states: *Every body continues in its state of rest or of uniform motion in a straight line unless it is compelled to change that state by forces acting on it.* This law leads us to the definition of force as that which changes or tends to change the condition of rest or of steady linear motion of a body.

Newton's Second Law of motion states: *The rate of change of momentum of a body is directly proportional to the impressed force and it takes place in the direction in which the force acts.* That is,

Force $\propto$ rate of change of momentum

Now,

$$\text{momentum} = \text{mass} \times \text{velocity}$$

$\therefore$ Rate of change of

$$\text{momentum} = \text{mass} \times \text{rate of change of velocity}$$
$$= \text{mass} \times \text{acceleration}$$

$\therefore$ Force $\propto$ mass $\times$ acceleration, that is,

$$F \propto m \times a \qquad \textit{Eqn. } 1.1$$

where F is force

m is mass

and a is acceleration.

The proportionality equation 1.1 does not provide us a means for measuring force. Different systems for measurement of forces have been developed by converting the proportionality equation 1.1 into an equation in the form $F = kma$ (where k is a constant). Defining the unit of force as a force required to move unit mass with unit accelerating, constant of proportionality k may be made unity. This leads to the practical definition of **unit force** as the *force required to produce unit acceleration in a body of unit mass.*

1.1 SYSTEMS OF UNITS

Three different systems of units are most commonly used in science and engineering. The difference among them arises out of the difference in the units used to represent the three fundamental quantities, namely, length (l), mass (m) and time (t). The three systems are:

– Metre-Kilogramme-Second (**MKS**) system,

– Centimeter-Gramme-Second (**CGS**) system, and

– Foot-Pound-Second (**FPS**) system.

The units of length, mass and time used in the above-mentioned systems are denoted by the abbreviated names of the systems themselves.

However, presently the whole world is in the process of switching over to **SI system of units**. SI stands for *Systéme Internationale d' Unités* or *International System of Units* and in this book SI units are used throughout.

As in **MKS** system in **SI** system also, the fundamental units are kilogramme for mass, metre for length and second for time. The difference between **MKS** system and **SI** system arises from the diference in mainly selecting unit of force. The unit for force is Newton (denoted by N) in **SI** units while in **MKS** system, it is Kilogramme-weight (kg-wt). Newton is the amount of force which causes one kilogramme mass to move with an acceleration of 1 m/sec^2. Kilogramme-weight (kg-wt) is the force required to move a mass of one kilogramme with an acceleration equal to gravitational acceleration. Gravitational acceleration g is equal to 9.81 m/sec^2 near the earth's surface, but in all problems invariably encountered by engineers, variation in g is negligible. Hence

1 kg-wt = 9.81 **N** *Eqn.* 1.2

It may be noted that in practice kg-wt is called kg (kilogramme) only.

The following SI prefixes are used when quantities are too big or too small.

Multiplying Factor	Prefix	Symbol
10^{12}	tera	**T**
10^{9}	giga	**G**
10^{6}	mega	**M**
10^{3}	kilo	**k**
10^{0}	–	–
10^{-3}	milli	**m**
10^{-6}	micro	**μ**
10^{-9}	nano	**n**
10^{-12}	pico	**p**
10^{-15}	femto	**f**
10^{-18}	atto	**a**

1.2 CHARACTERISTICS OF A FORCE

A force is completely defined only when the following four characteristics are specified:

- Magnitude,
- Point of application,
- Line of action, and
- Direction.

In Fig. 1.1, *AB* is a ladder kept against a wall. At point *C*, a person weighing 600 N is standing. The force applied by the person on the ladder has the following characteristics:

- Magnitude is 600 N,
- The point of application is at *C* which is 2m from the floor along the ladder,
- The line of action is vertical, and
- The direction is downward.

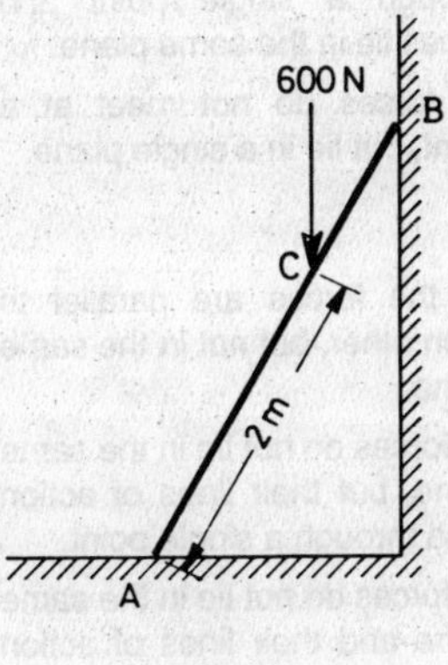

Fig. 1.1

Note that the magnitude of the force is written near the arrow. The line of the arrow shows the line of application, and the arrow head represents the point of application and the direction of the force.

1.3 SCALAR AND VECTOR QUANTITIES

A quantity is said to be **scalar** if it is completely defined by its magnitude alone. Examples are length, area and time.

A quantity is said to be **vector** if it is completely defined only when its magnitude and direction are specified. Hence, force is a vector. The other examples of vectors are velocity and acceleration.

1.4 SYSTEM OF FORCES

When several forces of different magnitude and direction act upon a body, they constitute a *system of forces*. If all the forces in a system lie in a single plane, it is called a *coplanar force system*. If the line of action of all the forces in a system pass through a single point it is called a *concurrent force system*. In a system of *parallel forces* all the forces are parallel to each other. If the line of action of all forces lie along a single line then it is called a *collinear force system*. Various system of forces, their characteristics and examples are given in Table 1.1.

Table 1.1 System of Forces

Force System	*Characteristic*	*Examples*
Collinear forces	Line of action of all the forces act along the same line.	Forces on a rope in a tug of war
Coplanar parallel forces	All forces are parallel to each other and lie in a single plane.	System of forces acting on a beam subjected to vertical loads (including reactions)
Coplanar like parallel forces	All forces are parallel to each other, lie in a single plane and are acting in the same direction.	Weight of a stationary train on a rail when the track is straight
Coplanar concurrent forces	Line of action of all forces pass through a single point and forces lie in the same plane.	Forces on a rod resting against a wall
Coplanar nonconcurrent forces	All forces do not meet at a point, but lie in a single plane.	Forces on a ladder resting against a wall when a person stands on a rung which is not at its centre of gravity
Non-coplanar parallel forces	All the forces are parallel to each other, but not in the same plane.	The weight of benches in a class room
Non-coplanar concurrent forces	All forces do not lie in the same plane, but their lines of action pass through a single point.	A tripod carrying a camera
Non-coplanar non-concurrent forces	All forces do not lie in the same plane and their lines of action do not pass through a single point.	Forces acting on a moving bus

1.5 IDEALISATIONS IN MECHANICS

A number of ideal conditions are assumed to exist when applying the principles of mechanics to practical problems. In fact, without such assumptions it is not feasible to arrive at applicable solutions. The experience of many centuries has shown that such assumptions–idealisations–do not bring down the accuracy of the results of analysis below the optimum level required by engineers to deal with practical problems. One such idealisation is **continuum**.

A body consists of several particles. It is a well-known fact that each particle can be sub-divided into molecules, atoms and electrons. It is not feasible to solve any engineering problem by treating a body as a conglomeration of such discrete particles. The body is assumed to consist of a continuous distribution of matter. In other words, the body is treated as a **continuum**. The concept of continuum of bodies enables us to treat the bodies as rigid and this simplifies the problems in engineering mechanics.

A **rigid body** may be defined as a body in which the relative positions of any two particles do not change under the action of the forces. For example, consider the body shown in Fig. 1.2. Many engineering problems are solved by assuming that the distance between *A* and *B* [Fig. 1.2(a)] is the same as between *A′* and *B′* [Fig. 1.2(b)] where *A′* and *B′* are the altered positions of the two particles *A* and *B* on the body after application of the forces P_1, P_2 and P_3.

Similarly, some objects are treated as particles. A **particle** may be defined as an object which has only mass, and no size. Such a body cannot exist theoretically, but when dealing with problems involving distances considerably larger when compared to the size of the body, the size of the body can be neglected without sacrificing accuracy. The examples are:

- A bomber aeroplane is a particle for a gunner operating from ground.
- A ship in mid-sea is a particle when studying its relative motion from a control tower in a port.
- In the study of movement of the earth in celestial sphere, earth is treated as a particle.

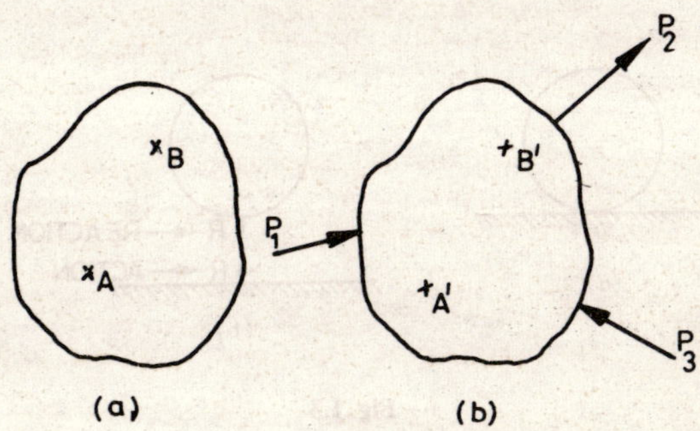

Fig. 1.2

Point force is yet another idealisation very commonly used in engineering mechanics. In Fig. 1.1, the weight of the man standing on the ladder is shown as a force applied at the point *C*. Actually the man cannot apply his weight through a single point. There is certain area of contact which is, however, small compared to the other dimensions involved in the problem. Therefore, not much accuracy is lost by treating it as a point force and thereby simplifying the problem.

There are many similar idealisations assumed in engineering mechanics to find the solution for practical problems, however, without sacrificing the optimum accuracy required. They relate to support conditions of bodies and about the system of forces. They are explained in the text alongwith the relevant topics.

1.6 FUNDAMENTAL LAWS OF MECHANICS

The following are considered fundamental laws in classical mechanics:

- Newton's first law,
- Newton's second law,
- Newton's third law,
- Law of transmissibility of forces, and
- Parallelogram law of forces.

The first two laws have been explained in the begining of this chapter. The others are explained in the following paragraphs.

Newton's Third Law states: *For every action there is an equal and opposite reaction.* Assume two bodies are in contact with each other and one body applies a force R on the other. At the point of contact, the second body develops a reactive force which is equal in magnitude to R, and acts in the same line as that of R, but opposite in direction.

In Fig. 1.3 the action of the ball and the reaction of the floor are shown. In Fig. 1.4 the action of the ladder on the wall and the floor and the reactions of the wall and the floor are illustrated.

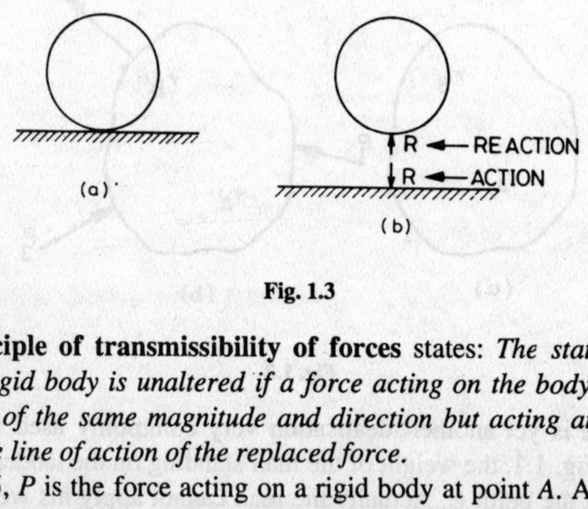

Fig. 1.3

The **Principle of transmissibility of forces** states: *The state of rest or of motion of a rigid body is unaltered if a force acting on the body is replaced by another force of the same magnitude and direction but acting anywhere on the body along the line of action of the replaced force.*

In Fig. 1.5, P is the force acting on a rigid body at point A. According to the principle of transmissibility of forces, this force has the same effect on the body as a force P applied at point B.

Reiteration of the following two points is considered necessary in this context.

1. In Engineering Mechanics we deal with only rigid bodies. If deformation of the body is to be considered in a problem, the law of transmissibility of forces will not hold good.
2. By transmission of the force, only the state of the body is unaltered, but not the internal stresses which may develop in the body.

   Hence the law of transmissibility of forces can be applied only to problems in which rigid bodies are involved.

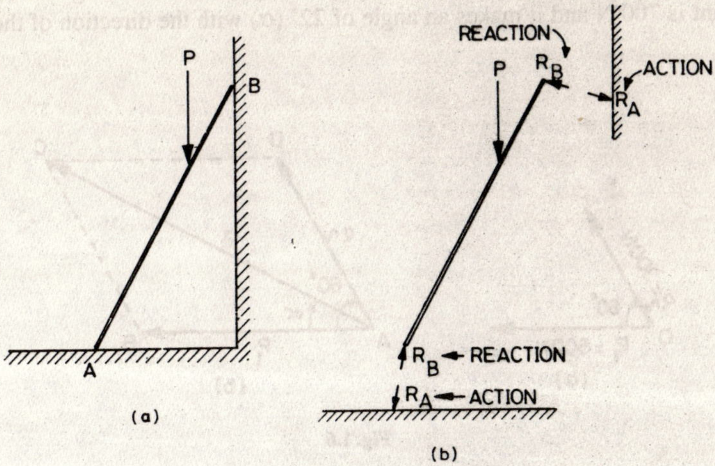

Fig. 1.4

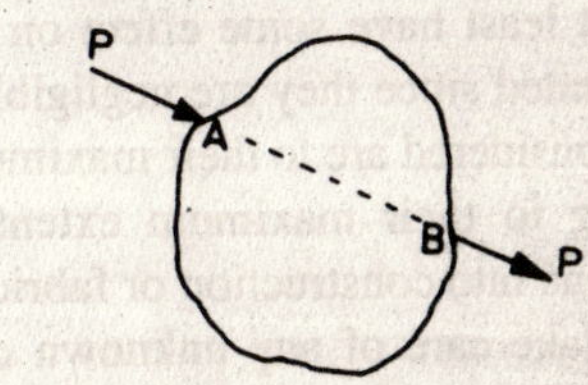

Fig. 1.5

The **Parallelogram law of forces** was formulated based on experimental results. Though Stevinus employed it in 1586, the credit of presenting it as a law goes to Varignon and Newton (1687). The parallelogram law of forces enables us to determine the single force (called Resultant) in magnitude and direction which can replace the two forces acting at a point with the same effect as that of the two forces. The law states: *If two forces acting simultaneously on a body at a point are represented in magnitude and direction by the two adjacent sides of a parallelogram, their resultant is represented in magnitude and direction by the diagonal of the* parallelogram which passes through the point *of intersection of the two sides representing the forces.*

In Fig. 1.6(a), two forces $P_1 = 500$ N and $P_2 = 300$ N are acting at point O, the angle between them being 60°. The parallelogram $ABCD$ shown in Fig. 1.6(b) is constructed such that the side AB represents P_1 and AD represents P_2. Then, according to the parallelogram law of forces, the diagonal AC which passes through the intersection of AB and AD represents the resultant of the two forces P_1

and P_2 in magnitude and the direction. In this problem, the magnitude of the resultant is 700 N and it makes an angle of 22° (α) with the direction of the force P_1.

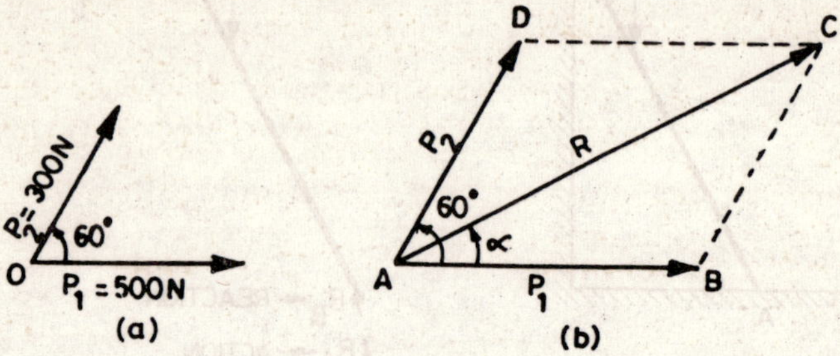

Fig. 1.6

1.7 ACCURACY OF CALCULATIONS

As explained in 1.6, a number of ideal conditions are assumed to exist while applying the principles of mechanics to practical engineering problems. These assumptions theoretically at least have some effect on the accuracy of the final results, but they are not counted since they are negligibly small. Further, in engineering analyses all loads considered are to their maximum extent. The possibility of some of them occurring to their maximum extent is remote. Then, while translating design calculations into construction or fabrication, an appropriate factor of safety is applied to take care of any unknown conditions. In view of all these, there is no point in noting down the calculations and answers beyond four digits. It is found that 0.2% accuracy, which is slide rule accuracy, is more than sufficient though electronic calculators give accuracy upto eight figures.

CHAPTER 2

Coplanar Concurrent Forces

If all the forces in a system lie in a single plane and pass through a single point, then the system constitutes a coplanar concurrent force system. In this chapter composition, resolution and equilibrium of such force systems are explained with application to many engineering problems.

2.1 COMPOSITION OF FORCES

It is possible to find a single force which will have the same effect as that of a number of forces acting on a body. Such a single force is called **Resultant** force and the process of finding out the resultant force is called **composition of forces.**

The fundamental principle of finding the resultant of a force is the parallelogram law of forces which is explained in Chapter 1. According to this law, if P_1 and P_2 are the two forces acting on a body as shown in Fig. 2.1, the resultant force can be obtained by constructing a parallelogram ABCD in which the sides AB and AD represent the forces P_1 and P_2 in magnitude and direction. Then, the diagonal AC represents the resultant R in magnitude and direction.

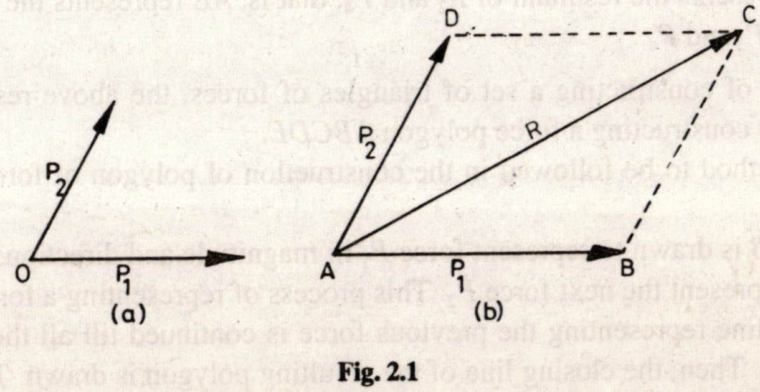

Fig. 2.1

It may be observed that the same result can be obtained by constructing the triangle ABC instead of constructing the parallelogram $ABCD$.

Line AB is drawn to represent the force P_1 and BC to represent P_2. The closing line AC of triangle ABC represents the resultant force in magnitude, line of action and direction.

Thus, **Triangle Law of Forces** can be stated as: *If two forces acting simultaneously on a body are represented by the sides of a triangle taken in order, their resultant is represented by the closing side of the triangle taken in the opposite order.*

If more than two forces are acting on a body, two forces at a time can be combined by the triangle law of forces and finally the resultant of all the forces acting on the body can be obtained.

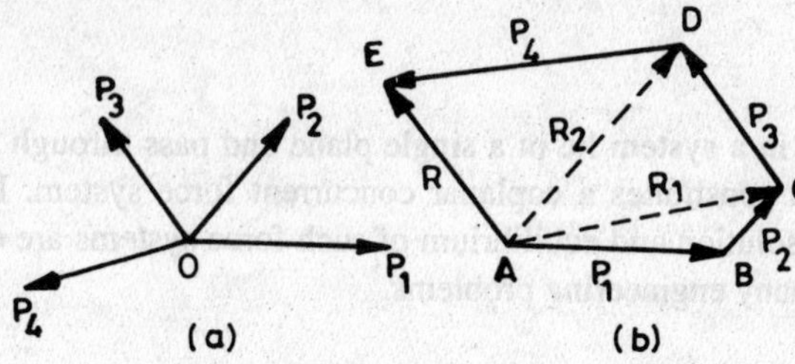

Fig. 2.2

In Fig. 2.2(a) a system of concurrent forces acting on a body are shown. In Fig. 2.2(b),

AB represents P_1 and *BC* represents P_2

As per triangle law of forces, *AC* represents the resultant of P_1 and P_2, say, R_1. Now *CD* is drawn to represent P_3. Then, in triangle *ACD*, *AC* represents R_1, *CD* represents P_3 and therefore

AD represents the resultant of R_1 and P_3, that is, *AD* represents the resultant of P_1, P_2 and P_3, say, R_2. Similarly, If *DE* is drawn to represent P_4 in triangle *ADE*,

AD represents R_2, and *DE* represents P_4.

AE represents the resultant of R_2 and P_4, that is, *AE* represents the resultant of P_1, P_2, P_3 and P_4.

Instead of constructing a set of triangles of forces, the above result may be obtained by constructing a force polygon *ABCDE*.

The method to be followed in the construction of polygon of forces is given below:

Line *AB* is drawn to represent force P_1 in magnitude and direction. Line *BC* is drawn to represent the next force P_2. This process of representing a force from the end of the line representing the previous force is continued till all the forces are represented. Then, the closing line of the resulting polygon is drawn. This closing line represents the resultant in magnitude as well as in direction. It may be noted that the resultant is directed towards the last point.

Thus, the **Polygon Law of Forces** may be stated as: *If a number of concurrent forces acting simultaneously on a body, are represented in magnitude and direction by the sides of a polygon, taken in order, then the resultant is represented in magnitude and direction by the closing side of the polygon, taken in opposite order.*

2.2 ANALYTICAL METHOD

The graphical method of finding the resultant using parallelogram law of forces, triangle law of forces and polygon law of forces gives a clear picture of the work being carried out. However, the main disadvantage is that one needs drawing aids like pencil, scale, drawing sheet, etc. Hence, many a times, it is advantageous to go for analytical method. Therefore there is need for the analytical method.

Consider two forces P_1 and P_2 acting on a body as shown in Fig. 2.3(a). Let the angle between the two forces be θ. The diagonal *AC* of the parallelogram *ABCD* shown in Fig. 2.3(b) represents the resultant. Drop perpendicular *CE* to *AB*.

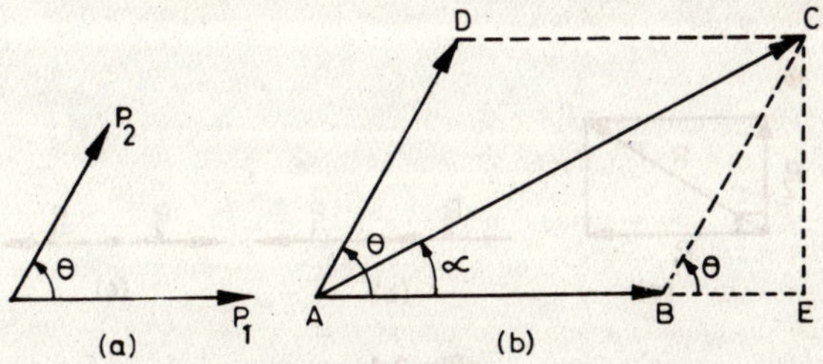

Fig. 2.3

Now the resultant *R* of P_1 and P_2 is given by:

$$R = AC$$

$$= \sqrt{AE^2 + CE^2}$$

$$= \sqrt{(AB + BE)^2 + CE^2}$$

But $$AB = P_1$$

$$BE = BC\cos\theta = P_2\cos\theta$$

and $$CE = BC\sin\theta = P_2\sin\theta$$

∴ $$R = \sqrt{(P_1 + P_2\cos\theta)^2 + (P_2\sin\theta)^2}$$

$$= \sqrt{P_1^2 + 2P_1P_2\cos\theta + P_2^2\cos^2\theta + P_2^2\sin^2\theta}$$

$$= \sqrt{P_1^2 + 2P_1P_2\cos\theta + P_2^2} \qquad Eqn.2.1$$

The inclination of the resultant to the direction of the force P_1 is given by:

$$\tan\alpha = \frac{CE}{AE} = \frac{P_2\sin\theta}{P_1 + P_2\cos\theta}$$

$$\alpha = \tan^{-1}\frac{P_2 \sin\theta}{P_1 + P_2\cos\theta} \qquad \textit{Eqn.2.2}$$

Particular cases:

1. When $\theta = 90°$ [Fig. 2.4(a)] $R = \sqrt{P_1^2 + P_2^2}$
2. When $\theta = 0°$ [Fig. 2.4(b)] $R = \sqrt{P_1^2 + 2P_1P_2 + P_2^2} = P_1 + P_2$
3. When $\theta = 180°$ [Fig. 2.4(c)] $R = \sqrt{P_1^2 - 2P_1P_2 + P_2^2} = P_1 - P_2$

It is clear from the cases shown in Fig. 2.4(b) and 2.4(c) that when the forces acting on a body are collinear, their resultant is equal to the algebraic sum of the forces.

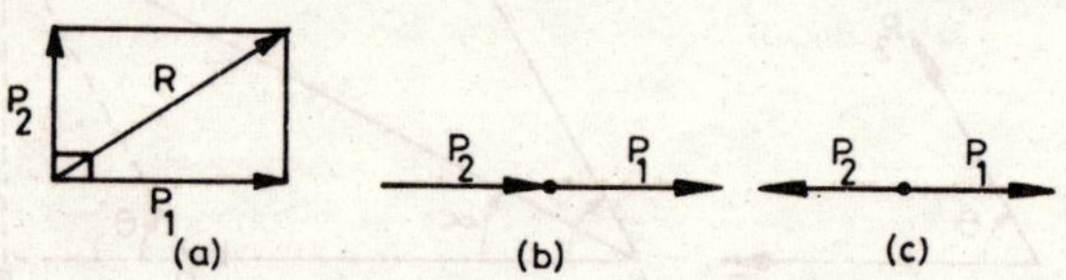

Fig. 2.4

2.3 RESOLUTION OF FORCES

The **Resolution** of forces is exactly the opposite process of composition of forces. It is the process of finding a number of component forces which will have the same effect on the body as the given single force. Exactly the opposite process of composition can be employed to get the resolved component of a given force.

In Fig. 2.5(a) the given force P is resolved into two components making angles α and β with P.

In Fig. 2.5(b) the force P is resolved into its rectangular components P_X and P_Y.

In Fig. 2.5(c) the force P is resolved into four components P_1, P_2, P_3 and P_4.

Note that all component forces act at the same point as the given force. Resolution of a force into its rectangular components is more useful for the analysis.

Ex. 2.1: The guy wire of an electric pole shown in Fig. 2.6(a) makes 30° to the pole and is subjected to 20 KN force. Find the vertical and horizontal components of the force.

From the triangle of forces,
vertical component P_V of P is given by:

$$P_V = P\cos 30° = 20\cos 30°$$

$$\mathbf{P_V = 17.321\ kN(down\ word)} \qquad \textit{Ans.}$$

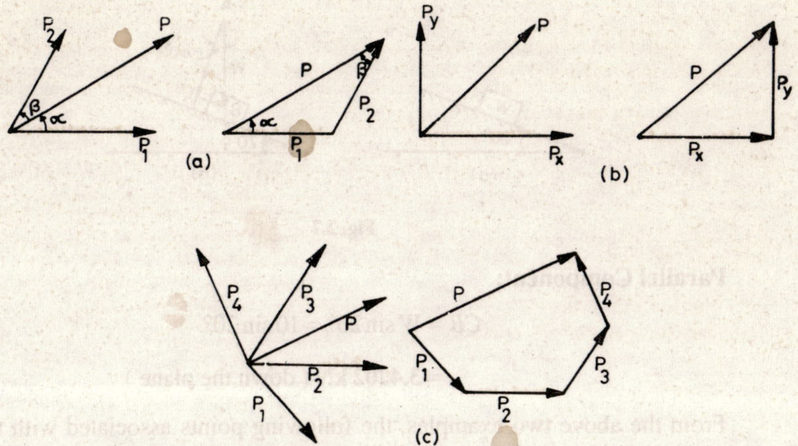

Fig. 2.5

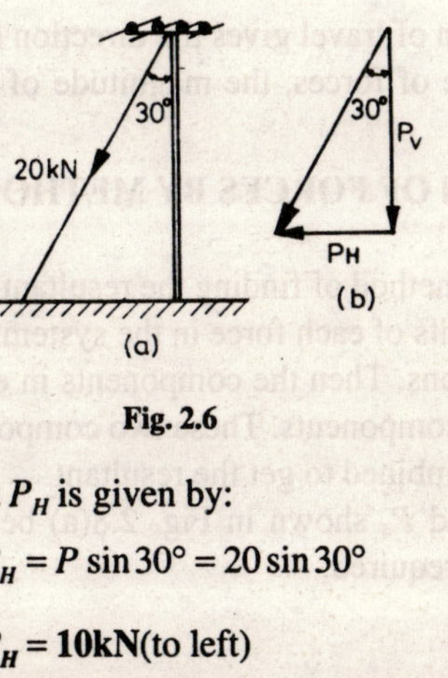

Fig. 2.6

and the horizontal component P_H is given by:

$$P_H = P \sin 30° = 20 \sin 30°$$

$$P_H = \mathbf{10kN}\text{(to left)} \qquad \textit{Ans.}$$

Ex: 2.2: A block weighing W = 10 kN is resting on an inclined plane as shown in Fig. 2.7(a). Determine its components normal to, and parallel to the inclined plane. The plane makes an angle of 20° with the horizontal. Hence the normal to the plane makes an angle of 20° with the vertical.

In Fig. 2.7(b), if *AB* represents the given force to some scale, *AC* represents the normal component and *CB* represents component parallel to the plane.

From Δ *ABC*:

Normal Component:

$$AC = W \cos 20° = 10 \cos 20°$$

$$= \mathbf{9.3969\ kN} \text{ (thrust on the plane)} \qquad \textit{Ans.}$$

Fig. 2.7

Parallel Component:

$$CB = W \sin 20° = 10 \sin 20°$$

$$= \mathbf{3.4202\ kN}\ (\text{ down the plane }) \qquad Ans.$$

From the above two examples, the following points associated with the analytical method of finding the components of a force may be noted:

1. Imagine that the arrow drawn represents given force to some scale.
2. Travel from tail of the force line to arrow head of the force in the direction of coordinates.
3. Then the direction of travel gives the direction of component forces.
4. From the triangle of forces, the magnitude of the components can be calculated.

2.4 COMPOSITION OF FORCES BY METHOD OF RESOLUTION

This is an analytical method of finding the resultant of a multiples of forces. In this method the components of each force in the system are first found in two mutually perpendicular directions. Then the components in each direction are algebraically added to get the two components. These two component forces which are mutually perpendicular, are combined to get the resultant.

Let P_1, P_2, P_3 and P_4 shown in Fig. 2.8(a) be the system of four forces the resultant of which is required.

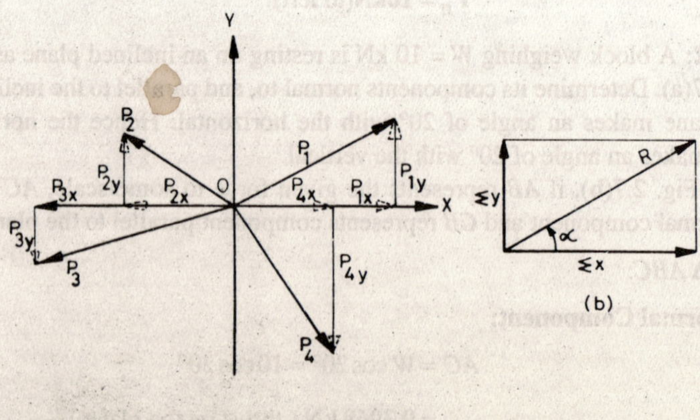

Fig. 2.8

The procedure to get the resultant is given below:

Step 1: Find the components of all the forces in X and Y directions. Thus, P_{1X}, P_{2X}, P_{3X}, P_{4X}, P_{1Y}, P_{2Y}, P_{3Y} and P_{4Y} are obtained.

Step 2: Find the algebraic sum of the component forces in X and Y directions.

$$\sum X = P_{1X} + P_{2X} + P_{3X} + P_{4X}$$

$$\sum Y = P_{1Y} + P_{2Y} + P_{3Y} + P_{4Y}$$

(*Note*: In the above case P_{2X}, P_{3X}, P_{3Y} and P_{4Y} have negative values.)

Step 3: Now the system of forces is equal to two mutually perpendicular forces, namely, $\sum X$ and $\sum Y$ as shown in Fig. 2.8(b). Since these two forces are perpendicular, the parallelogram of forces becomes a rectangle. Hence the resultant R is given by:

$$R = \sqrt{(\sum X)^2 + (\sum Y)^2} \qquad \textit{Eqn.2.3}$$

and its inclination to X axis is given by:

$$\alpha = \tan^{-1}\left(\frac{\sum Y}{\sum X}\right) \qquad \textit{Eqn.2.4}$$

Note:

$$R\cos\alpha = \sum X \qquad \textit{Eqn.2.5}$$

and $$R\sin\alpha = \sum Y \qquad \textit{Eqn.2.6}$$

i.e. $\sum X$ and $\sum Y$ are the X and Y components of the resultant.

Ex: 2.3: Determine the resultant of the three forces acting on a hook shown in Fig. 2.9(a).

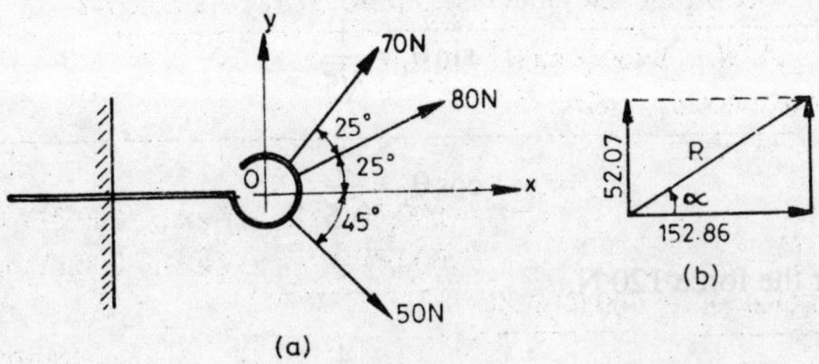

Fig. 2.9

Force	x component	y component
70N	45.00	53.62
80N	72.50	33.81
50N	35.36	–35.36
	$\sum X = 152.86$	$\sum Y = 52.07$

$$R = \sqrt{152.86^2 + 52.07^2}$$

i.e. $\quad R = \mathbf{161.48\ N}$ *Ans*.

$$\alpha = \tan^{-1}\frac{52.07}{152.86}$$

i.e. $\quad \alpha = \mathbf{18.81°}$ as shown in Fig.2.9(b). *Ans*.

Ex. 2.4: A system of four forces acting on a body is as shown in Fig. 2.10(a). Determine the resultant.

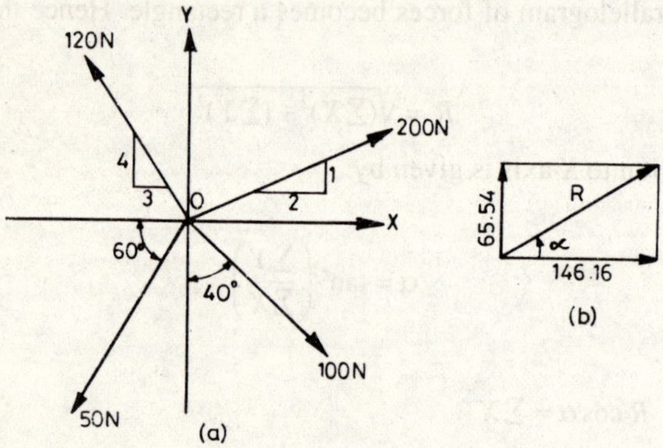

Fig. 2.10(a, b)

If θ_1 is the inclination of the 200 N force to x axis,

$$\tan\theta_1 = \frac{1}{2}$$

$$\sin\theta_1 = \frac{1}{\sqrt{5}}$$

$$\cos\theta_1 = \frac{2}{\sqrt{5}}$$

Similarly for the force 120 N,

$$\tan\theta_2 = \frac{4}{3}$$

$$\sin\theta_2 = \frac{4}{5}$$

$$\cos\theta_2 = \frac{3}{5}$$

$$\text{Now, } \Sigma X = 200 \times \frac{2}{\sqrt{5}} - 120 \times \frac{3}{5} - 50 \cos 60° + 100 \sin 40°$$

$$= 146.16 \text{ N.}$$

$$\Sigma Y = 200 \times \frac{1}{\sqrt{5}} + 120 \times \frac{4}{5} - 50 \times \sin 60° - 100 \cos 40°$$

$$= 65.54 \text{ N}$$

$$R = \sqrt{146.16^2 + 65.54^2}$$

$$\mathbf{R = 160.18\ N} \qquad \textit{Ans.}$$

$$\alpha = \tan^{-1} \frac{65.54}{146.16} = \mathbf{24.15°} \text{ as shown in Fig. 2.10(b)} \qquad \textit{Ans.}$$

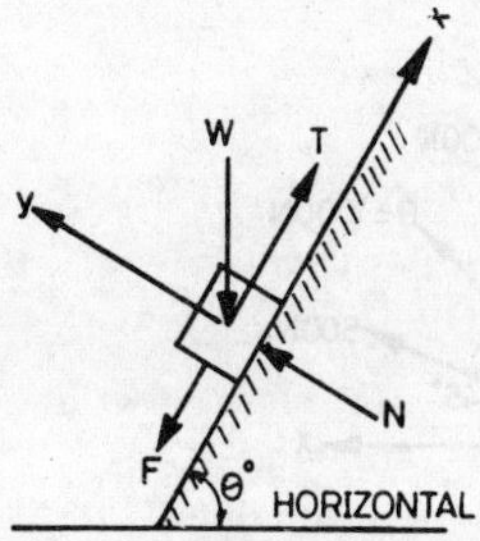

Fig. 2.11

Ex. 2.5: A system of forces acting on a body resting on an inclined plane is as shown in Fig. 2.11. Determine the resultant force if $\theta = 60°$ and if $W = 1000$ N; $N = 500$ N; $F = 100$ N; and $T = 1200$ N.

In this problem, note that selecting X and Y axes parallel to the plane and perpendicular to the plane is convenient.

$$\Sigma X = T - F - W \sin\theta$$

$$= 1200 - 100 - 1000 \sin 60°$$

$$= 233.97\text{N}$$

$$\Sigma Y = N - W \cos 60° = 500 - 1000 \cos 60° = 0.$$

∴ **Resultant is a force of 233.97 N directed up the plane.** *Ans.*

Ex. 2.6: Two forces acting on a body are 500 N and 1000 N as shown in Fig. 2.12(a). Determine the third force F such that the resultant of all the three forces is 1000 N directed at 45° to x axis.

Let the third force F make an angle θ with x axis.

Then,

$$R\cos\alpha = \sum X$$

i.e. $1000 \cos 45° = 500 \cos 30° + 1000 \sin 30° + F \cos\theta$

$\therefore \quad F\cos\theta = -255.9$

and $R\sin\alpha = \sum Y$

$$1000\sin 45° = 500\sin 30° + 1000\cos 30 + F\sin\theta$$

$\therefore \quad F\sin\theta = -408.91$

$\therefore \quad F = \sqrt{225.9^2 + 408.91^2}$

i.e, $F = \mathbf{467.2\ N}$ *Ans.*

$\theta = \tan^{-1}\left(\dfrac{408.91}{225.9}\right) = \mathbf{61.08°}$ as shown in Fig. 2.12(b). *Ans.*

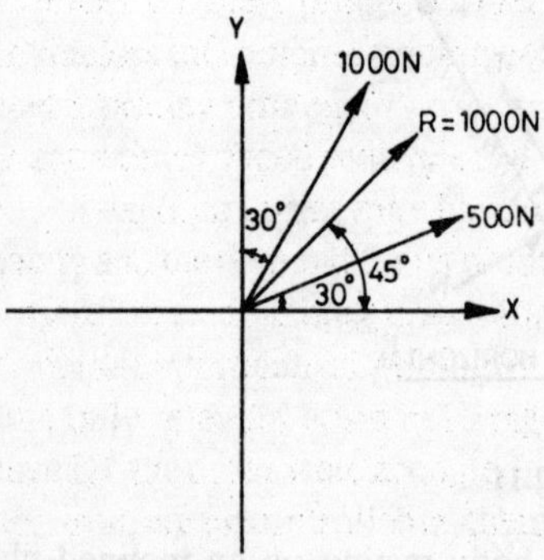

Fig. 2.12(a)

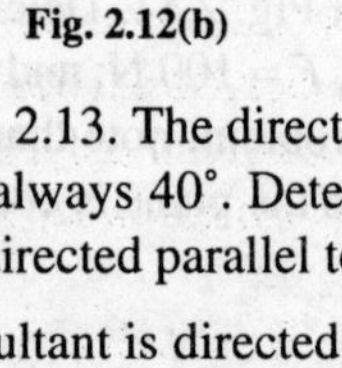

Fig. 2.12(b)

Ex. 2.7: Three forces acting at a point are shown in Fig. 2.13. The direction of the 300 N forces may vary, but the angle between them is always 40°. Determine the value of θ for which the resultant of the three forces is directed parallel to *b-b*.

Let the *X* and *Y* axes be as shown in Fig. 2.13. If the resultant is directed along the *X* axis, its component in *Y* direction is zero.

i.e, $0 = \sum Y$

$$= 300\sin\theta + 300\sin(40+\theta) - 500\sin 30°$$

$$\therefore \quad \sin\theta + \sin(40+\theta) = \frac{500\sin 30°}{300}$$

$= 0.8333 \quad \therefore \quad \sin\theta + \sin(40+\theta) = 0.8333$

$$2\sin\left(\frac{40+\theta+\theta}{2}\right)\times\cos\left(\frac{40+\theta-\theta}{2}\right)=0.8333$$

$$2\sin(20+\theta)\times\cos(20)=0.8333$$

$\therefore \quad \boldsymbol{\theta = 6.35°}$ *Ans.*

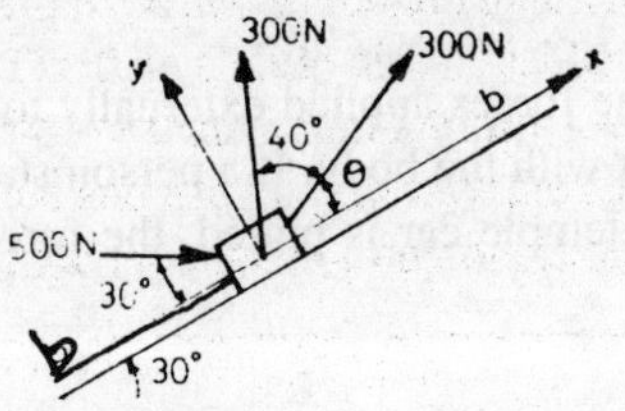

Fig. 2.13

2.5 EQUILIBRIANT

According to Newton's second law of motion a body starts moving with uniform acceleration if it is acted upon by a force. Hence, if another force equal in magnitude and opposite in direction to the force causing the motion, is applied to the body, the body comes to rest. Such a force which is equal and opposite to a given force is called Equilibriant. To find the equilibriant of a force system, the resultant of the forces is found as described earlier and the equilibriant is shown in the opposite direction to that of the resultant. For example, in Ex. 2.5 the equilibriant of the system is a force of magnitude 233.97 N directed down the plane.

2.6 EQUILIBRIUM OF A BODY

A body is said to be in equilibrium when it is at rest or continues to be in steady linear motion. According to Newton's law of motion, it means that the resultant of all the forces acting on a body in equilibrium, is zero. In graphical terms, it means that the force polygon closes.

The resultant R of a system of concurrent forces is zero only when the following conditions are satisfied:

$$\Sigma X = 0 \qquad \textit{Eqn.2.7}$$

$$\Sigma Y = 0 \qquad \textit{Eqn.2.8}$$

It may be observed that only one of the above two conditions is not sufficient. For example $\Sigma X = 0$ means that $R\cos\alpha = 0$. This will ensure that the resultant R cannot exist in any direction except in y direction ($\alpha = 90°$). Hence the condition $\Sigma Y = 0$ also should be satisfied to ensure that the resultant R does not exist, that is, equilibrium condition exists.

2.7 TYPES OF FORCES ON A BODY

While applying equilibrium conditions to a body it is essential that all forces acting on the body should be considered. The various forces acting on a body may be grouped as:

- Applied forces
- Non-applied forces

Applied Forces

Applied forces are the forces applied externally to a body. Each of the forces has got a point of contact with the body. If a person stands on a ladder, his weight is an applied force. If a temple car is pulled, the force in the rope is an applied force for the car.

Non-Applied Forces

There are two types of non-applied forces: (a) Self weight and (b) Reactions.

Self Weight – Every object is subjected to gravitational acceleration and hence has got a self-weight.

$$W = mg$$

where m is mass of the body and

g is gravitational acceleration (9.81 m/sec^2 near the earth surface)

Self-weight always acts in vertically downward direction. When analysing equilibrium conditions of a body, self-weight is treated as acting through the centre of gravity of the body. If self-weight is very small, it may be neglected.

Reactions – These are self-adjusting forces developed by the other bodies which come in contact with the body under consideration. According to Newton's third law of motion, the reactions are equal and opposite to the actions. The reactions adjust themselves to bring the body to equilibrium.

If the surface of contact is smooth, the direction of the reaction is normal to the surface of contact. If the surface of contact is not smooth, apart from normal reaction, there will be frictional reaction also. Hence the resultant reaction will not be normal to the surface of contact.

2.8 FREE BODY DIAGRAM

In many problems, it is essential to isolate the body under consideration from the other bodies in contact and draw all the forces acting on the body. For this, first the body is drawn and then applied forces, self-weight and the reactions at the points of contact with other bodies are drawn. Such a *diagram of the body in which the body under consideration is freed from all the contact surfaces and all the forces acting on it (including reactions at contact surfaces) are drawn, is called* **a Free Body Diagram (FBD)**. Free Body Diagrams (FBD) are shown for a few typical cases in Table 2.1.

Table 2.1 Free Body Diagrams (FBD) for a few typical cases

Reacting bodies	FBD required for	FBD
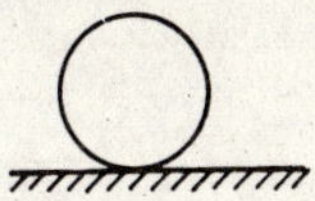	**Ball**	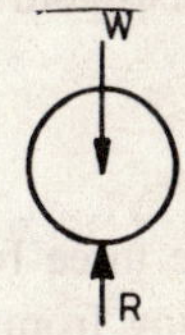
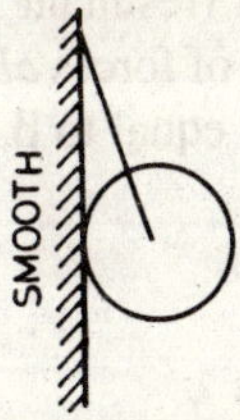	**Ball**	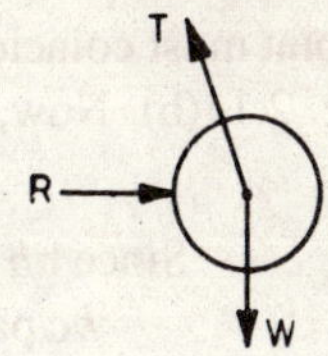
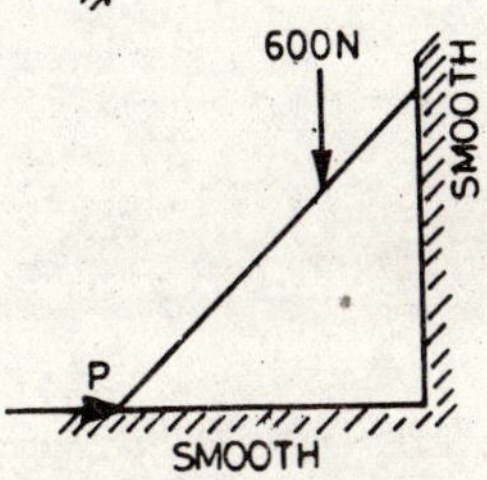	**Ladder**	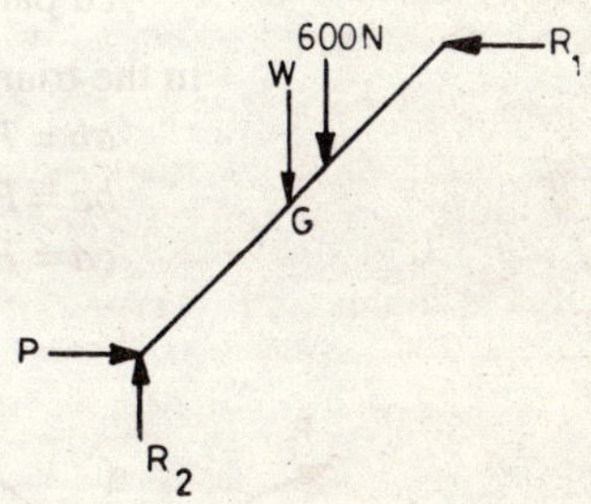
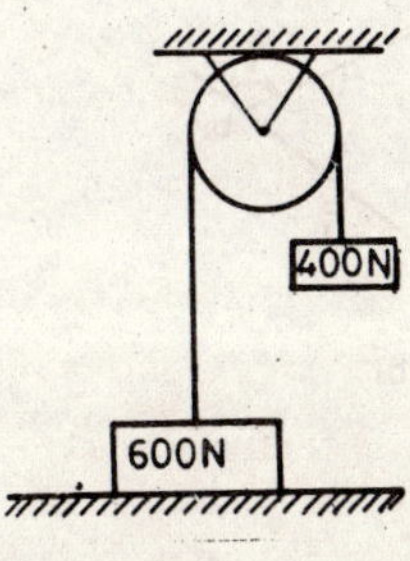	**Block weighing 600 N**	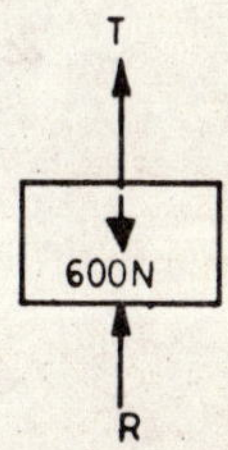
Fig		**Fig**

2.9 LAMI'S THEOREM

If a body is in equilibrium under the action of a number of forces, it may be

analysed using the equations 2.7 and 2.8. However, if the body is in equilibrium under the action of only three coplanar concurrent forces, Lami's theorem can be used.

Lami's theorem states : *If a body is in equilibrium under the action of three forces, each force is proportional to the sine of the angle between the other two forces.*

Thus, for the system of forces shown in Fig. 2.14(a).

$$\frac{P_1}{\sin\alpha} = \frac{P_2}{\sin\beta} = \frac{P_3}{\sin\gamma}$$

Proof: Draw the three forces P_1, P_2 and P_3 one after the other in direction and magnitude starting from point *a*. Since the body is in euqilibrium (resultant is zero), the last point must coincide with *a*. Thus, it results in a triangle of forces *abc* as shown in Fig. 2.14(b). Now, the external angles at *a*, *b* and *c* are equal to β, γ and α.

Since *ab* parallel to P_1,
bc parallel to P_2, and
ca parallel to P_3

in the triangle of forces *abc*
$ab = P_1$,
$bc = P_2$, and
$ca = P_3$.

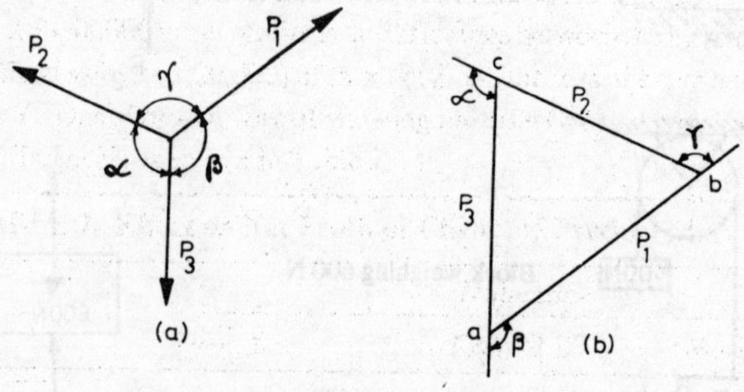

Fig. 2.14

Applying sine rule for the triangle *abc*,

$$\frac{ab}{\sin(180-\alpha)} = \frac{bc}{\sin(180-\beta)} = \frac{ca}{\sin(180-\gamma)}$$

$$i.e. \quad \frac{P_1}{\sin\alpha} = \frac{P_2}{\sin\beta} = \frac{P_3}{\sin\gamma}$$

Note: While determining the direction of the reaction on a body note that if the body is in equilibrium under the action of only three coplanar forces, those three forces must be concurrent. Proof of this statement is given in Chapter 3. However, this concept is used in this chapter also.

Ex. 2.8: A sphere of weight 100 N is tied to a smooth wall by a string as shown in Fig. 2.15(a). Find the tension T in the string and reaction R of the wall.

Free Body Diagram of the sphere is as shown in Fig. 2.15(b). Applying Lami's theorem to the system of forces,

$$\frac{T}{\sin 90^\circ} = \frac{R}{\sin(180-15)} = \frac{100}{\sin(90-15)}$$

$$\therefore \quad \boldsymbol{T = 103.53\ \text{N}} \qquad \textit{Ans.}$$

$$\boldsymbol{R = 26.79\text{N}} \qquad \textit{Ans.}$$

The above problem may be solved using equations of equilibrium also.

Taking horizontal direction as X axis and vertical direction as Y axis,

$\sum Y = 0$ gives

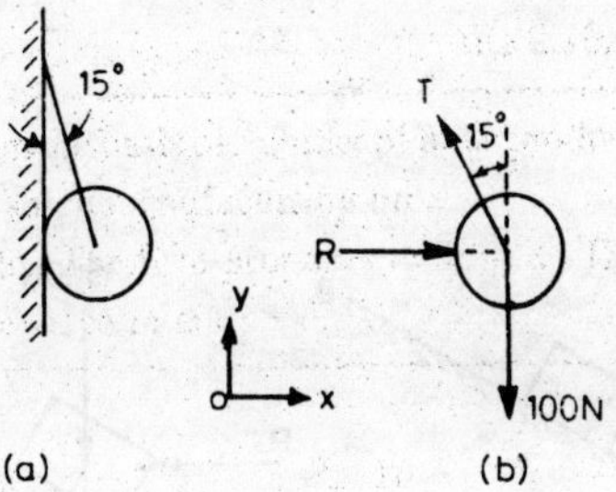

Fig. 2.15

$T \cos 15° - 100 = 0$ *Ans.*

$\mathbf{T = 103.53\ N}$

$\sum X = 0$ gives

$R - T \sin 15° = 0$

$\mathbf{R = 26.79\ N}$ *Ans.*

Notes:

1. The string can have only tension in it (it can pull a body), but cannot have compression in it (cannot push a body).
2. The wall reaction is a push, but cannot be a pull on the body
3. If the magnitude of reaction comes out to be negative, then assumed direction of reaction is wrong. It is acting exactly in the opposite to the assumed direction. However, the magnitude will be the same. Hence no further analysis is required. This advantage is not there in using Lami's equation. Hence, it is advisable for beginners to use equations of equilibrium, instead of Lami's theorem even if the body is in equilibrium under the action of only three forces.

Ex. 2.9: Determine the horizontal force P to be applied to a block of weight 1500 N to hold it in position on a smooth inclined plane AB which makes an angle of 30° with the horizontal [Fig. 2.16(a)].

The body is in equilibrium under the action of applied force P, self-weight 1500 N and normal reaction R from the plane. Since R, which is normal to the plane, makes 30° with the vertical (or 60° with the horizontal),

$$\Sigma Y = 0$$

$$R\cos 30° - 1500 = 0$$

$$\mathbf{R = 1732.06N}$$ *Ans.*

$$\Sigma X = 0$$

$$P - R\sin 30° = 0$$

$$P = R\sin 30°$$

$$\mathbf{P = 866.03N}$$ *Ans.*

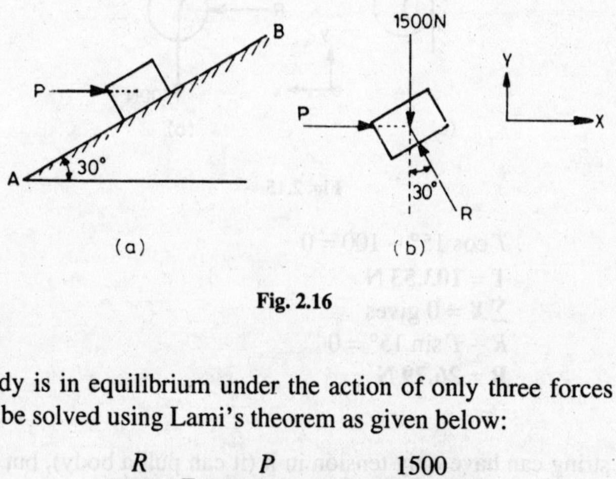

Fig. 2.16

Note:

Since the body is in equilibrium under the action of only three forces the above problem can be solved using Lami's theorem as given below:

$$\frac{R}{\sin 90°} = \frac{P}{\sin(180-30)} = \frac{1500}{\sin(90+30)}$$

$$\mathbf{R = 1732.06} \text{ and } \mathbf{P = 866.03}.$$ *Ans.*

Ex. 2.10: A roller of weight 10 kN rests on a smooth horizontal floor and is connected to the floor by the bar AC as shown in Fig. 2.17(a). Determine the force in the bar AC and reaction from floor if the roller is subjected to a horizontal force of 5 kN and an inclined force of 7 kN as shown in the figure.

A bar can develop a tensile force or a compressive force. Let the force developed be a compressive force S (push on the cylinder). Free Body Diagram of the cylinder is as shown in Fig. 2.17(b).

Since there are more than three forces in the system, Lami's equations cannot be applied. Consider the components in horizontal and vertical directions.

$$\Sigma H = 0$$

$$S\cos 30° + 5 - 7\cos 45° = 0$$

$$S = \frac{7\cos 45° - 5}{\cos 30°} = -0.058\text{kN}$$

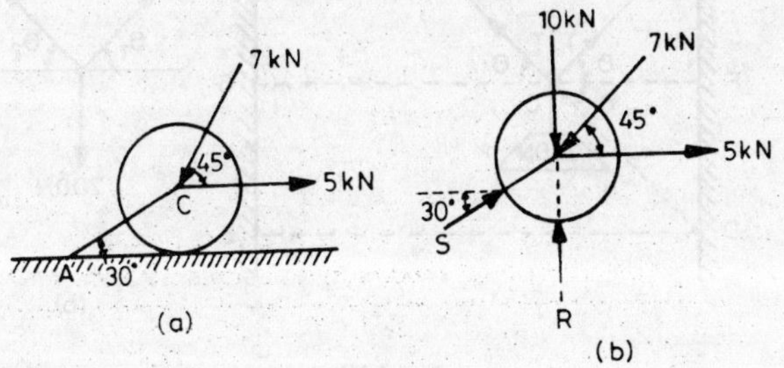

Fig. 2.17

Since the value of S is negative the force exerted by the bar is not a push, but it is pull (tensile force in bar) of magnitude 0.058 kN. Ans.

$$\Sigma V = 0$$

$$R - 10 - 7\sin 45° + S\sin 30° = 0$$

$$R = 10 + 7\sin 45° - S\sin 30°$$

$$= 10 + 7\sin 45° - (-0.058)\sin 30°$$

$$\mathbf{R = 14.979kN} \qquad \textit{Ans.}$$

Ex. 2.11: A cord *ACB* 5m long is attached at points *A* and *B* to two vertical walls 3m apart as shown in Fig. 2.18(a). A pully *C* of negligible radius carries a suspended load of 200 N and is free to roll without friction along the cord. Determine the position of equilibrium, as defined by the distance *X*, that the pulley will assume and also the tensile force in the cord.

The pulley *C* is in equilibrium under the action of tensile forces in *CA* and *CB* and vertical downward load 200 N. The tensile forces in segment *CA* and *CB* are the same since the pulley is frictionless. Now consider the equilibrium of pulley *C* shown in Fig. 2.18(b).

$$\Sigma H = 0$$

$$T\cos\theta_1 - T\cos\theta_2 = 0$$

$\therefore \quad \theta_1 = \theta_2$, say, equal to θ

Now, let *BC* be extended to *D*.

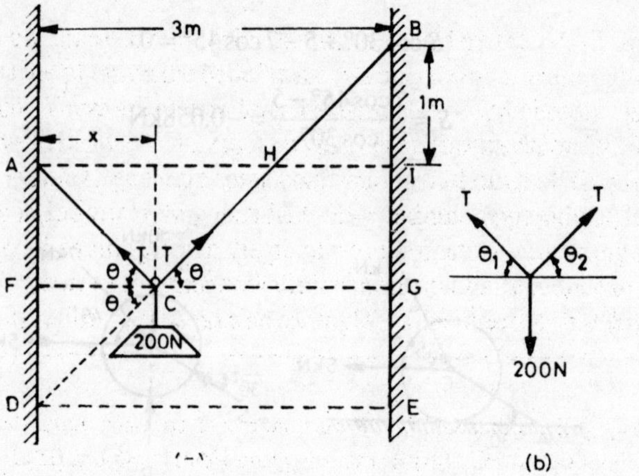

Fig. 2.18

$\Delta\, CFD = \Delta CFA$

$\therefore \quad CD = AC$

$$BD = BC + CD = BC + AC = \text{length of chord} = 5\text{m}$$

$$\text{DE} = 3\text{m}$$

$\therefore \quad BE = 4\text{m}$

As $\Delta\, BHI$ is similar to $\Delta\, BDE$

$$\mathbf{HI = \frac{BI}{BE} \times DE = \frac{1}{4} \times 3 = 0.75}$$

$$\therefore \quad \mathbf{AH = 3 - 0.75 = 2.25}$$

$\therefore \quad$ **x = 1.125m. Since AH = 2x** *Ans.*

At C,

$$\Sigma V = 0$$

$$2 \times T \sin\theta = 200$$

$$2 \times T \times \frac{4}{5} = 200$$

$\therefore \quad$ $\mathbf{T = 125\ N}$ *Ans.*

Ex. 2.12: A roller of radius r = 300 mm and weight 2000 N is to be pulled over a curb of height 150 mm [Fig. 2.19(a)] by a horizontal force P applied to the end of a string wound tightly around the circumference of the roller. Find the magnitude of P required to start the roller move over the curb. What is the least pull P through the centre of the wheel to just turn the roller over the curb?

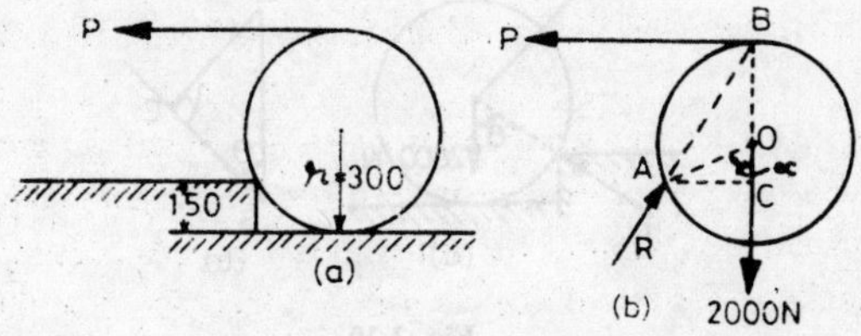

Fig. 2.19

When the roller is about to turn over the curb, the contact with the floor is lost and hence there is no reaction from the floor. The reaction R from the curb must pass through the intersection of P and the line of action of self weight, since the body is in equilibrium under the action of only three forces (all the three forces must be concurrent).

Referring to Fig. 2.19(b),

$$\cos\alpha = \frac{OC}{AO} = \frac{(300-150)}{300} = \frac{1}{2}$$

$$\therefore \quad \alpha = 60°$$

Now in ΔAOB, $\angle OAB = \angle OBA$ since $OA = OB$

but $\angle OAB + \angle OBA = \alpha$

$$2\angle OBA = 60°$$

$$\angle OBA = 30°$$

i.e. the reaction makes 30° with the vertical

$$\Sigma V = 0$$

$$R\cos 30° - 2000 = 0$$

$$R = 2309.40\text{N}$$

$$\Sigma H = 0$$

$$P - R\sin 30° = 0$$

$$P = 2309.40 \times \sin 30°$$

$$\mathbf{P = 1154.70\ N} \qquad \textit{Ans.}$$

Least force through the centre of wheel:

Now the reaction from the curb must pass through the centre of the wheel since the other two forces pass through that point. Its inclination to vertical is $\theta = 60°$.

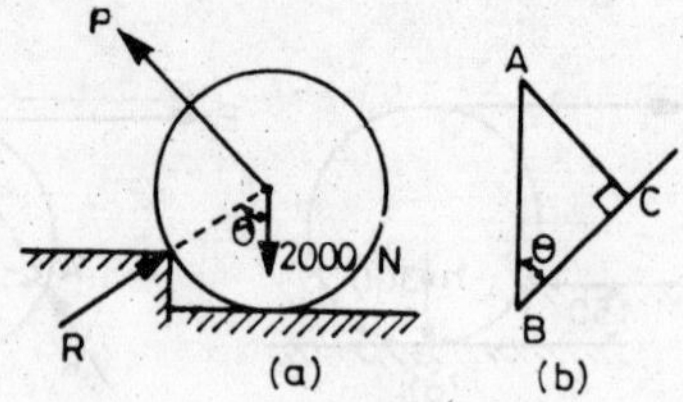

Fig. 2.20

If the triangle of forces ABC is constructed [Fig. 2.20(b)], representing self-weight by AB, reaction R by BC and pull P by AC, it may be observed that AC to be least, it should be perpendicular to BC. In other words, P makes 90° with the line of action of R.

From triangle of forces ABC, we get

$$P = AC = AB \sin \theta$$

$$= 2000 \sin 60°$$

$$\mathbf{P = 1732.05\ N} \qquad \textit{Ans.}$$

2.10 EQUILIBRIUM OF CONNECTED BODIES

When two or more bodies are in contact with one another, the system of forces appears as though it is a nonconcurrent forces system. However, when each body is considered separately, in many situations it turns out to be a set of concurrent force system. In such instances, first, the body subjected to only two unknown forces is to be analysed followed by the analysis of other connected body/bodies. Such examples are illustrated below.

Ex. 2.13: A system of connected flexible cables shown in Fig. 2.21(a) is supporting two vertical forces 200 N and 250 N at points B and D. Determine the forces in various segments of the cable.

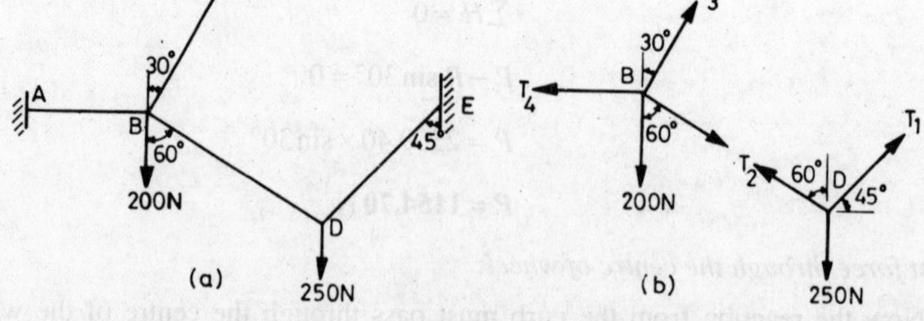

Fig. 2.21

Free body diagrams of the points B and D are shown in Fig. 2.21(b). Let the forces in the members be as shown in the figure.

Applying Lami's theorem to the system of forces at point D,

$$\frac{T_1}{\sin 120°} = \frac{T_2}{\sin 135°} = \frac{250}{\sin 105°}$$

$$\therefore \quad \mathbf{T_1 = 224.14\,N} \qquad \textit{Ans.}$$

$$\mathbf{T_2 = 183.01\,N} \qquad \textit{Ans.}$$

consider the system of forces acting at B.

$$\Sigma V = 0$$

$$T_3 \cos 30° - 200 - T_2 \cos 60° = 0$$

$$T_3 = \frac{200 + 183.01 \cos 60°}{\cos 30°}$$

$$\mathbf{T_3 = 336.60\,N} \qquad \textit{Ans.}$$

$$\Sigma H = 0$$

$$T_4 - T_2 \sin 60° - T_3 \sin 30° = 0$$

$$T_4 = 183.01 \times \sin 60° + 336.60 \sin 30°$$

$$\mathbf{T_4 = 255.81\,N.} \qquad \textit{Ans.}$$

Ex. 2.14: A rope AB, 4.5 m long is connected at two points A and B at the same level 4 m apart. A load of 1500 N is suspended from a point C on the rope 1.5 m from A as shown in Fig. 2.22(a). What load connected at a point D on the rope, 1 m from B will be necessary to keep the position CD level?

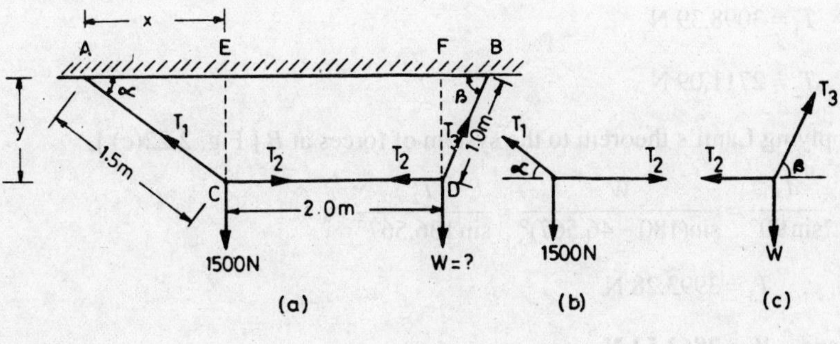

Fig. 2.22

Drop perpendiculars CE and DF on AB.

Let $\quad CE = y$, and

$\quad AE = x$

From $\Delta\, AEC$

$$x^2 + y^2 = 1.5^2$$
$$= 2.25 \qquad ...(1)$$

Now, $AB = 4$ m

and $AC + CD + BD = 4.5$ m

i.e. $CD = 4.5 - 1.5 - 1.0$

$= 2.0$ m

$\therefore$ $EF = 2.0$ m

$\therefore$
$$BF = AB - (AE + EF)$$
$$= 4 - (x + 2.0)$$
$$= 2 - x \qquad ...(2)$$

From $\Delta\, BFD$,

$$BF^2 + DF^2 = 1^2$$
$$(2 - x)^2 + y^2 = 1 \qquad ...(3)$$

From (1) and (3)

$x^2 - (2 - x)^2 = 1.25$

i.e. $x^2 - 4 + 4x - x^2 = 1.25$

$x = 1.3125$ m

$$\therefore \quad \alpha = \cos^{-1}\left(\frac{1.3125}{1.5}\right) = 28.955°$$

$$\beta = \cos^{-1}\left(\frac{2 - 1.3125}{1}\right) = 46.567°$$

Applying Lami's theorem to the system of forces acting at point C [Fig. 2.22(b)],

$$\frac{T_1}{\sin 90°} = \frac{T_2}{\sin 118.955°} = \frac{1500}{\sin(180 - 28.955)°}$$

$T_1 = 3098.39$ N

$T_2 = 2711.09$ N

Applying Lami's theorem to the system of forces at B [Fig. 2.22(c)],

$$\frac{T_3}{\sin 90°} = \frac{W}{\sin(180 - 46.567)°} = \frac{T_2}{\sin 136.567°}$$

$\therefore$ $T_3 = 3993.28$ N

and $W =$ **2863.53 N** *Ans.*

Ex. 2.15: A wire rope is fixed at two points A and D as shown in Fig. 2.23(a). Two weights 20 kN and 30 kN are attached to it at B and C, respectively. The weights rest with portions *AB* and *BC* inclined at angles 30° and 50° respectively, to the vertical as shown in figure. Find the tension in the wire in segments *AB*, *BC* and *CD* and also the inclination of the segments *CD* to vertical.

Applying Lami's theorem for the system of forces at *B* [Fig. 2.23(b)],

$$\frac{T_1}{\sin 50°} = \frac{T_2}{\sin 150°} = \frac{20}{\sin 160°}$$

$\mathbf{T_1 = 44.79\ kN}$ *Ans.*

$\mathbf{T_2 = 29.24\ kN}$ *Ans.*

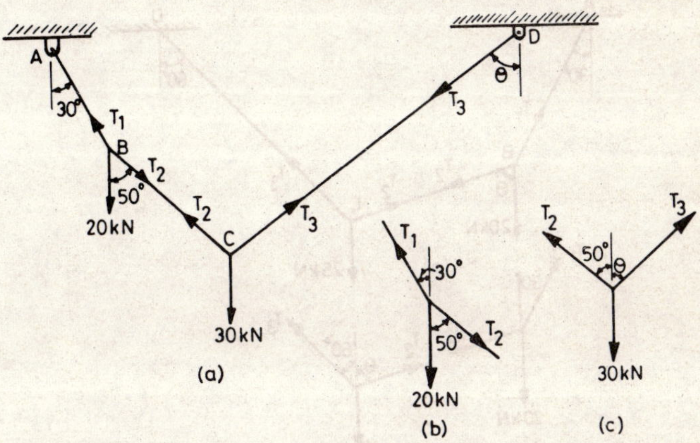

Fig. 2.23

Writing equations of equilibrium for the system of forces at *C* [Fig 2.23(c)], we get:

$$T_3 \sin\theta = T_2 \sin 50°$$

$$= 22.4\text{kN} \qquad \ldots(1)$$

and

$$T_3 \cos\theta = 30 - T_2 \cos 50°$$

$$= 11.20 \qquad \ldots(2)$$

From (1) and (2), $\tan\theta = 1.998687$

$\therefore \quad \mathbf{\theta = 63.422°}$ *Ans.*

and $\mathbf{T_3 = 25.045\ kN.}$ *Ans.*

Ex. 2.15: A wire is fixed at two points *A* and *D* as shown in Fig. 2.24(a). Two weights 20 kN and 25 kN are supported at *B* and *C*, respectively. When equilibrium is reached it is found that inclination of *AB* is 30° and that of *CD* is 60° to the vertical. Determine the tension in the segments *AB*, *BC* and *CD* of the rope and also the inclination of *BC* to the vertical.

Writing down equations of equilibrium for the system of forces at joints *B* and *C* [Fig. 2.24(b) and (c)], we have:

$$T_2 \sin\theta = T_1 \sin 30° \quad ...(1)$$

$$T_2 \cos\theta = T_1 \cos 30° - 20 \quad ...(2)$$

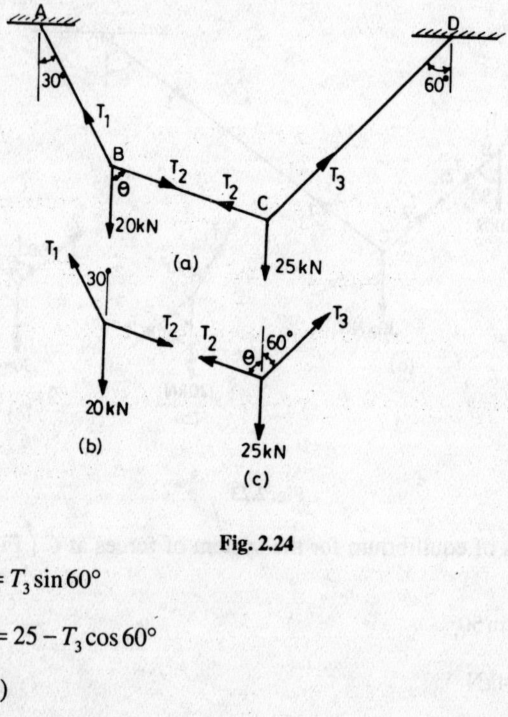

Fig. 2.24

$$T_2 \sin\theta = T_3 \sin 60° \quad ...(3)$$

$$T_2 \cos\theta = 25 - T_3 \cos 60° \quad ...(4)$$

From (1) and (3)

$$T_1 \sin 30° = T_3 \sin 60°$$

$$\therefore \quad T_1 = \sqrt{3}\, T_3 \quad ...(5)$$

From (2) and (4),

$$T_1 \cos 30° - 20 = 25 - T_3 \cos 60°$$

$$\sqrt{3}\, T_3 \frac{\sqrt{3}}{2} + T_3 \frac{1}{2} = 45$$

$$\mathbf{T_3 = 22.5\ kN}$$ *Ans.*

From (5), $\mathbf{T_1 = 38.97\ kN}$ *Ans.*

From (1), $T_2 \sin\theta = 19.48$

From (2), $T_2, \cos\theta = 13.75$

$\therefore \quad \tan\theta = 1.4223007$

$$\theta = \mathbf{54.8875^\circ} \qquad \textit{Ans.}$$

$$T_2 = \mathbf{23.91 kN} \qquad \textit{Ans.}$$

Ex. 2.16: A 600 N cylinder is supported by the frame *BCD* as shown in Fig. 2.25(a). The frame is hinged at *D*. Determine the reactions at *A*, *B*, *C* and *D*.
Free Body diagram of sphere and frame are shown in Figs. 2.25(b) and 2.25(c), respectively.

Considering the sphere,

$$\Sigma V = 0$$

$$\mathbf{R_B = 600\ N} \qquad \text{...(1)} \quad \textit{Ans.}$$

$$\Sigma H = 0$$

$$R_A = R_C \qquad \text{...(2)}$$

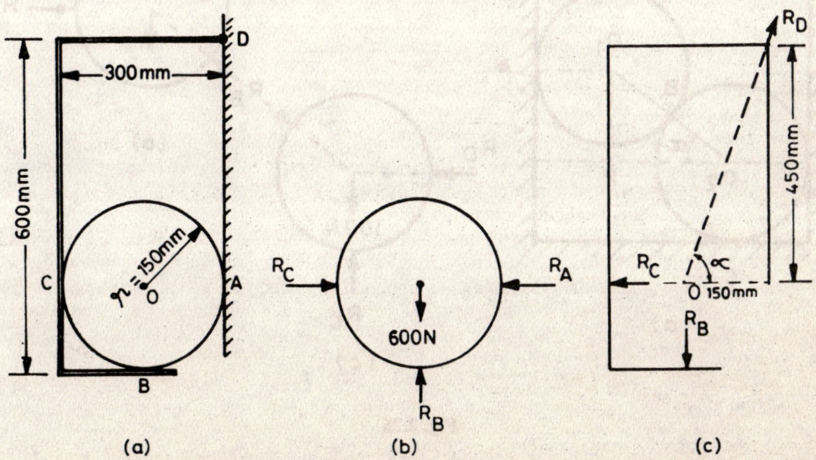

Fig. 2.25

As the frame is in equilibrium under the action of three forces only, they must be concurrent forces. In other words, reaction at *D* has line of action along *OD*. Hence, its inclination to horizontal is given by:

$$\tan\alpha = \frac{450}{150} = 3$$

$$\alpha = 71.56505°$$

$$\Sigma V = 0$$

$$R_D \sin\alpha = R_B = 600$$

$\therefore$ $\mathbf{R_D = 632.456\ N}$ *Ans.*

$$\Sigma H = 0$$

$$R_C = R_D \cos\alpha$$

$\therefore$ $\mathbf{R_C = 200\ N}$ *Ans.*

From (2), $\therefore$ $\mathbf{R_A = 200\ N.}$ *Ans.*

Ex. 2.17: Two smooth spheres each of radius 100 mm and weight 100 N, rest in a horizontal channel having vertical walls, the distance between which is 360 mm. Find the reactions at the points of contacts *A*, *B*, *C* and *D* shown in Fig. 2.26(a).

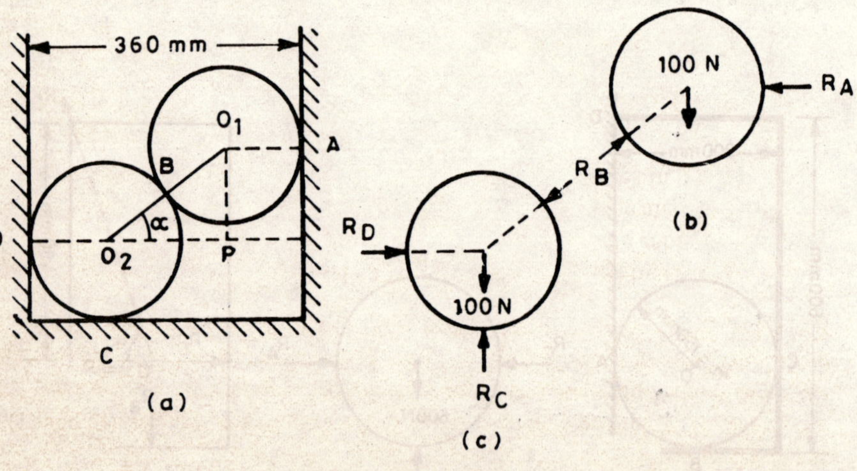

Fig. 2.26

Let O_1 and O_2 be the centres of the first and second spheres. Drop perpendicular O_1P to the horizontal line through O_2. Fig. 2.26(b) and 2.26(c) show free body diagram of the sphere 1 and 2, respectively. Since the surface of contact are smooth, reaction of *B* is in the radial direction, i.e. in the direction O_1O_2. Let it make angle α with the horizontal. Then,

$$\cos\alpha = \frac{O_2P}{O_1O_2} = \frac{360 - O_1A - O_2D}{O_1B + BO_2}$$

$$= \frac{360 - 100 - 100}{100 + 100} = 0.8$$

$\therefore \quad \sin\alpha = 0.6.$

Consider sphere No. 1.

$$\Sigma V = 0$$

$$R_B \times 0.6 = 100$$

$$\mathbf{R_B = 166.67N} \qquad \textit{Ans.}$$

$$\Sigma H = 0$$

$$R_A = R_B \times 0.8$$

$$\therefore \quad \mathbf{R_A = 133.33\ N} \qquad \textit{Ans.}$$

Consider sphere No. 2

$$\Sigma V = 0$$

$$R_C = 100 + R_B \times 0.6$$

$$\therefore \quad \mathbf{R_C = 200\ N} \qquad \textit{Ans.}$$

$$\Sigma H = 0$$

$$R_D = R_B \times 0.8$$

$$\therefore \quad \mathbf{R_D = 133.33\ N} \qquad \textit{Ans.}$$

Ex. 2.18: Two cylinders, *A* of weight 4000 N and *B* of weight 2000 N rest on smooth inclines as shown in Fig. 2.27(a). They are connected by a bar of negligible weight hinged to each cylinder at its geometric centre by smooth pins. Find the force *P* to be applied as shown in the figure such that it will hold the system in the given position.

Fig. 2.27(b) and 2.27(c) show the free body diagram of the two cylinders. Applying Lami's theorem to the system of forces on cylinder *A*, we get:

$$\frac{C}{\sin 60°} = \frac{4000}{\sin(60 + 90 - 15)}$$

$$C = 4898.98\ \text{N}$$

Consider cylinder *B*.
Summation of the forces parallel to the inclined plane (45° to horizontal) = 0, gives:

$$P \cos 15° + 2000 \cos 45° - C \cos 60° = 0$$

$$\therefore \quad P = \frac{4898.98 \cos 60° - 2000 \cos 45°}{\cos 15°}$$

$$P = \mathbf{1071.8\ N}$$ *Ans.*

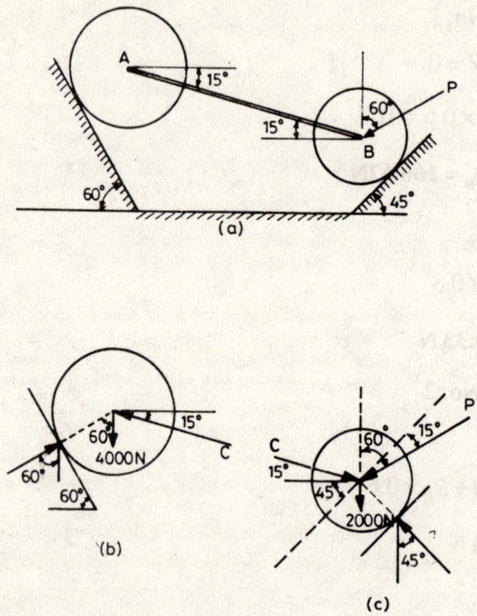

Fig. 2.27

PROBLEMS FOR EXERCISE

2.1 The resultant of two forces, one of which is double the other is 260 N. If the direction of the larger force is reversed and the other remains unaltered, the resultant reduces to 180 N. Determine the magnitude of the forces and the angle between the forces.

[*Ans.* $P_1 = 100$ N; $P_2 = 200$ N; $\theta = 63.8961°$]

2.2 A body is subjected to the three forces as shown in Fig. 2.28. If possible determine the direction of the force *F* so that the resultant is in *X* direction, when (a) $F = 5000$ N, (b) $F = 3000$ N.

[*Ans.* (a) 36.87°; (b) Not possible]

2.3 The frictionless pulley *A* shown in Fig. 2.29 is supported by two bars *AB* and *AC* which are hinged at *B* and *C* to a vertical wall. The flexible cable *DG* hinged at *D*, goes over the pulley and supports a load of 20 kN at *G*. The angles between the various members are shown in the figure. Determine the forces in the bars *AB* and *AC*. Neglect the size of the pulley.

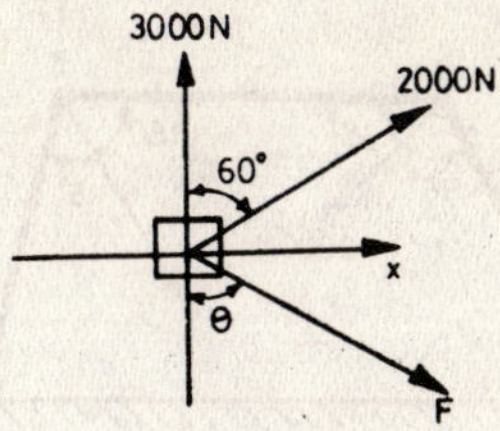

Fig. 2.28

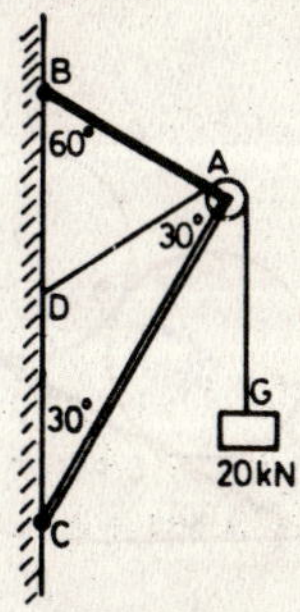

Fig. 2.29

[*Ans.* $F_{AB} = 0$; $F_{AC} = 34.641$ kN. Hint: Force in AD = 20 kN, tensile]

2.4 A cord supported at *A* and *B* carries a load of 10 kN at *D* and a load of *W* at *C* as shown in Fig. 2.30. Find the value of *W* so that *CD* remains horizontal.

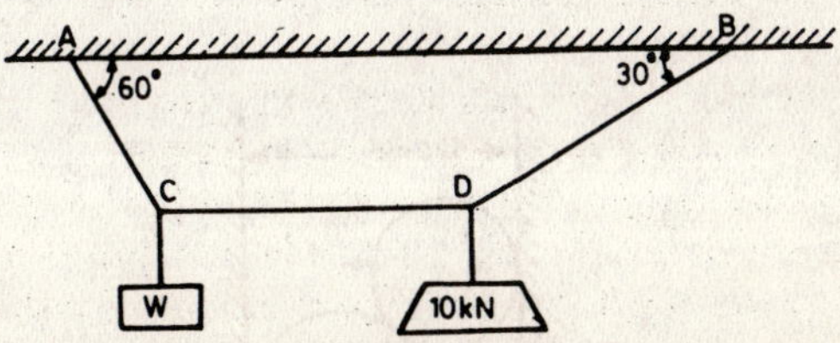

Fig. 2.30

[*Ans.* W = 30 kN]

2.5 Three bars, hinged at *A* and *D* and pinned at *B* and *C* as shown in Fig. 2.31 form a four-linked mechanism. Determine the value of *P* that will prevent movement of bars.

[*Ans.* P = 3047.2 N]

2.6 Two identical rollers, each of weights 100 N are supported by an inclined plane and a vertical wall as shown in Fig. 2.32. Assuming smooth surfaces, find the reactions induced at the points of supports *A*, *B*, *C* and *D*.

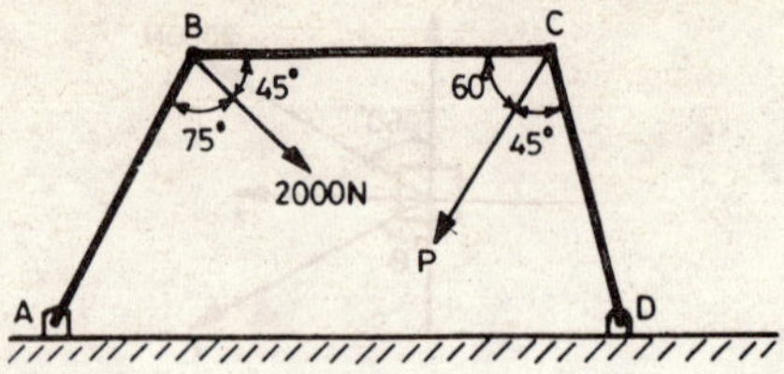

Fig. 2.31

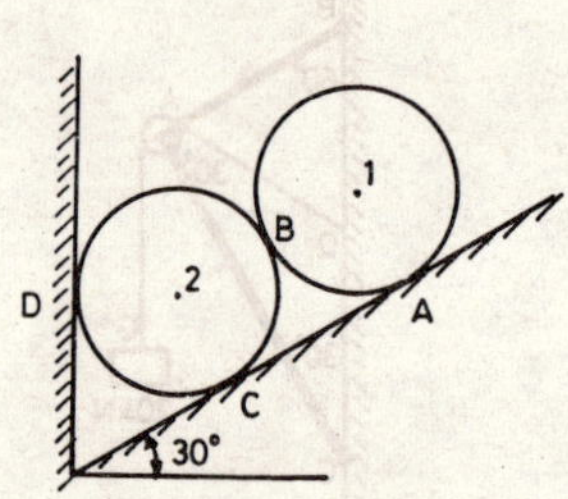

Fig. 2.32

[*Ans.* $R_A = 86.67$ N; $R_B = 50.00$ N; $R_C = 144.4$ N; $R_D = 115.52$ N]

2.7 Two cylinders of diameters 100 mm and 50 mm, weighing 200 N and 50 N, respectively are placed in a trough as shown in Fig. 2.33. Neglecting friction, find the reactions at contact surfaces 1, 2, 3, and 4.

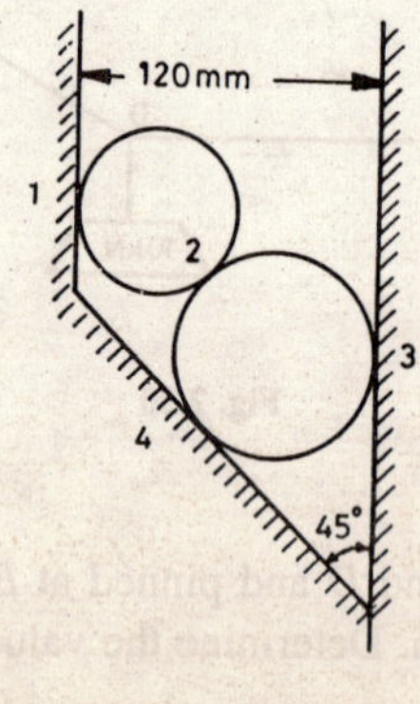

Fig. 2.33

[*Ans.* $R_1 = 37.5$ N; $R_2 = 62.5$ N; $R_3 = 287.5$ N; $R_4 = 353.5$ N]

2.8 Three cylinders are piled in a rectangular ditch as shown in Fig. 2.34. Neglecting friction, determine the reaction between cylinder *A* and the vertical wall. Weights and radii of the cylinders are as given below:

Cylinder	Weight	Radius
A	75 N	100 mm
B	200 N	150 mm
C	100 N	125 mm

[*Ans.* $R = 226.86$ N]

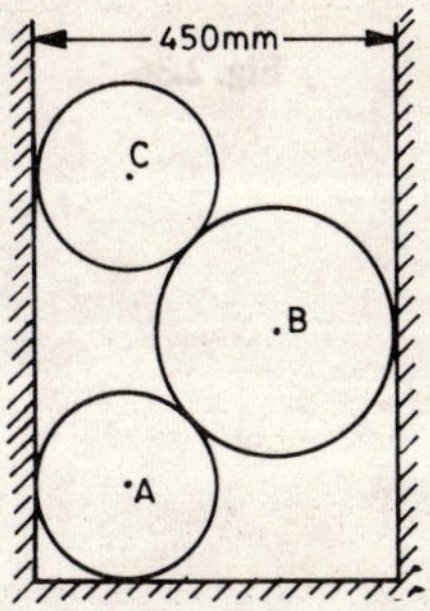

Fig. 2.34

2.9 Three spheres *A*, *B* and *C* having their diameters 500 mm, 500mm and 800 mm, respectively, are placed in a trench with smooth side walls and floor as shown in Fig. 2.35. The centre-to-centre distance of spheres *A* and *B* is 600 mm. The weights of the cylinders *A*, *B* and *C* are 4 kN, 4 kN and 8 kN respectively. Determine the reactions at *P*, *Q*, *R* and *S*.

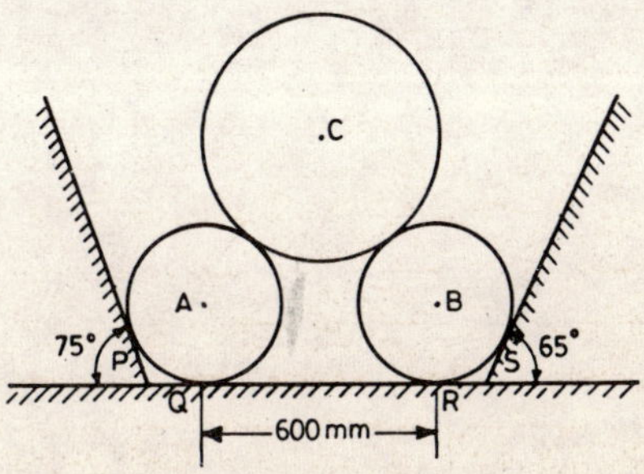

Fig. 2.35

[*Ans.* $R_P = 2.1545$ kN; $R_Q = 7.4424$ kN; $R_R = 7.0296$ kN; $R_S = 2.2962$ kN].

2.10 Three uniform, homogeneous and smooth spheres *A*, *B* and *C* weighing 300 N, 600 N and 300 N, respectively and having diameters 800 mm, 1200 mm and 800 mm, respectively, are placed in a trench as shown in Fig. 2.36. Determine the reactions at the contact points *P*, *Q*, *R* and *S*.

[*Ans.* $R_P = 61.24$ N; $R_Q = 631.58$ N; $R_R = 1095.21$ N; $R_S = 290.43$ N].

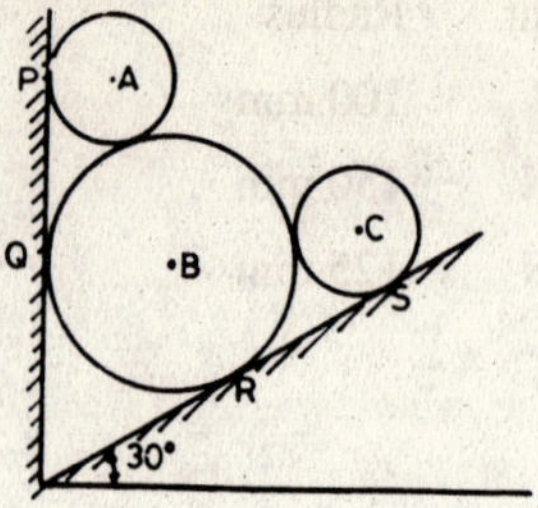

Fig. 2.36

CHAPTER 3

Coplanar Nonconcurrent Forces

If all the forces in a system lie in the same plane and the lines of action of all the forces do not pass through a single point, the system is said to be *coplanar non-concurrent force system.* Three such systems are shown in Fig. 3.1.

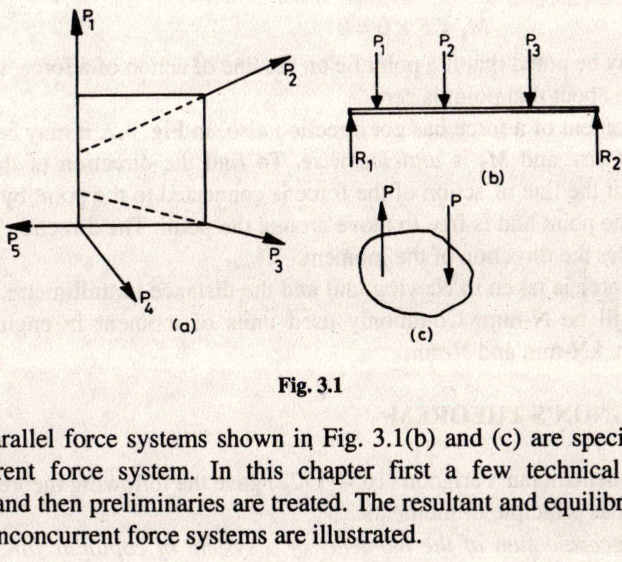

Fig. 3.1

The parallel force systems shown in Fig. 3.1(b) and (c) are special cases of nonconcurrent force system. In this chapter first a few technical terms are explained and then preliminaries are treated. The resultant and equilibrium conditions of nonconcurrent force systems are illustrated.

3.1 MOMENT OF A FORCE

Moment of a force about a point is the measure of its rotational effect. *Moment* is defined as the product of the magnitude of the force and the perpendicular distance of the point from the line of action of the force. The point about which the moment is considered is called *moment centre* and the perpendicular distance of the point from the line of action of the force is called *moment arm.* Referring to Fig. 3.2, if d_1 is the perpendicular distance of point 1 from the line of action of force F, the moment of F about point 1 is given by

$$M_1 = F\,d_1$$

Similarly, moment about point 2 is given by

$$M_2, = F\,d_2$$

If the moment centre 3 lies on the line of action of the force F, the moment arm is zero and hence,

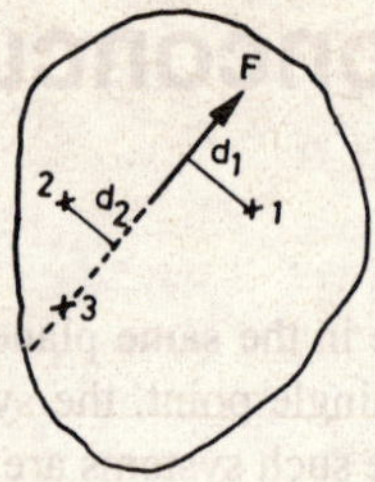

Fig. 3.2

$$M_3 = F \times 0 = 0$$

Thus, it may be noted that if a point lie on the line of action of a force, the moment of the force about that point is zero.

The moment of a force has got direction also. In Fig. 3.2, it may be noted that M_1 is *clockwise* and M_2 is *anticlockwise*. To find the direction of the moment, imagine that the line of action of the force is connected to the point by a rigid rod pinned at the point and is free to move around the point. The direction of the rotation indicates the direction of the moment.

If the force is taken in Newton unit and the distance in millimetre, the unit of moment will be N-mm. Commonly used units of moment in engineering are kN-m, N-m, kN-mm and N-mm.

3.2 VARIGNON'S THEOREM

French mathematician Varignon (1654-1722) gave the following theorem which is also known as principle of moments:

The algebraic sum of the moments of a system of coplanar forces about a moment centre in their plane is equal to the moment of their resultant force about the same moment centre.

Proof: Referring to Fig. 3.3, let R be the resultant of forces P_1 and P_2 and B the moment centre. Let d, d_1 and d_2 be the moment arms of the forces, R, P_1 and P_2, respectively from the moment centre B. Then in this case, we have to prove that:

$$Rd = P_1\,d_1 + P_2\,d_2$$

Join AB and consider it as y axis and draw x axis at right angles to it at A [Fig. 3.3 (b)]. Denoting by θ the angle that R makes with x axis and noting that the same angle is formed by perpendicular to R at B with AB_1, we can write:

$$Rd = R \times AB \cos\theta$$

$$= AB \times (R \cos\theta)$$

$$= AB \times R_x \qquad \textit{Eqn}.3.1$$

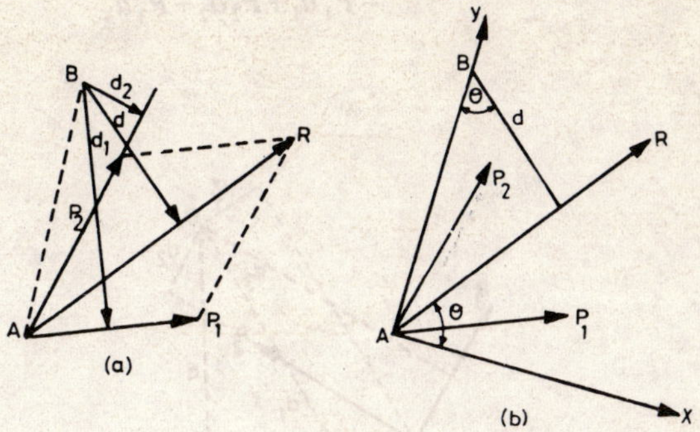

Fig. 3.3

where R_x denotes the component of R in x direction.

Similarly, if P_{1x} and P_{2x} are the components of P_1 and P_2, in x direction, respectively, then

$$P_1 d_1 = AB\, P_{1x} \qquad Eqn.3.2$$

$$\text{and} \quad P_2 d_2 = AB\, P_{2x} \qquad Eqn.3.3$$

From Eqns. 3.2 and 3.3,

$$P_1 d_1 + P_2 d_2 = AB\,(P_{1x} + P_{2x})$$

$$= AB \times R_x \qquad Eqn.3.4$$

Since, the sum of x components of individual forces is equal to the x component of the resultant R from Eqn. 3.1 and 3.4, we can conclude:

$$Rd = P_1\, d_1 + P_2\, d_2$$

If a system of forces consists of more than two forces, the above result can be extended as given below:

Let P_1, P_2, P_3 and P_4 be four concurrent forces and R be their resultant. Let d_1, d_2, d_3, d_4 and a be the distances of line of action of forces P_1, P_2, P_3, P_4 and R, respectively from the moment centre O.

If R_1 is the resultant of P_1 and P_2 and its distance from O is a_1, then applying Varignon's theorem:

$$R_1\, a_1 = P_1\, d_1 + P_2\, d_2$$

If R_2 is the resultant of R_1 and P_3 (and hence of P_1, P_2 and P_3) and its distance from O is a_2, then from Varignon's theorem:

$$R_2 a_2 = R_1 a_1 + P_3 d_3$$

$$= P_1 d_1 + P_2 d_2 + P_3 d_3$$

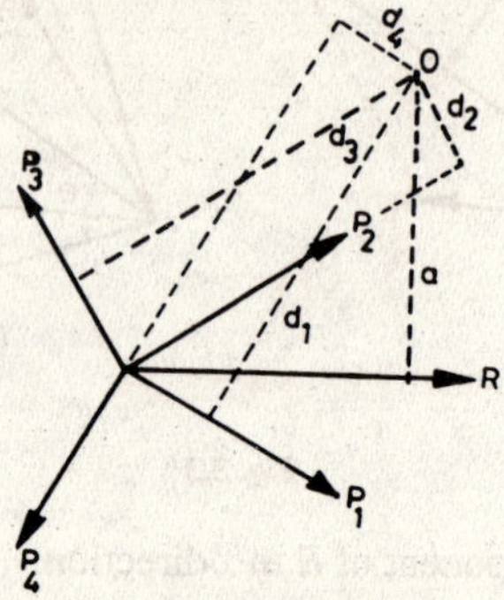

Fig. 3.4

Now considering R_2 and P_4, we can write:

$$Ra = R_2 a_2 + P_4 d_4$$

Since R is the resultant of R_2 and P_4 (i.e. P_1, P_2, P_3 and P_4).

$$\therefore \quad Ra = P_1 d_1 + P_2 d_2 + P_3 d_3 + P_4 d_4$$

Thus, the moment of the resultant of a number of forces about a moment centre is equal to the sum of the moments of its component forces about the same moment centre.

Ex. 3.1: Find the moment of 100 N force acting at B about point A as shown in Fig. 3.5.

100 N force may be resolved into its horizontal components as 100 cos 60° and vertical component 100 sin 60°. From Varignon's theorem, moment of 100 N force about the point A is equal to sum of the moments of its components about A.

Taking clockwise moment as positive,

$$M_A = 100 \cos 60° \times 500 - 100 \sin 60° \times 400$$

$$= 25,000 - 34,641.02$$

$$= -9641.02 \text{ N-mm}$$

= 9641.016 N-mm Anticlockwise *Ans.*

Ex. 3.2: What will be the y intercept of the 5000 N force if its moment about A is 8000 N-m in Fig. 3.6

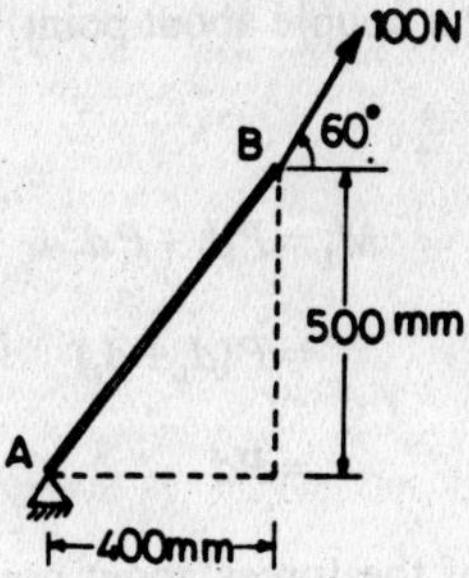

Fig. 3.5

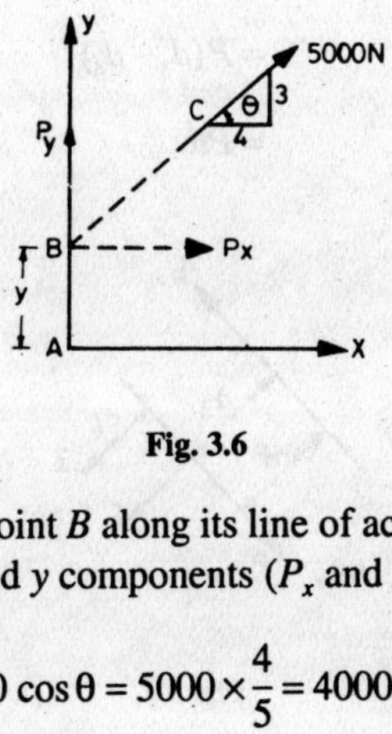

Fig. 3.6

5000 N force is shifted to a point *B* along its line of action (law of transmissibility) and it is resolved into its *x* and *y* components (P_x and P_y as shown in Fig. 3.6).

$$P_x = 5000 \cos\theta = 5000 \times \frac{4}{5} = 4000 \text{ N}, \quad \text{and}$$

$$P_y = 5000 \sin\theta = 5000 \times \frac{3}{5} = 3000 \text{ N}.$$

By Varignon's therorem, moment of 5000 N force about *A* is equal to moment of its component about the same point.

$$8000 = 4000 \times y + 3000 \times 0$$

$$\therefore \quad y = 2\text{m} \qquad \textit{Ans.}$$

3.3 COUPLE

Two parallel forces equal in magnitude and opposite in direction and separated by a definite distance are said to form a couple. The sum of the forces forming a couple in any direction is zero, which means the translatory effect of the couple is zero.

An interesting property can be observed if we consider rotational effect of a couple about any point. Let the magnitude of the forces forming the couple be P and the perpendicular distance between the two forces be d. Consider the moment of the two forces constituting a couple about point 1 as shown in Fig. 3.7 (a). Let the moment be M_1 then,

$$M_1 = Pd_1 + Pd_2$$

$$= P(d_1 + d_2)$$

$$= Pd$$

Now, consider the moment of the forces about point 2 which is outside the two forces as shown in Fig. 3.7 (b). Let M_2 be the moment.
Then,

$$M_2 = Pd_3 - Pd_4$$

$$= P(d_3 - d_4)$$

$$= Pd$$

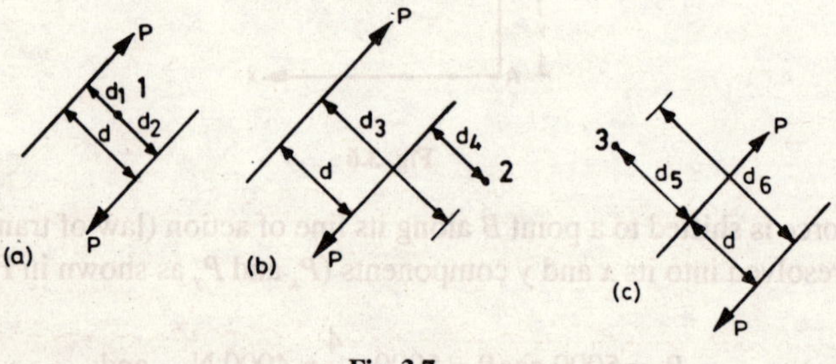

Fig. 3.7

Thus, moment of a couple about any point is the same. Now we can list the following characteristics of a couple:

– A couple consists of a pair of equal and opposite parallel forces which are separated by a definite distance;

– The translatory effect of a couple on the body is zero;

– The rotational effect (moment) of a couple about any point is a constant and it is equal to the product of the magnitude of the force and the perpendicular distance between the two forces.

Since the only effect of a couple is a moment and this moment is the same about any point, the effect of a couple is unchanged if:

– The couple is rotated through any angle;

– The couple is shifted to any other position;

– The couple is replaced by another pair of forces whose rotational effect is the same.

3.4 RESOLUTION OF A FORCE INTO A FORCE AND A COUPLE

It will be advantageous to resolve a force acting at a point on a body into a force acting at some other suitable point on the body and a couple. In Fig. 3.8 (a), P is a force acting on a body at A.

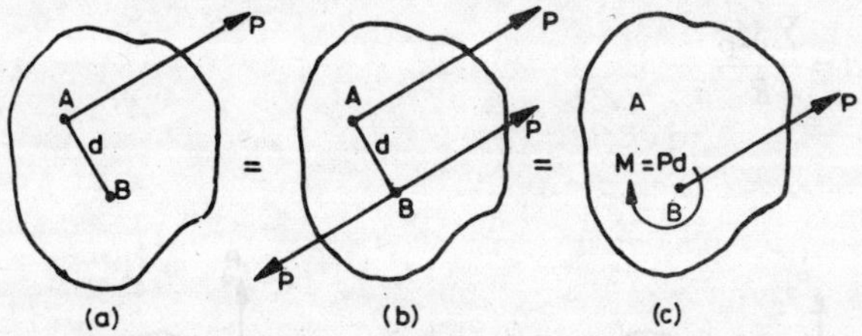

Fig. 3.8

Now it can be shown that P at A may be resolved into a force P at B and a couple of magnitude $M = P \times d$, where d is the perpendicular distance of B from the line of action of P through A.

By applying equal and opposite forces P at B the system of forces is not disturbed. Hence the system of forces in Fig. 3.8(b) is the same as the system given in Fig. 3.8(a). Now the original force P at A and the opposite force P at B form a couple of magnitude Pd. The system in Fig. 3.8(b) can be replaced by the system shown in Fig. 3.8(c). Thus, the given force P at A is replaced by a force P at B and a moment Pd.

3.5 RESULTANT OF FORCE SYSTEMS

The resultant of force system is the one which will have the same rotational and translation effect as the given system of forces. It may be a single force, a pure moment or a force and a moment.

Let P_1, P_2 and P_3 [Fig. 3.9 (a)] constitute a system of forces acting on a body. Each force can be replaced by a force of the same magnitude and acting in the same direction at point O and a moment about O. Thus, the given system in Fig. 3.9(a) is equal to the system shown in Fig. 3.9 (b) where ΣM_O is the algebraic sum of the moments of the given forces about O.

At O, the concurrent forces P_1, P_2 and P_3 can be combined as usual to get the resultant force R. Now the resultant of the given system is equal to a force R at O and a moment ΣM_O as shown in Fig. 3.9 (c).

The force R and moment ΣM_O shown in Fig. 3.9 (c) can be replaced by a single force R acting at a distance d from O such that the moment produced by this force R is equal to ΣM_O [Ref. 3.9 (d)].

Thus, we get a single force R acting at a distance d from the point O which gives the same effect as the constituent forces of the systems. Thus, the resultant of the given forces may be reduced to a single force.

Mathematically,

$$\left.\begin{aligned} R &= \sqrt{(\Sigma x)^2 + (\Sigma y)^2} \\ \tan\alpha &= \frac{\Sigma y}{\Sigma x} \end{aligned}\right\} \qquad \text{Eqn. 3.5}$$

$$d = \frac{\Sigma M_O}{R}$$

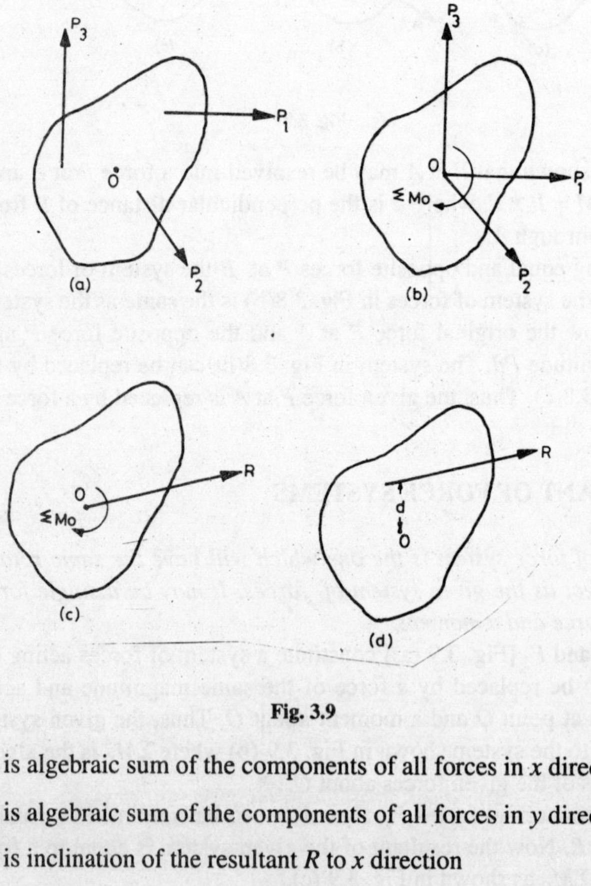

Fig. 3.9

where, Σx is algebraic sum of the components of all forces in x direction

Σy is algebraic sum of the components of all forces in y direction

α is inclination of the resultant R to x direction

ΣM_O is algebraic sum of the moments of all the forces about a point O

d is distance of the resultant R from the point O.

Note:

R is marked at distance d such that it produces the same direction of moment about point O as $\sum M_O$.

Sometimes the values of $\sum x$ and $\sum y$ may come out to be zero, but $\sum M_O$ may exist. This means that the resultant of the system gets reduced to a pure couple.

3.6 x AND y INTERCEPTS OF RESULTANT

In some situations we may be interested in finding only the distance of R along x or y axis, that is x and y intercepts.

Let d be the distance of the resultant from O and α be its inclination to x axis (Fig. 3.10). Then the intercepts are given by:

$$x = \frac{d}{\sin \alpha} \qquad y = \frac{d}{\cos \alpha} \qquad \textit{Eqn.}3.6$$

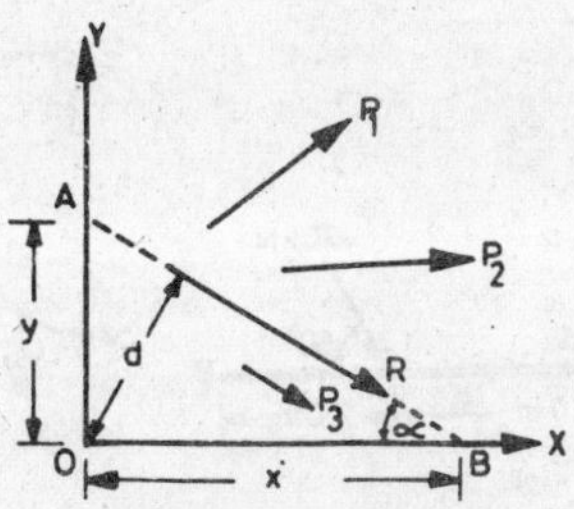

Fig. 3.10

Another method of finding the intercepts is as follows:

Let $R_x = \Sigma x$ and $R_y = \Sigma y$ be the components of the resultant R in x and y directions. Considering the moment of R about O as the sum of moments of its components about A (Varignon's theorem) we get (ref. Fig. 3.11).

$$Rd = \Sigma M_O$$

$$R_x \times 0 + R_y\, x = \Sigma M_O$$

$$\therefore \quad x = \frac{\Sigma M_O}{R_y} = \frac{\Sigma M_O}{\Sigma y} \qquad \textit{Eqn.}3.7$$

Similarly, resolving the resultant into its components at B, it can be shown that:

$$y = \frac{\Sigma M_O}{R_x} = \frac{\Sigma M_O}{\Sigma x} \qquad \textit{Eqn.}3.8$$

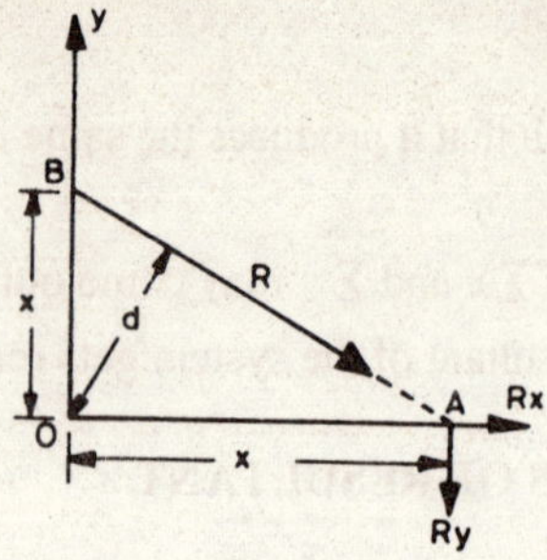

Fig. 3.11

Ex. 3.3: A system of loads acting on a beam is shown in Fig. 3.12 (a). Determine the resultant of the loads.

Taking horizontal direction as x axis and the vertical direction as y axis.

$$\Sigma x = 20 \cos 60° = 10 \text{ kN}$$

$$\Sigma y = 20 + 30 + 20 \sin 60°$$

$$= 67.3205$$

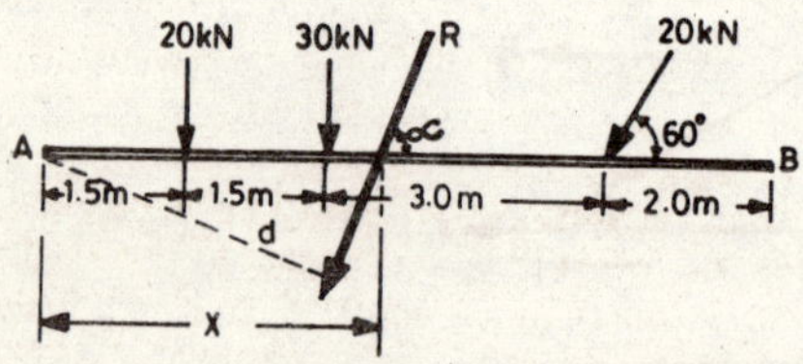

Fig. 3.12 (a)

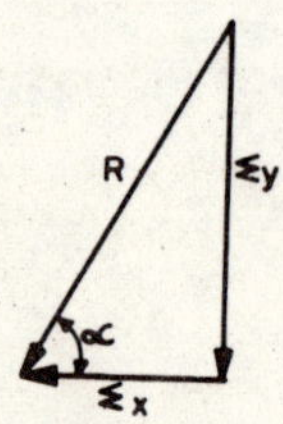

Fig. 3.12 (b)

$$\therefore \quad R = \sqrt{(\Sigma x)^2 + (\Sigma y)^2}$$

$$= \sqrt{10^2 + (67.3205)^2}$$

i.e. $\mathbf{R = 68.0592 \text{ kN}}$ *Ans.*

$$\tan \alpha = \frac{\Sigma y}{\Sigma x} = 6.7321$$

$\therefore \quad \alpha = \mathbf{81.55°}$ *Ans.*

Now taking moment about A,

$$\Sigma M_A = 20\times 1.5 + 30\times 3.0 + 20\sin 60°\times 6$$

$$= 223.9231 \text{ kN-m}$$

$\therefore$ The distance of the resultant from point O is given by:

$$d = \frac{\Sigma M_A}{R} = \frac{223.9231}{68.0592} = 3.290 \text{ m}$$

$$\therefore \quad x = \frac{d}{\sin\alpha} = \frac{3.290}{\sin 81.55°}$$

$$\mathbf{x = 3.326\ m} \qquad \textit{Ans.}$$

The value of x intercept may be obtained using Eqn. 3.7 also. Thus,

$$x = \frac{\Sigma M_O}{R_y} = \frac{\Sigma M_O}{\Sigma y} = \frac{223.9231}{67.3205} = \mathbf{3.326\ m} \qquad \textit{Ans.}$$

Ex. 3.4: Find the resultant of the force system shown in Fig. 3.13 (a) acting on a lamina of equilateral triangular shape.

$$\Sigma x = 80 - 100\cos 60° - 120\cos 30°$$

$$= -73.92 \text{ N}$$

(Negative sign shows that R_x acts from right to left)

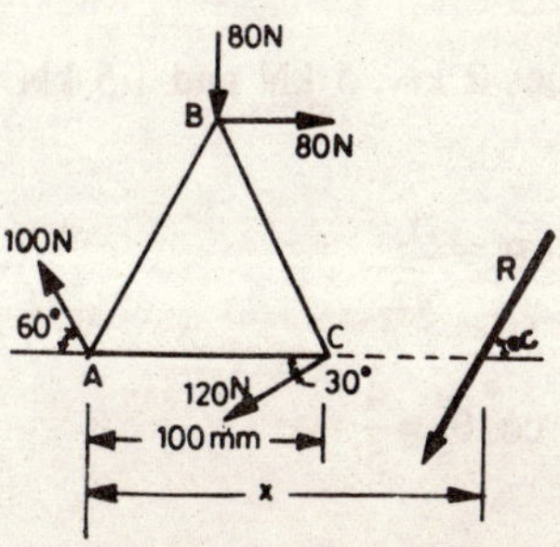

Fig. 3.13 (a)

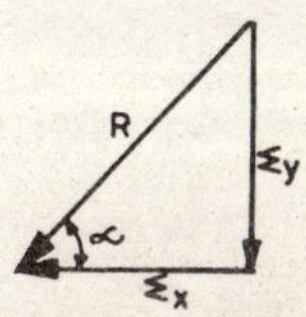

Fig. 3.13 (b)

$$R_x = 73.92 \text{ N}$$

$$\Sigma y = 80 + 120\sin 30° - 100\sin 60°$$

$$R_y = 53.40 \text{ N}$$

$$\therefore \quad R = \sqrt{73.92^2 + 53.40^2}$$

$$\mathbf{R = 91.19\ N} \qquad \textit{Ans.}$$

$$\tan \alpha = \frac{\Sigma y}{\Sigma x} = \frac{53.40}{73.92}$$

$$\alpha = \mathbf{35.84°} \qquad \textit{Ans.}$$

Let x be the distance from A at which the resultant cuts AC. Then taking A as moment centre,

$$53.40x = 80 \times 100 \sin 60° + 80 \times 50 + 120 \sin 30° \times 100$$

$$x = \mathbf{317.008\ m} \text{ to the right of } A \text{ as shown in Fig.3.13(a).} \qquad \textit{Ans.}$$

Ex. 3.5: Find the resultant of a set of coplanar forces acting on a lamina as shown in Fig. 3.14 (a). Each square has side of 10 mm.

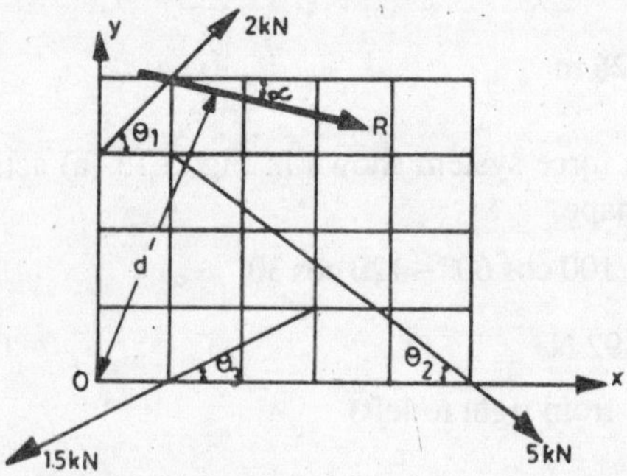

Fig. 3.14 (a)

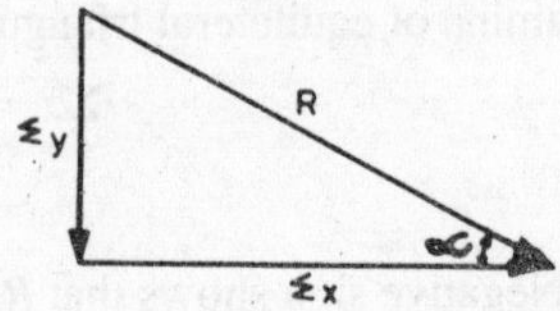

Fig. 3.14 (b)

If θ_1, θ_2 and θ_3 are the slopes of the forces 2 kN, 5 kN and 1.5 kN forces with respect to x axis, then

$$\sin \theta_1 = \cos \theta_1 = \frac{1}{\sqrt{2}}$$

$$\sin \theta_2 = \frac{3}{5}, \quad \cos \theta_2 = \frac{4}{5}$$

$$\sin \theta_3 = \frac{1}{\sqrt{5}}, \quad \cos \theta_3 = \frac{2}{\sqrt{5}}$$

$$\Sigma x = 2 \times \frac{1}{\sqrt{2}} + 5 \times \frac{4}{5} - 1.5 \times \frac{2}{\sqrt{5}} = 4.0726 \text{ N}$$

$$\Sigma y = -2 \times \frac{1}{\sqrt{2}} + 5 \times \frac{3}{5} + 1.5 \times \frac{1}{\sqrt{5}} = 2.26 \text{ N}$$

$$\boldsymbol{R} = \sqrt{(\Sigma x)^2 + (\Sigma y)^2} = \mathbf{4.66\ N} \qquad \textit{Ans.}$$

$$\tan \alpha = \frac{2.26}{4.0726}$$

$\therefore \quad \alpha = \mathbf{28.99°}$ *Ans.*

Distance d of the resultant from O is given by

$$Rd = \Sigma M_O$$

$$4.66d = 2\times\frac{1}{\sqrt{2}}\times 30 + 5\times\frac{3}{5}\times 50 + 1.5\times\frac{1}{\sqrt{5}}\times 10$$

$$= 199.13$$

$d = \mathbf{42.77mm}$ as show in Fig. 3.14 (a) *Ans.*

Note:

To find moment of forces about O, if 2 kN force is resolved at it's intersection with y axis and if 5 kN and 3 kN forces are resolved at their intersection with x axis, and then Varignon theorem is used.

Ex. 3.6: The system of forces acting on a bell crank is shown in Fig. 3.15 (a). Determine the magnitude, direction and the point of application of the resultant.

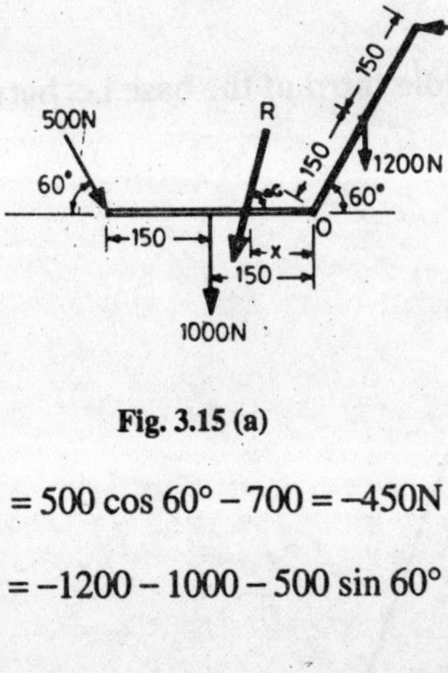

Fig. 3.15 (a)

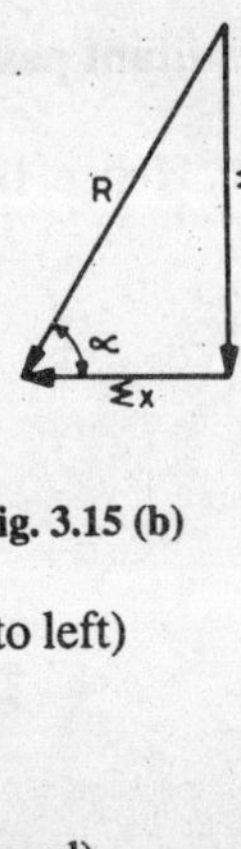

Fig. 3.15 (b)

$R_x = \Sigma x = 500 \cos 60° - 700 = -450\text{N} = 450\text{ N}$ (from right to left)

$R_y = \Sigma y = -1200 - 1000 - 500 \sin 60° = -2633.01\text{ N}$

$= 2633.01\text{ N}$ (downward)

$$\therefore \quad R = \sqrt{450^2 + (2633.01)^2}$$

$R = \mathbf{2671.19\ N}$ *Ans.*

$$\tan\alpha = \frac{2633.01}{450}$$

$\alpha = \mathbf{80.30°}$, as shown in Fig. 3.15*b* *Ans.*

Let the point of application of the resultant be at a distance x from the point O along the horizontal arm. Then,

$$x \times 2633.01 = 500 \sin 60° \times 300 + 1000 \times 150$$

$$- 1200 \times 150 \cos 60° + 700 \times 300 \sin 60°$$

$x = $ **141.195 mm**, as shown in Fig.3.15(a) *Ans*.

Ex. 3.7: Various forces to be considered for the stability analysis of a dam are shown in the Fig. 3.16. The dam is safe if the resultant force passes through middle third of the base. Verify whether the dam is safe.

$$\Sigma x = 500 \text{ kN}$$

$$\Sigma y = -1120 + 120 - 420$$

$$= -1420 \text{ kN}$$

$$= 1420 \text{ kN (downward)}$$

Let x be the distance from O where the resultant cuts the base.

$$\text{Then } x\Sigma y = \Sigma M_O$$

$$x \times 1420 = 500 \times 4 + 1120 \times 2 - 120 \times 4 + 420 \times 5$$

$$\therefore \quad x = 4.126 \text{ m}$$

The resultant passes through the middle third of the base i.e. between $\frac{7}{3}$ m and $2 \times \frac{7}{3}$ m. Hence, the dam is safe. *Ans.*

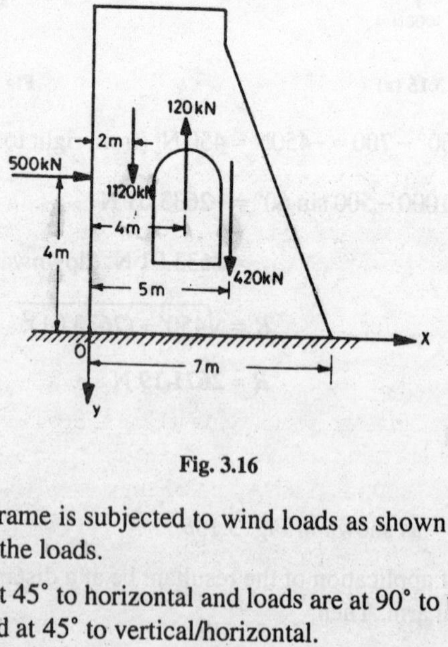

Fig. 3.16

Ex. 3.8: A building frame is subjected to wind loads as shown in Fig. 3.17. Determine the resultant of the loads.

The roof is inclined at 45° to horizontal and loads are at 90° to the roof. Hence, the loads are also inclined at 45° to vertical/horizontal.

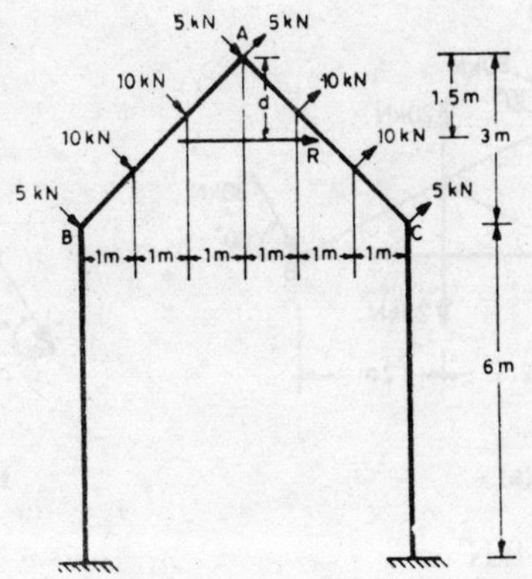

Fig. 3.17

Now,

$$\Sigma x = (5+10+10+5+5+10+10+5)\cos 45°$$

$$= 60 \times \frac{1}{\sqrt{2}} = 42.4264 \text{ kN}$$

$$\Sigma y = -(5+10+10+5)\sin 45° + (5+10+10+5)\sin 45°$$

$$= 0$$

$\therefore \quad R = \Sigma x = 42.4264$ kN

and its direction is horizontal.

Let R be at a distance d from the ridge A.

Then, $\quad Rd = \Sigma M_A$

$$60 \times \frac{1}{\sqrt{2}} \times d = 5 \times \frac{3}{\sqrt{2}} + 10 \times \frac{2}{\sqrt{2}} + 10 \times \frac{1}{\sqrt{2}} + 10 \times \frac{1}{\sqrt{2}} + 10 \times \frac{2}{\sqrt{2}} \times 5 \times \frac{3}{\sqrt{2}}$$

$d = 1.5$m

$\therefore$ **Resultant is a horizontal force of magnitude 42.4264 kN at 1.5 m below A.**

Ans.

Ex. 3.9: Determine the magnitude, direction and line of action of the equilibriant of the given set of coplanar forces acting on a planar structure shown in Fig. 3.18.

The two 40 kN forces acting on the smooth pulley may be replaced by a pair of 40 kN forces acting at centre of pulley C and parallel to the given forces, since the sum of moments of the two given forces about C is zero.

Now, $\Sigma x = 20\cos 45° - 30\cos 60° - 50\cos 30° + 40\cos 20° - 40\sin 30°$

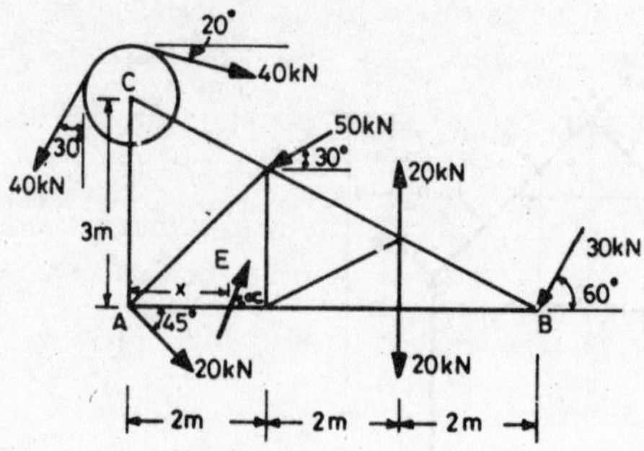

Fig. 3.18 (a)

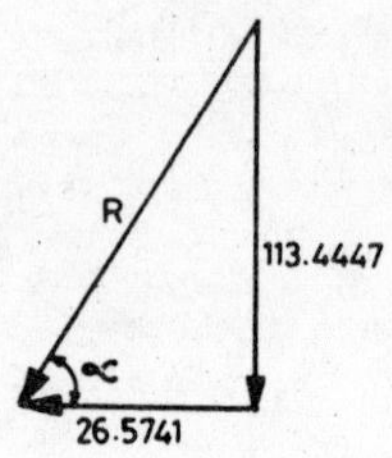

Fig. 3.18 (b)

$$\Sigma x = -26.5714\text{kN}$$

$= 26.5714$ kN (from right to left, because of the negative sign)

$$\Sigma y = -20 \sin 45° - 20 + 20 - 30 \sin 60° - 50 \sin 30°$$

$$-40 \sin 20° - 40 \cos 30°$$

$$= -113.4447 \text{ kN}$$

$= 113.4447$ kN downward.

$$R = \sqrt{(\Sigma x)^2 + (\Sigma y)^2}$$

$$= 116.515 \text{ kN}$$

$$\tan \alpha = \frac{113.4447}{26.5714}$$

$\alpha = 76.82°$ as shown in Fig.3.18(b)

Let the resultant intersect *AB* at a distance *x* from *A*. Then,

$$x \Sigma y = \Sigma M_A$$

$$x \times 113.4447 = 20 \times 4 - 20 \times 4 + (30 \sin 60°) \times 6 + (50 \sin 30°) \times 2 - (50 \cos 30°) \times 2$$

$$+ (40 \cos 20°) \times 3 - (40 \sin 30°) \times 3$$

$$x = 1.516 \text{ m}$$

The equilibriant is equal and opposite to the resultant. Hence, it is as shown in Fig. 3.18 (a) in which **E = 116.515 kN** **α = 76.82°** **and** **x = 1.516 m** *Ans.*

3.7 EQUILIBRIUM CONDITIONS

A body is said to be in equilibrium when it does not have any translatory or rotatory motion in any direction. This means that when the body is in equilibrium, the following two simultaneous conditions are to be satisfied:

(1) The algebraic sum of the components of the forces along each of the three mutually perpendicular directions, is zero.

(2) The algebraic sum of the components of the moments acting on the body about each of the three mutually perpendicular axes, is zero.

If x, y and z are the three mutually perpendicular direction, the first condition results into:

$$\Sigma x = 0 \qquad \Sigma y = 0 \qquad \text{and} \quad \Sigma z = 0$$

and the second condition results into:

$$\Sigma M_x = 0 \qquad \Sigma M_y = 0 \qquad \text{and} \quad \Sigma M_z = 0$$

In the case of coplanar forces (acting on x–y plane), $\Sigma z = 0$, $\Sigma M_x = 0$, $\Sigma M_y = 0$ and z axis reduces to a point. If A is a point in $x - y$ plane, M_z reduced to M_A. Hence, the necessary and sufficient conditions for equilibrium of coplanar force system are:

$$\Sigma x = 0$$

$$\Sigma y = 0 \qquad \textit{Eqn.}\,3.9$$

$$\text{and} \quad \Sigma M_A = 0$$

Referring to Fig. 3.19, A, B and C are three points in the x–y plane which are not collinear. Any system of forces acting on a body may be replaced by a single force and a couple. Let R be such resultant force.

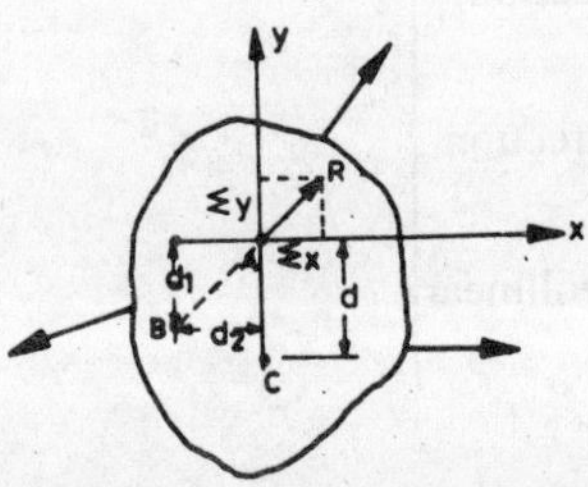

Fig. 3.19

If the condition $\Sigma M_A = 0$ is satisfied, we can conclude that the resultant R should be a force passing through A. If $\Sigma M_B = 0$ is satisfied, the line of action of the resultant passes through B, i.e. AB is the line of action. Now, if $\Sigma M_C = 0$ is also satisfied, the resultant should be zero since the line of action of the resultant (line AB) cannot pass through C also. Thus if A, B, C are not collinear points, the following conditions are necessary and sufficient conditions of equilibrium.

$$\left.\begin{aligned} \Sigma M_A &= 0 \\ \Sigma M_B &= 0 \\ \Sigma M_C &= 0 \end{aligned}\right\} \qquad \textit{Eqn.}\ 3.10$$

The equilibrium Equations 3.10 are not independent of equilibrium Equations 3.9. Two of them are common to the two sets. Referring to Fig. 3.19, R can be resolved into its components perpendicular to and parallel to AC.

Then,

$$\Sigma M_C = (\Sigma x)d$$

$$\therefore \quad \text{if} \quad \Sigma M_C = 0, (\Sigma x)d = 0 \quad \text{or} \quad \Sigma x = 0$$

Now,

$$\Sigma M_B = 0$$

$$\text{i.e.} \quad (\Sigma x)d_1 + (\Sigma y)d_2 = 0$$

$$\text{But} \quad \Sigma x = 0$$

$$\therefore \quad (\Sigma y)d_2 = 0$$

Since ABC are not colinear, $d_2 \neq 0$

$$\therefore \quad \Sigma y = 0$$

Thus, $\Sigma M_C = 0$ is identical to $\Sigma X = 0$ and $\Sigma M_B = 0$ is identical to $\Sigma y = 0$. Hence any one of the following sets may be used as equations of equilibrium:

(1) $\Sigma X = 0; \Sigma Y = 0; \Sigma M_A = 0$

(2) Line AB is not in y direction,
$\Sigma Y = 0; \Sigma M_A = 0; \Sigma M_B = 0$

(3) If line AB is not in x direction
$\Sigma X = 0; \Sigma M_A = 0; \Sigma M_B = 0$

(4) If A, B and C are non-collinear
$\Sigma M_A = 0; \Sigma M_B = 0; \Sigma M_C = 0$

Eqn. 3.11

It can be proved that *if a system is in equilibrium under the action of three forces, those three forces must be concurrent.*

Let P_1, P_2 and P_3 be the forces acting on a body as shown in Fig. 3.20 and let P_1 and P_2 intersect at A. Then applying moment equilibrium condltion about A, we get,

$$\Sigma M_A = 0$$

$$\text{i.e} \quad P_3\, d = 0$$

where d is distance of line of action of P_3 from A.

Since P_3 is not zero,

$$d = 0,$$

i.e. P_3 also must pass through A.

Hence, the proposition is proved.

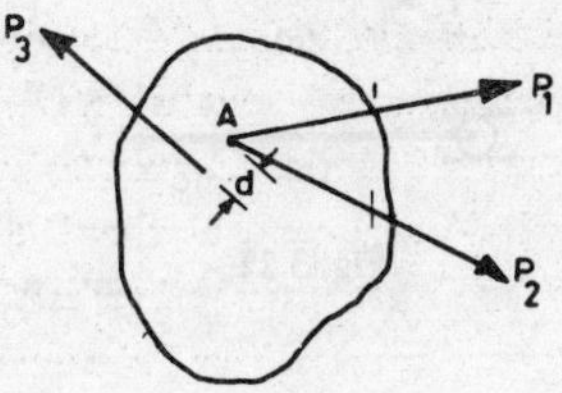

Fig. 3.20

3.8 REACTIONS AT SUPPORTS OF BEAMS

A **beam** may be defined as a structural element which has one dimension (length) considerably larger than the other two dimensions, namely, breadth and depth, and is supported at a few points. It is usually loaded in vertical direction. Due to applied loads reactions develop at supports, and the system of forces consisting of applied loads and reactions keep the beam in equilibrium. The nature of reaction depends upon the type of supports.

Types of Supports

Various types of supports and reactions developed are listed below:
Simple Support – If the beam rests simply on a support it is called a simple-support. In such case the reaction at the support is at right angles to the support and the beam is free to move in the direction of its axis and also it is free to rotate about the support (Fig. 3.21)

Fig. 3.21

Roller Support – In this case, beam end is supported on rollers. In such cases, reaction is normal to the support since rollers can be treated as frinctionless. [Fig. 3.22 (a)]. Many mechanical components are having roller supports which roll between guides. In such cases reaction will be normal to the guides, in both the direction (Fig. 3.22). At roller support beam is free to move along the support. It can rotate about the support also.

Hinged Support – At a hinged end, a beam cannot move in any direction. However, it can rotate about the support (Fig. 3.23). Hence the support will not develop any resisting moment, but it can develop reaction in any direction to keep the end sationary. The reaction R can be split into its horizontal and vertical components for the purpose of analysis.
Thus, in Fig. 3.23,

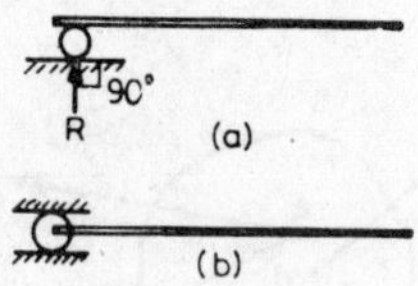

Fig. 3.22

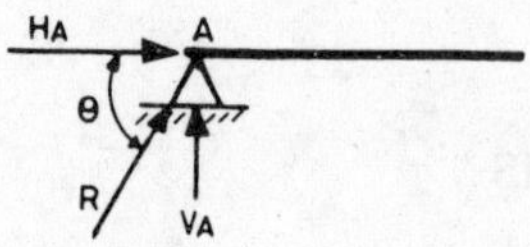

Fig. 3.23

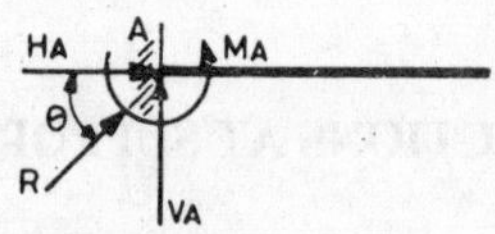

Fig. 3.24

$$V_A = R \sin \theta$$

$$H_A = R \cos \theta$$

Fixed Support – At such supports, the beam end is not free to translate or rotate. Translation is prevented by developing support reaction in any required direction.

Referring to Fig. 3.24 the support reaction R which is at an angle θ to x axis may be represented by H_A and V_A, where

$$V_A = R \sin \theta$$

$$H_A = R \cos \theta$$

Rotation is prevented by developing support moment M_A as shown in Fig. 3.24. Thus at fixed support A, there are three reactions H_A, V_A and M_A.

Type of Beams

Simply Supported Beam – When both ends of a beam are simply supported it is called simply supported beam (Fig. 3.25). Such a beam can support load in the direction normal to its axis.

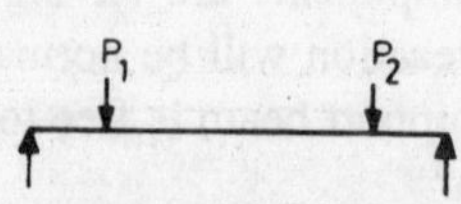

Fig. 3.25

Beam with one End Hinged and the Other on Rollers – If one end of a beam is hinged and other end is on rollers, the beam can resist load in any direction (*see* Fig. 3.26).

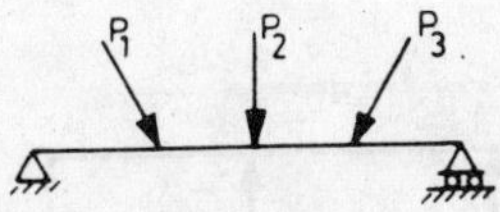

Fig. 3.26

Over-hanging Beam – If a beam is projecting beyond the support, it is called an over-hanging beam (Fig. 3.27). The overhang may be only on one side as in Fig. 3.27 (a) or may be on both sides as in Fig. 3.27 (b).

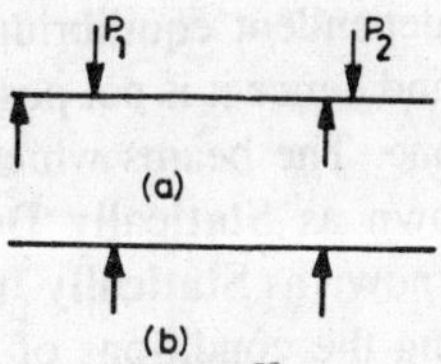

Fig. 3.27

Cantilever Beam – If a beam is fixed at one end and is free at the other end, it is called cantilever beam (Fig. 3.28).

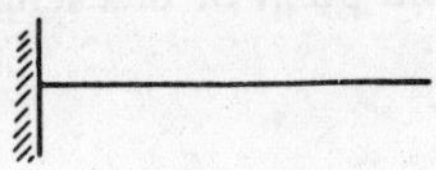

Fig. 3.28

Propped Cantilever – It is a beam with one end fixed and the other end simply supported (Fig. 3.29).

Fig. 3.29

Both Ends Hinged – In these beams both ends will be having hinged supports (Fig. 3.30)

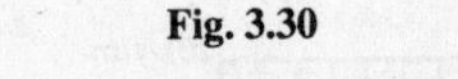

Fig. 3.30

Continuous Beam – A beam is said to be continuous, if it is supported at more than two points (Fig. 3.31).

Fig. 3.31

In the case of simply supported beams, beams with one end hinged and the other on rollers, cantilever and over-hanging beams, it is possible to determine the reactions for given loadings by using the Equations of equilirium 3.11 only. In the other cases, the number of independent equilibrium equations are less than the number of unknown reactions and hence it is not possible to analyse them by using equilibrium equations 3.11 alone. The beams which can be analysed using only equilibrium equations are known as **Statically Determinate** beams and those which cannot be analysed are known as **Statically Indeterminate** beams. The latter beams can be analysed using the conditions of continuity in deformations in addition to equilibrium equations. Such cases will not be treated in this book.

Types of Loading

Usual types of loadings on the beams are discussed here.

Concentrated Loads – If a load is acting on a beam over a very small length, it is approximated as acting at the mid point of that length and is represented by an arrow as shown in Fig. 3.32.

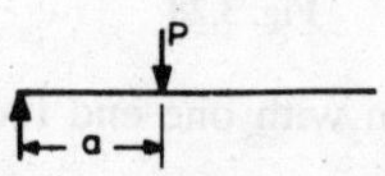

Fig. 3.32

Uniformly Distributed Load (*UDL*) – Over considerably long distance such load has got uniform intensity. It is represented as shown in Fig. 3.33 (a) or as in (b). For finding reaction, this load may be assumed as total load acting at the centre of gravity of the loading (middle of the loaded length). For example, in the beam shown in Fig. 3.33, the given load may be replaced by a 20 × 4 = 80 kN concentrated load acting at a distance 2 m from the left support.

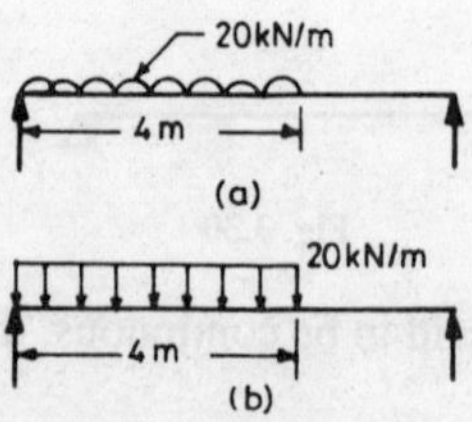

Fig. 3.33

Uniformly Varying Load – The load shown in Fig. 3.34 varies uniformly from *C* to *D*. Its intensity is zero at *C* and is 15 kN/m at *D*. In the load diagram, the ordinate represents the load intensity and the abscissa represents the position of load on the beam.

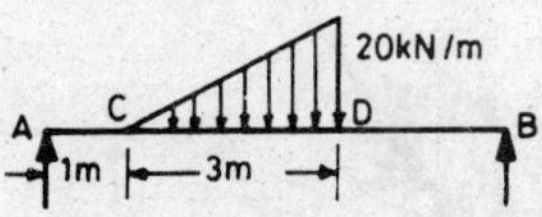

Fig. 3.34

Hence the area of the triangle represents the total load and the centroid of the triangle represents the centre of gravity of the load. Thus, total load in this case is $\frac{1}{2} \times 3 \times 20 = 30$ kN and the centre of gravity of this loading is at $\frac{1}{3} \times 3 = 1$ m from *D*, i.e. $1 + 3 - 1 = 3$ m from *A*. For finding the reactions, we can assume that the given load is equivalent to 30 kN acting at 3 m from *A*.

General Loading – Fig. 3.35 shows a general loading. Here the ordinate represents the intensity of loading and abscissa represents position of the load on the beam. For simplicity in analysis such loadings are replaced by a set of equivalent concentrated loads.

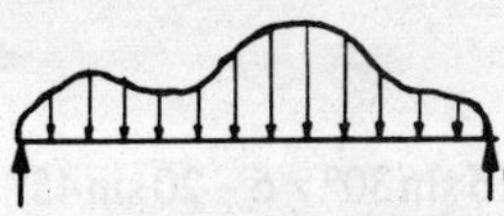

Fig. 3.35

External Moment – A beam may be subjected to external moment at certain points. In Fig. 3.36, the beam is subjected to clockwise moment of 30 kN-m at a distance of 2 m from the left support.

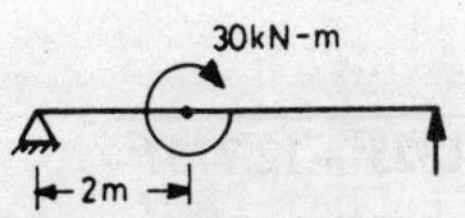

Fig. 3.36

Analysis of Beams

A beam is in equilibrium under the action of the loads and the reactions. Hence the equations of equilibrium may be written down for the system of forces consisting of reactions and the loads. Solutions of these equations give the unknown reactions.

Ex. 3.10: The beam *AB* of span 12 m shown in Fig. 3.37 (a) is hinged at *A* and is on rollers at *B*. Determine the reactions at *A* and *B* for the loading shown in the figure.

At A the reaction can be in any direction. Let this reaction be represented by its components V_A and H_A as shown in Fig. 3.37(b). At B the reaction is in vertical direction only. The beam is in equilibrium under the action of system of forces shown in Fig. 3.37 (b).

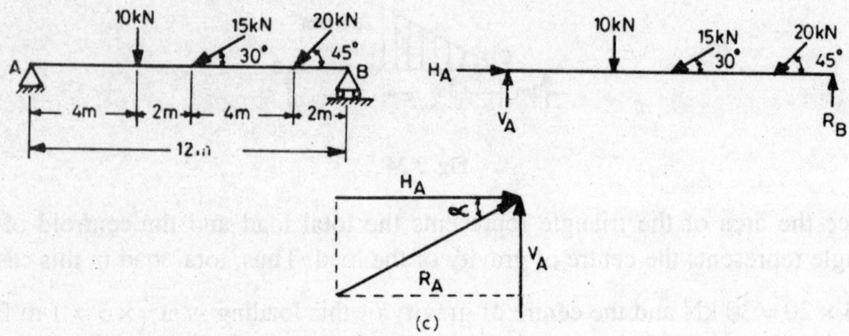

Fig. 3.37

Now, $\Sigma H = 0$

$$H_A - 15\cos 30° - 20\cos 45° = 0$$

$$H_A = 27.1325 \text{ kN}$$

$$\Sigma M_A = 0$$

$$R_B \times 12 - 10 \times 4 - 15\sin 30° \times 6 - 20\sin 45° \times 10 = 0$$

$$R_B = \mathbf{18.8684\ kN}$$ *Ans.*

$$\Sigma V = 0$$

$$V_A + 18.8684 - 10 - 15\sin 30° - 20\sin 45° = 0$$

$\therefore$ $V_A = 12.7737$ kN

$\therefore$ $R_A = \sqrt{H_A^2 + V_A^2} = \sqrt{27.1325^2 + 12.7737^2}$

i.e. $R_A = \mathbf{29.989\ kN}$ *Ans.*

and $\alpha = \tan^{-1}\dfrac{(12.7737)}{(27.1325)} = \mathbf{25.21°}$ *Ans.*

Ex. 3.11: Find the reactions at supports A and B in the beam AB shown in Fig. 3.38 (a)

The reaction at B will be at right angles to the support, i.e. at 60° to horizontal as shown in the figure. Let the components of the reactions at A be H_A and V_A. Then

$\Sigma M_A = 0$ gives

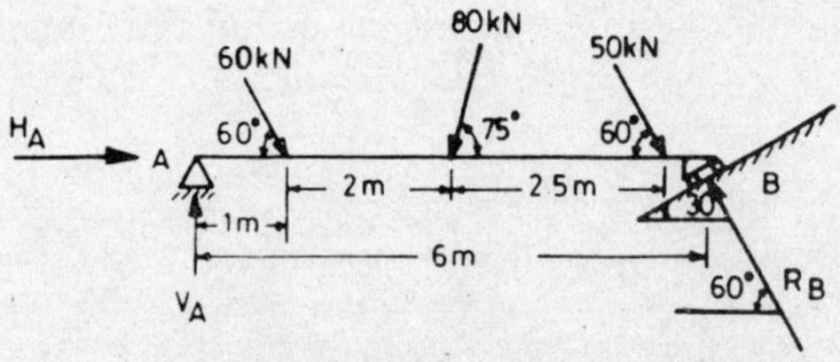

Fig. 3.38 (a)

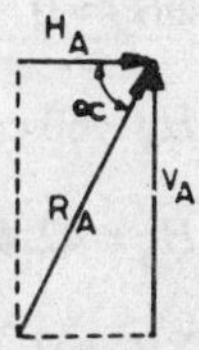

Fig. 3.38 (b)

$$R_B \sin 60° \times 6 - 60 \sin 60° \times 1 - 80 \times \sin 75° \times 3 - 50 \times \sin 60° \times 5.5 = 0$$

$\therefore \quad$ $R_B = \mathbf{100.4475\ kN}$ *Ans.*

$$\Sigma H = 0$$

$$H_A + 60 \cos 60° - 80 \cos 75° + 50 \cos 60° - R_B \cos 60° = 0$$

$$H_A = -60 \cos 60° + 80 \cos 75° - 50 \cos 60° + 100.4475 \cos 60°$$

$$= 15.9293 \text{ kN}$$

$$\Sigma V = 0$$

$$V_A + R_B \sin 60° - 60 \sin 60° - 80 \sin 75° - 50 \sin 60° = 0$$

$$V_A = -100.4475 \sin 60° + 60 \sin 60° + 80 \sin 75°$$

$$+50 \sin 60° = 85.5468 \text{ kN}$$

$\therefore \quad R_A = \sqrt{15.9293^2 + 85.5468^2}$

i.e $R_A = \mathbf{87.0172\ kN}$ *Ans.*

$$\alpha = \tan^{-1} \frac{85.5468}{15.9293}$$

i.e. $\alpha = \mathbf{79.45°}$, as shown in Fig. 3.38 (b) *Ans.*

Ex. 3.12: Find the reactions at supports A and B of the loaded beam shown in Fig. 3.39(a).

The reaction at A is vertical. Let H_B and V_B be the components of the reaction at B.

$$\Sigma M_B = 0$$

$$R_A \times 9 - 20 \times 7 - 30 \times 4 \times 5 - 60 \sin 45° \times 2 = 0$$

$\therefore \quad R_A = \mathbf{91.6503\ kN}$ *Ans.*

$\Sigma H_A = 0$

$H_B - 60 \cos 45° = 0$

$\therefore \quad H_B = \mathbf{42.4264\ kN}$ *Ans.*

$\Sigma V_A = 0$

$91.6503 + V_B - 20 - 30 \times 4 - 60 \sin 45° = 0$

$V_B = \mathbf{90.7761\ kN}$ *Ans.*

$\therefore \quad R_B = \sqrt{42.4264^2 + 90.7761^2}$

$R_B = \mathbf{100.2013\ kN}$ *Ans.*

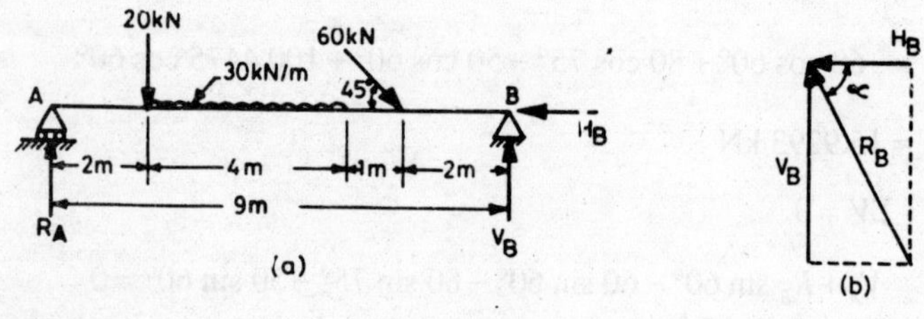

Fig. 3.39 (a, b)

$$\alpha = \tan^{-1} \frac{90.7761}{42.4264}$$

$\alpha = \mathbf{64.95°}$, as shown in Fig. 3.37 (b) *Ans.*

Ex. 3.13: The cantilever shown in Fig. 3.40 is fixed at A and is free at B. Determine the reactions when it is loaded as shown in the figure.

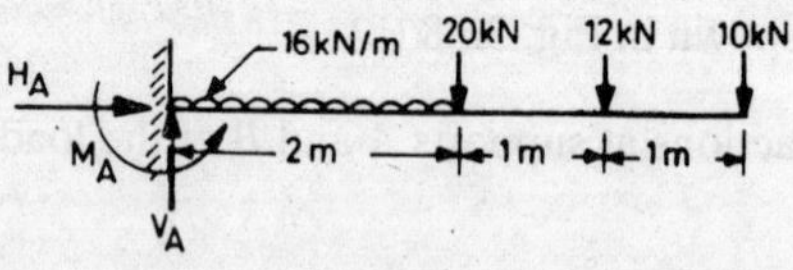

Fig. 3.40

Let the reactions at A be H_A, V_A and M_A as shown in the figure

Now, $\Sigma H = 0$

$H_A = \mathbf{0}$ *Ans.*

$\Sigma V = 0$

$$V_A - 16 \times 2 - 20 - 12 - 10 = 0$$

$\therefore$ $\mathbf{V_A = 74\ kN}$ *Ans.* $\Sigma M = 0$

$$M_A - 16 \times 2 \times 1 - 20 \times 2 - 12 \times 3 - 10 \times 4 = 0$$

$\mathbf{M_A = 148\ kN\text{-}m}$ *Ans.*

Ex. 3.14: Determine the reactions at supports *A* and *B* of the overhanging beam shown in Fig, 3.41.

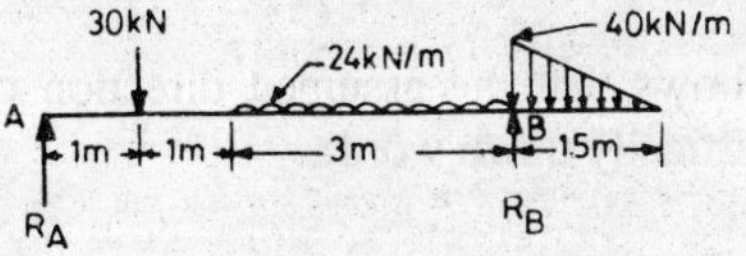

Fig. 3.41

$$\Sigma M_A = 0$$

$$R_B \times 5 - 30 \times 1 - 24 \times 3 \times 3.5 - \frac{1}{2} \times 40 \times 1.5 \times 5.5 = 0$$

$\mathbf{R_B = 89.4\ kN}$ *Ans.*

$$\Sigma V = 0$$

$$R_A - 30 - 24 \times 3 - \frac{1}{2} \times 40 \times 1.5 + R_B = 0$$

$$R_A = 30 + 72 + 30 - 89.4$$

$\mathbf{R_A = 42.6\ kN}$ *Ans.*

Ex. 3.15: Determine the reactions at *A* and *B* of the over-hanging beam shown in Fig. 3.42 (a).

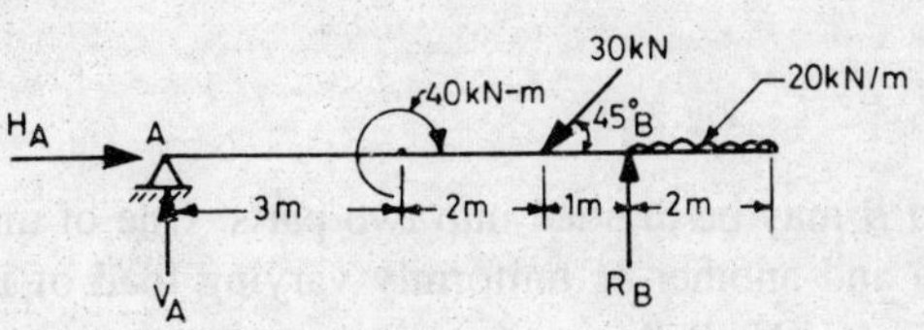

Fig. 3.42 (a)

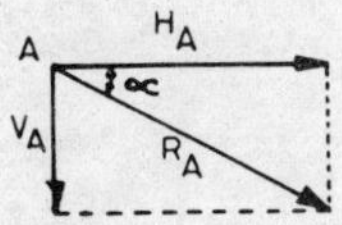

Fig. 3.42 (b)

$$\Sigma M_A = 0$$

$$R_B \times 6 - 40 - 30 \sin 45° \times 5 - 20 \times 2 \times 7 = 0$$

$\mathbf{R_B = 71.0110\ kN}$ *Ans.*

$\Sigma H = 0$

$H_A = 30\cos 45°$

$= 21.2132$ kN

$\Sigma V = 0$

$V_A - 30\sin 45° + R_B - 20 \times 2 = 0$

$V_A = 30\sin 45° - R_B + 40$

$V_A = -9.7978$ kN

(Negative sign shows that the assumed direction of V_A is wrong. In other words, V_A is acting vertically down word).

$R_A = \sqrt{V_A^2 + H_A^2}$

$R_A = 23.3666$ kN *Ans.*

$\alpha = \tan^{-1}\dfrac{V_A}{H_A}$

$\alpha =$ **24.79°**, as shown in Fig. 3.40 (b) *Ans.*

Ex. 3.16 : Find the reactions at supports *A* and *B* of the loaded beam shown in Fig. 3.43.

Since all loads are vertical, the horizontal component of reaction at *B* is zero, i.e., the reaction R_B is vertical.

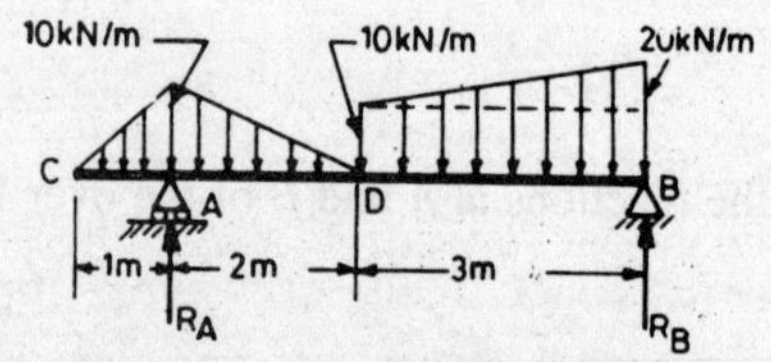

Fig. 3.43

The load between *D* and *B* may be divided into two parts: One of uniformly distributed load of 10 kN/m and another of uniformly varying load of intensity zero at *D* and 10 kN (20–10 = 10 kN) at *B*.

$\Sigma M_A = 0$

$$R_B \times 5 + \frac{1}{2} \times 1 \times 10 \times \frac{1}{3} - \frac{1}{2} \times 2 \times 10 \times \frac{1}{3} \times 2 - 10 \times 3 \times 3.5 - \frac{1}{2} \times 3 \times 10 \times 4 = 0$$

$$R_B = \mathbf{34.0kN}$$ *Ans.*

$$\Sigma V = 0$$

$$R_A + 34.0 - \frac{1}{2}\times 1\times 10 - \frac{1}{2}\times 2\times 10 - 10\times 3 - \frac{1}{2}\times 3\times 10 = 0$$

$$R_A = 5 + 10 + 30 + 15 - 34.0$$

i.e. $R_A = \mathbf{26.0kN}$ *Ans.*

Ex. 3.17: A beam *AB* 20 m long supported on two intermediate supports 12m apart, carries a uniformly distributed load of 6 kN/m and two concentrated loads of 30 kN at left end *A* and 50 kN at the right end *B* as shown in Fig. 3.44. How far away should the first support *C* be located from the end *A* so that the reactions at both the supports are equal?

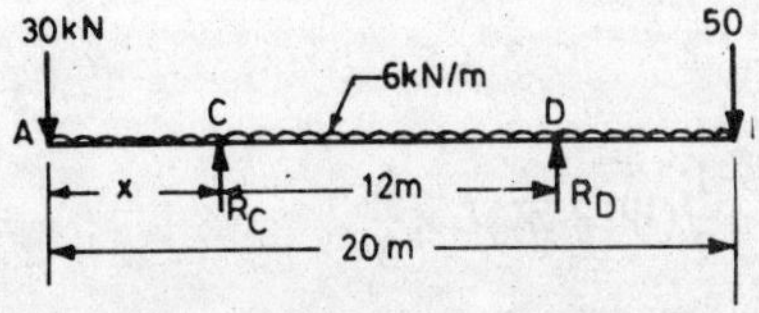

Fig. 3.44

Let the left support *C* be at a distance *x* metres from *A*.
Now, $R_C = R_D$ (given)

$$\Sigma V = 0$$

$$R_C + R_D - 30 - 6\times 20 - 50 = 0$$

or $2R_C = 30 + 120 + 50$ since $R_C = R_D$

or $R_C = 100\text{kN}$

$\therefore$ $R_D = 100 \text{ kN}$

$$\Sigma M_A = 0$$

$$100x + 100(12+x) - 6\times 20\times 10 - 50\times 20 = 0$$

$$200x = 1000$$

$$\mathbf{x = 5m}$$ *Ans.*

Ex. 3.18: A beam *AB* supports a uniformly distributed load of intensity W_1 and rests on soil which exerts a uniformly varying upward reaction as shown in Fig. 3.45. Determine W_2 and W_3 corresponding to equilibrium. Note that soil can exert

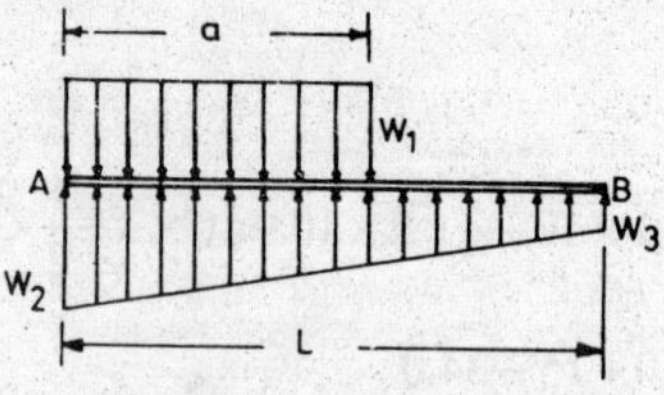

Fig. 3.45

only an upward reaction at any point on the beam. Hence, state for what range of values of $\frac{a}{L}$ the results obtained are valid.

$$\Sigma V = 0$$

$$a W_1 = \frac{W_2 + W_3}{2} L \qquad \ldots(1)$$

$$\Sigma M_A = 0$$

$$a W_1 \frac{a}{2} = W_3 L \times \frac{L}{2} + \frac{1}{2}(W_2 - W_3) L \times \frac{L}{3}$$

$$W_1 a^2 = W_3 L^2 + (W_2 - W_3)\frac{L^2}{3} \qquad \ldots(2)$$

$$\text{From (1), } W_2 = \frac{2a W_1}{L} - W_3 \qquad \ldots(3)$$

$$\text{From (2), } W_1 a^2 = W_3 L^2 + \left(\frac{2a W_1}{L} - W_3 - W_3\right)\frac{L^2}{3}$$

$$= \frac{2aL}{3} W_1 + W_3\left(1 - \frac{2}{3}\right) L^2$$

$$\therefore \quad W_1\left(a^2 - \frac{2aL}{3}\right) = W_3 \frac{L^2}{3}$$

$$W_3 = \left(\frac{3W_1}{L^2}\right)\left(a^2 - \frac{2aL}{3}\right) \qquad \ldots(4) \qquad \textit{Ans.}$$

Substituting this value of W_3 in (3) we get,

$$W_2 = \frac{2a W_1}{L} - \frac{3W_1}{L^2}\left(a^2 - \frac{2aL}{3}\right)$$

$$W_2 = \left(\frac{4a}{L} - \frac{3a^2}{L^2}\right) W_1 \qquad \ldots(5) \qquad \textit{Ans.}$$

For the above expressions to be valid W_2 and W_3 should not be negative. From (4),

$$0 = \frac{3W_1}{L^2}\left(a^2 - \frac{2aL}{3}\right)$$

$$\therefore \quad a = \frac{2L}{3} \quad \text{i.e.} \quad \frac{a}{L} = \frac{2}{3}$$

From (5), $0 = \left(\frac{4a}{L} - \frac{3a^2}{L^2}\right) W_1$

$$\text{i.e.} \quad \frac{4a}{L} - \frac{3a^2}{L^2} = 0$$

$$\text{i.e.} \quad \frac{a}{L} = \frac{4}{3}$$

Hence, the results obtained are valid for the range $\frac{a}{L} = \frac{2}{3}$ to $\frac{4}{3}$ only. However, there is no possibility of loading beyond $\frac{a}{L} = 1$. Hence, **the valid range of $\frac{a}{L}$ is from $\frac{2}{3}$ to 1.**

Ex. 3.19: Determine the reactions at A, B and D of the system shown in Fig. 3.46 (a). Neglect the self-weight of the members.

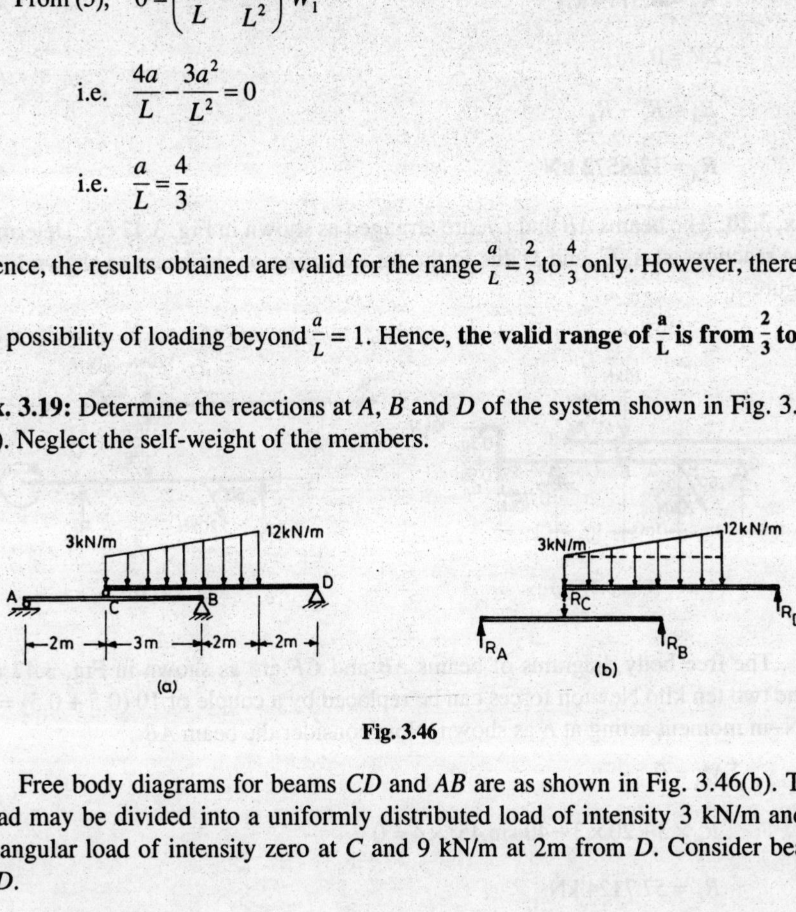

Fig. 3.46

Free body diagrams for beams CD and AB are as shown in Fig. 3.46(b). The load may be divided into a uniformly distributed load of intensity 3 kN/m and a triangular load of intensity zero at C and 9 kN/m at 2m from D. Consider beam CD.

$$\Sigma M_C = 0$$

$$R_D \times 7 - 3 \times 5 \times 2.5 - \frac{1}{2} \times 5 \times 9 \times \frac{2}{3} \times 5 = 0$$

$$\boldsymbol{R_D = 16.0714 \text{ kN}} \qquad \textit{Ans.} \qquad \Sigma V = 0$$

$$R_C + 16.0714 - 3 \times 5 - \frac{1}{2} \times 5 \times 9 = 0$$

$$R_C = 21.4286 \text{ kN}$$

Consider beam AB

$$\Sigma M_A = 0$$

$$R_B \times 5 - R_C \times 2 = 0$$

$$R_B = \frac{21.4286 \times 2}{5}$$

$$\mathbf{R_B = 8.5714\ kN} \qquad \textit{Ans.}$$

$$\Sigma V = 0$$

$$R_A = R_C - R_B$$

$$\mathbf{R_A = 12.8572\ kN} \qquad \textit{Ans.}$$

Ex. 3.20: The beams *AB* and *CF* are arranged as shown in Fig. 3.47 (a). Determine the reactions at *A*, *C*, and *D* due to the forces acting on the beam as shown in the figure.

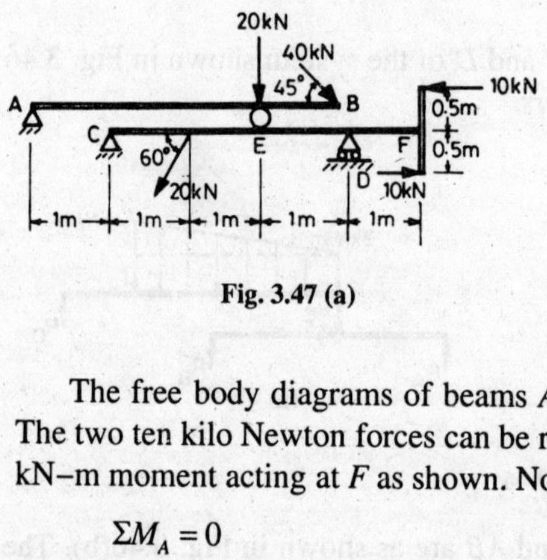

Fig. 3.47 (a)

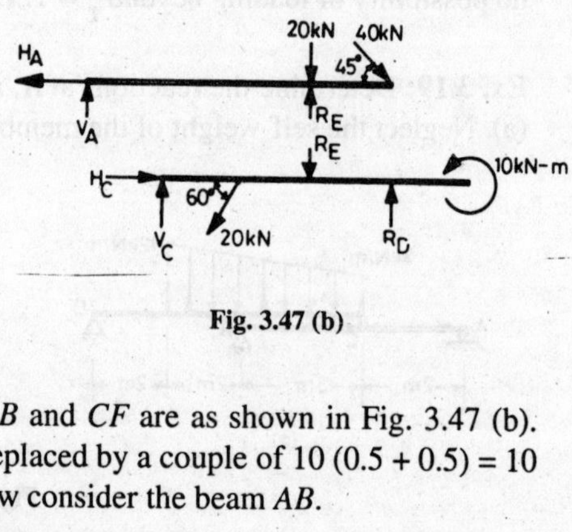

Fig. 3.47 (b)

The free body diagrams of beams *AB* and *CF* are as shown in Fig. 3.47 (b). The two ten kilo Newton forces can be replaced by a couple of 10 (0.5 + 0.5) = 10 kN–m moment acting at *F* as shown. Now consider the beam *AB*.

$$\Sigma M_A = 0$$

$$R_E \times 3 - 20 \times 3 - 40 \sin 45° \times 4 = 0$$

$$R_E = 57.7124 \text{ kN}$$

$$\Sigma H = 0$$

$$H_A = 40 \cos 45° = 28.2843 \text{ kN}$$

$$\Sigma V = 0$$

$$V_A + 57.7124 - 20 - 40 \sin 45° = 0$$

$$V_A = -9.4281$$

or $V_A = 9.4281$ kN, downwards.

$R_A = \sqrt{H_A^2 + V_A^2}$

i.e. $\mathbf{R_A = 29.8143\ kN}$ *Ans.*

$\alpha_1 = \tan^{-1}\dfrac{V_A}{H_A}$

$\alpha_1 = \mathbf{18.43°}$ as shown in Fig. 3.47 (c) *Ans.*

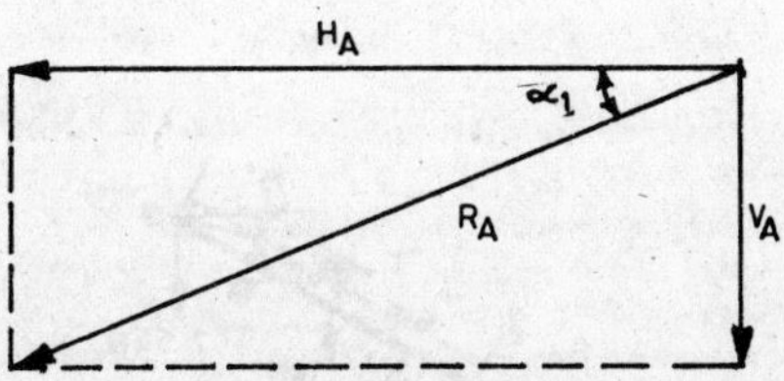

Fig. 3.47 (c)

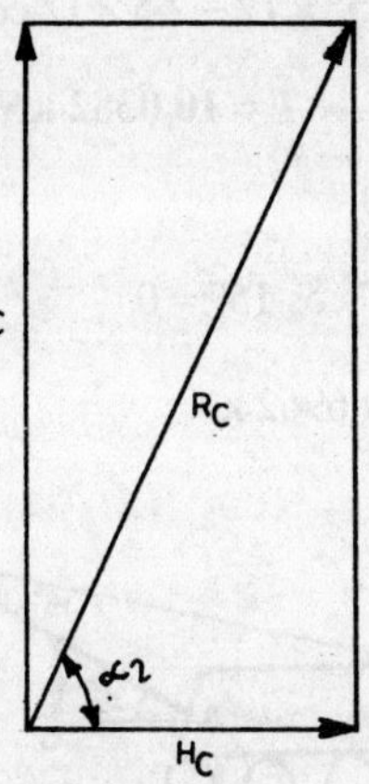

Fig. 3.47 (d)

Consider beam CF

$\Sigma M_C = 0$

$R_D \times 3 - 20 \sin 60° \times 1 - 57.7124 \times 2 + 10 = 0$

$\mathbf{R_D = 40.9151\ kN}$ *Ans.* $\Sigma H = 0$

$H_C - 20 \cos 60° = 0$

$H_C = 10$ kN

$\Sigma V = 0$

$V_C - 20 \sin 60° - 57.7124 + 40.9151 = 0$

$V_C = 34.1178$ kN

$R_C = \sqrt{V_C^2 + H_C^2}$

$\mathbf{R_C = 35.5531\ kN}$ *Ans.*

$\alpha_2 = \tan^{-1}\dfrac{V_C}{H_C}$ $\quad\alpha_2 = \mathbf{73.66°}$ as shown in Fig. 3.45 (d) *Ans.*

Ex. 3.21: The 12 m boom *AB* weighs 1 kN, the distance of the centre of gravity *G* being 6 m from *A*. For the position shown, determine the tension *T* in the cable and the reaction at *B* [Ref. Fig. 3.48 (a)].

The free body diagram of the boom is shown in Fig. 3.48 (b).

$$\Sigma M_A = 0$$

$$T \sin 15° \times 12 - 2.5 \times 12 \cos 30° - 1 \times 6 \cos 30° = 0$$

$$\mathbf{T = 10.0382\ kN}$$ *Ans.*

$$\Sigma H = 0$$

$$H_A - T \cos 15° = 0$$

$$H_A = 9.6962 \text{ kN}$$

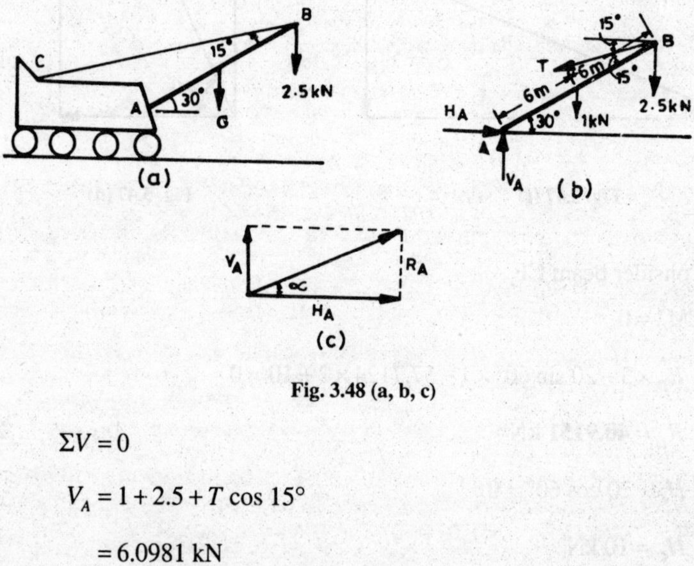

Fig. 3.48 (a, b, c)

$$\Sigma V = 0$$

$$V_A = 1 + 2.5 + T \cos 15°$$

$$= 6.0981 \text{ kN}$$

$$R_A = \sqrt{V_A^2 + H_A^2}$$

$$\mathbf{R_A = 11.4544\ kN}$$ *Ans.*

$$\alpha = \tan^{-1}\frac{6.0981}{9.6962}$$

$$= \mathbf{32.17°} \text{ as shown in Fig. 3.48 (c)}$$ *Ans.*

Ex. 3.22: A cable car used for carrying materials in a hydroelectric project is at rest on a track formed at an angle of 30° with the vertical. The gross weight of the car and its load is 60 kN and its centroid is at a point 800 mm from the track half way between the axles. The car is held by a cable as shown in Fig. 3.49. The axles of the car are at a distance 1.2m. Find the tension in the cables and reaction at each of the axles neglecting friction of the track.

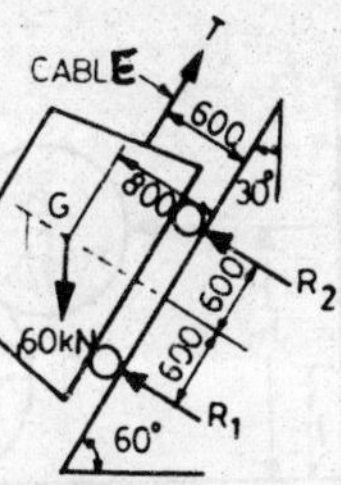

Fig. 3.49

Let T be the tension in the cable and the reaction at the pair of wheels be R_1 and R_2 as shown in Fig. 3.49.

Now, Σ of forces parallel to the track = 0, gives

$$T - 60 \sin 60° = 0$$

$$\mathbf{T = 51.9615\ kN} \qquad \textit{Ans.}$$

Taking moment equilibrium condition about upper axle point on track, we get:

$$R_1 \times 1200 + T \times 600 - 60 \sin 60° \times 800 - 60 \cos 60° \times 600 = 0$$

$$\mathbf{R_1 = 23.6603\ kN} \qquad \textit{Ans.}$$

Σ of forces normal to the plane = 0, gives:

$$R_1 + R_2 - 60 \cos 60° = 0$$

$$R_2 = 30 - 23.6603$$

$$\mathbf{R_2 = 6.3397\ kN} \qquad \textit{Ans.}$$

Ex. 3.23: A hollow right circular cylinder of radius 800 mm is open at both ends and rests on a smooth horizontal plane as shown in Fig. 3.50 (a). Inside the cylinder there are two spheres having weights 1 kN and 3 kN and radii 400 mm and 600 mm, respectively. The lower sphere also rests on the horizontal plane. Neglecting friction find the minimum weight W of the cylinder for which it will not tip over.

Join the centres of spheres O_1 and O_2 and drop $O_1 D$ perpendicular to horizontal through O_2.

Now, $O_1O_2 = 400 + 600 = 1000$ mm

$$O_2D = 1600 - 400 - 600 = 600 \text{ mm}$$

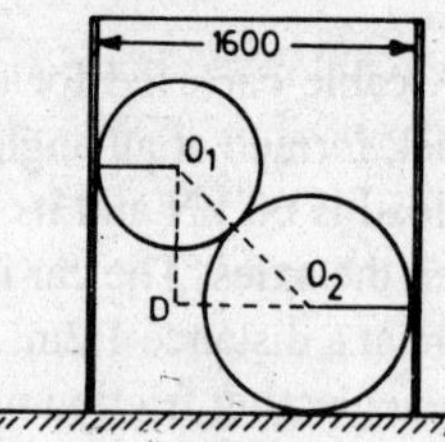

If α is the inclination of O_2O_1 to horizontal,

$$\cos\alpha = \frac{600}{1000} = 0.6$$

$$\therefore \quad \sin\alpha = 0.8$$

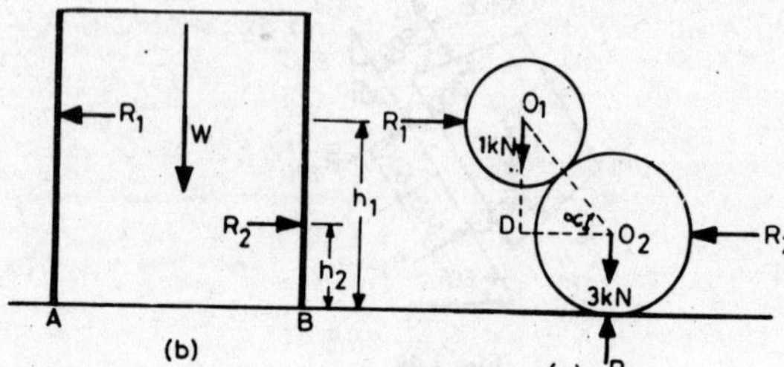

Fig. 3.50

Free body diagrams of cylinder and spheres are shown in Fig. 3.50 (b) and (c). Considering the equilibrium of the spheres,

$$\Sigma M_{O_2} = 0$$

$$R_1 \times O_1\,O_2 \sin\alpha - 1 \times O_2 D = 0$$

$$R_1 \times 1000 \times 0.8 = 1 \times 600$$

$$R_1 = 0.75 \text{ kN}$$

$$\Sigma H = 0$$

$$R_2 = R_1 = 0.75 \text{ kN}$$

$$\Sigma V = 0$$

$$R_3 = 1 + 3 = 4 \text{ kN}$$

Now consider the equilibrium of cylinder. When it is about to tip over *A*, there is no reaction from ground at *B*. The reaction will be only at *A*. $\Sigma M_A = 0$ gives,

$$R_1 h_1 - R_2 h_2 - W \times 800 = 0$$

$$R_1(h_1 - h_2) - W \times 800 = 0$$

Since $R_1 = R_2$

$$0.75 \times O_1 D = W \times 800$$

$$0.75 \times 1000 \times 0.8 = W \times 800$$

$\therefore$ $\quad$ $\mathbf{W = 0.75\ kN}$ *Ans.*

Ex. 3.24: A 500 *N* cylinder, 1 m in diameter is loaded between the cross pieces which make an angle of 60° with each other and are pinned at *C*. Determine the tension in the horizontal rope *DE* assuming a smooth floor [Ref. Fig. 3.51 (a)].

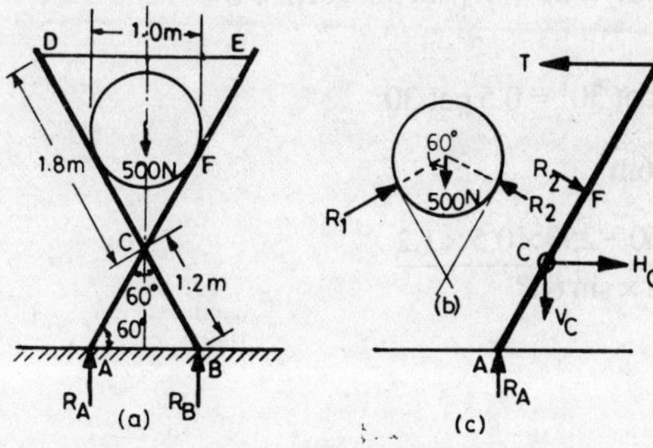

Fig. 3.51

Consider the equilibrium of the entire system.
Due to symmetry

$$R_A = R_B$$

$$\Sigma V = 0$$

$$R_A + R_B = 500$$

$$\therefore \quad R_A = 250 \text{ N}$$

and hence $\quad R_B = 250$ N

Now consider the equilibrium of the cylinder [Fig. 3.51(b)]. The members *AE* and *BE* make 60° with horizontal. Hence the reaction R_2 and R_1 make 60° with the vertical.

$$\Sigma H = 0$$

$$R_1 \sin 60° - R_2 \sin 60° = 0$$

$$\therefore \quad R_1 = R_2$$

$$\Sigma V = 0$$

$$R_1 \cos 60° + R_2 \cos 60° - 500 = 0$$

$$R_1 \times 0.5 + R_2 \times 0.5 = 500$$

i.e. $\quad R_1 = 500$ N since $R_1 = R_2$

$$\therefore \quad R_2 = 500 \text{ N}$$

Now consider the equilibrium of the member *AE*. The free body diagram is shown in Fig. 3.49(c).

$\Sigma M_C = 0$

$T \sin 60° \times 1.8 - R_2 \times CF - R_A \sin 30° \times 1.2 = 0$

Now $CF = OF \cot 30° = 0.5 \cot 30°$

$= 0.866\text{m}$

$\therefore \quad T = \dfrac{500 \times 0.8660 + 250 \times 0.5 \times 1.2}{1.8 \times \sin 60°}$

$= \mathbf{374.00\ N}$ *Ans.*

Note:
If the reactions on pin at *C* are required, the other two equilibrium equations, namely $\Sigma x = 0$ and $\Sigma y = 0$, are to be considered.

Ex. 3.25: The frame shown in Fig. 3.52(a) is supported by a hinge at *A* and a roller at *E*. Compute the horizontal and vertical components of the hinge forces at *B* and *C* as they act upon member *AC*.

Consider the equilibrium of the entire frame.

$\Sigma M_A = 0$

$R_E \times 5 - 1200 \times (1 + 3 + 1.5) = 0$

$R_E = 1320\ \text{N}$

$\Sigma H = 0$, gives

$H_A = 0$

$\Sigma V = 0$

$R_A + R_E - 1200 = 0$

$R_A = 1200 - 1320 = -120\ \text{N}$

$= 120\ \text{N}$ (The negative sign shows that R_A acts downwards)

Consider the equilibrium of the horizontal member *BDF* [Fig. 3.52 (c)].

$\Sigma M_D = 0$

$V_B \times 3 - 1200 \times 1.5 = 0$

$\mathbf{V_B = 600\ N}$ **(downward)** *Ans.*

Consider the equilibrium of the member *ABC*. [*FBD* in Fig. 3.52 (b)]

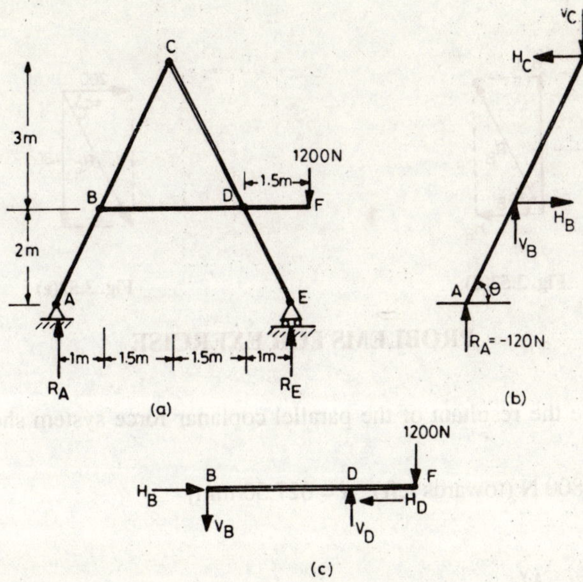

Fig. 3.52 (a, b, c)

$$\Sigma V = 0$$

$$V_C + 120 - 600 = 0$$

$\mathbf{V_C = 480\ N}$ **(downward)** *Ans.*

$$\Sigma H = 0$$

$$H_C = H_B$$

$$\Sigma M_C = 0$$

$$H_B \times 3 - 600 \times 1.5 + 120 \times 2.5 = 0$$

$\mathbf{H_B = 200\ N}$ **(towards right)** *Ans.*

$\therefore\quad \mathbf{H_C = 200\ N}$ **(towards left)** *Ans.*

$R_B = \sqrt{600^2 + 200^2} = \mathbf{632.45\ N}$ *Ans.*

$\alpha_B = \mathbf{71.56°}$ as shown in Fig. 3.52 (d) *Ans.*

$R_C = \sqrt{200^2 + 480^2} = \mathbf{520\ N}$ *Ans.*

$\alpha_C = \tan^{-1} \dfrac{480}{200} = \mathbf{67.38°}$ as shown in Fig. 3.52 (e) *Ans.*

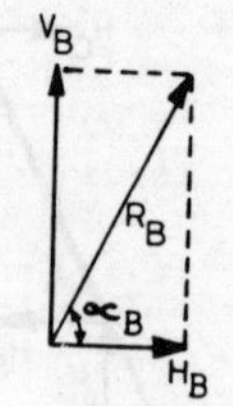

Fig. 3.52(d)

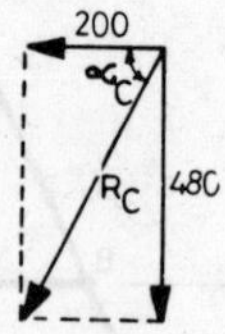

Fig. 3.52(e)

PROBLEMS FOR EXERCISE

3.1 Determine the resultant of the parallel coplanar force system shown in Fig. 3.53.

[*Ans.* $R = 800$ N (towards left); $d = 627.50$ mm]

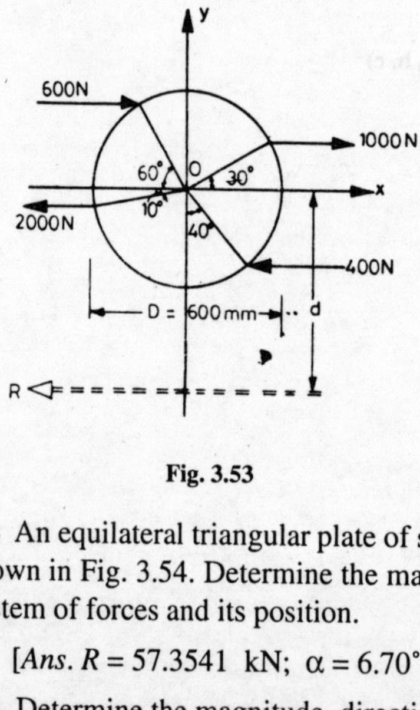

Fig. 3.53

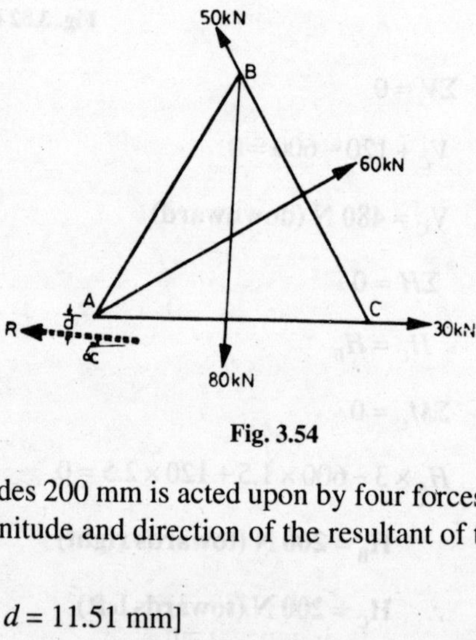

Fig. 3.54

3.2 An equilateral triangular plate of sides 200 mm is acted upon by four forces as shown in Fig. 3.54. Determine the magnitude and direction of the resultant of this system of forces and its position.

[*Ans.* $R = 57.3541$ kN; $\alpha = 6.70°$; $d = 11.51$ mm]

3.3 Determine the magnitude, direction and the line of application of the equilibriant of the set of forces shown in Fig. 3.55, which will keep the plane body *ABCDEFGH* in equilibrium.

[*Ans.* $E = 23.6517$ kN; $\alpha = 24.37°$; x from $A = 1.041$ m]

3.4 Determine the resultant of the four forces acting on a body as shown in Fig. 3.56.

[*Ans.* $R = 200$; $\alpha = 60°$; at a distance $y = 5.7679$ m below O]

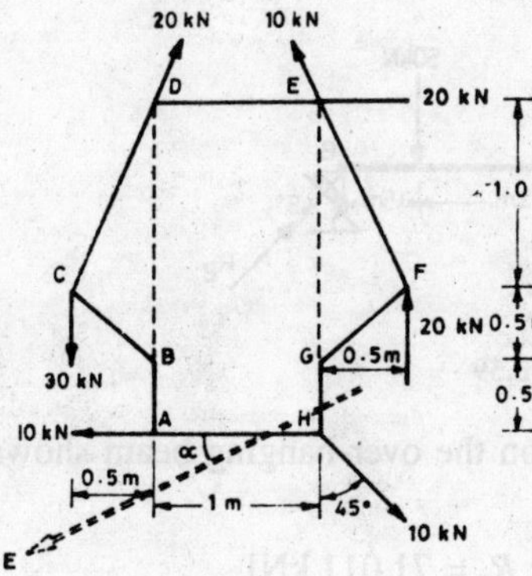

Fig. 3.55

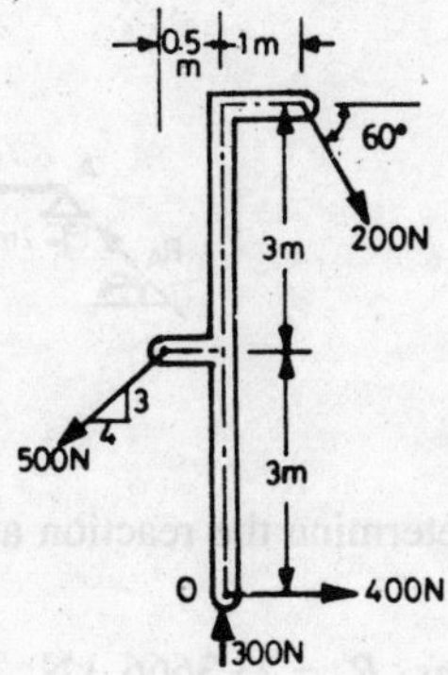

Fig. 3.56

3.5 A bracket is subjected to the system of forces and couples as shown in Fig. 3.57. Find the resultant of the system and the point of intersection of its line of action with (a) line *AB*, (b) line *BC* and (c) line *CD*.

[Ans. R = 485.4 N ; α = 34.50° ; Y_{BA} = 112.5 mm ; X_{BC} = 163.6 mm ; Y_{CD} = 93.75 mm]

3.6 Determine the resultant of the three forces acting on the dam section shown in Fig. 3.58 and locate its intersection with the base *AB*. For a safe design this intersection should occur within the middle third. Is it a safe design?

[*Ans.* Resultant intersects *AB* 3.333 m from *A*. It is a good design].

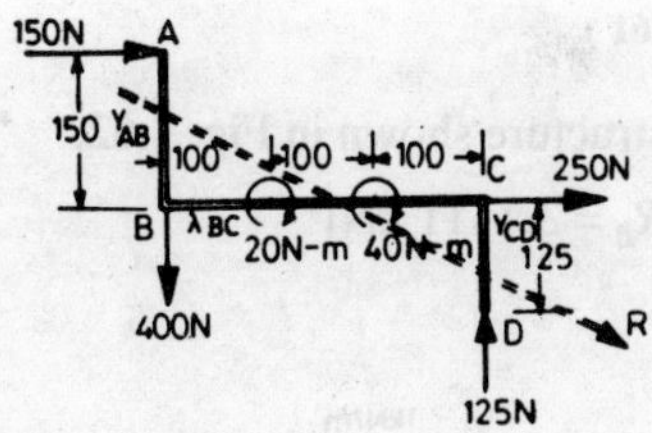

Fig. 3.57

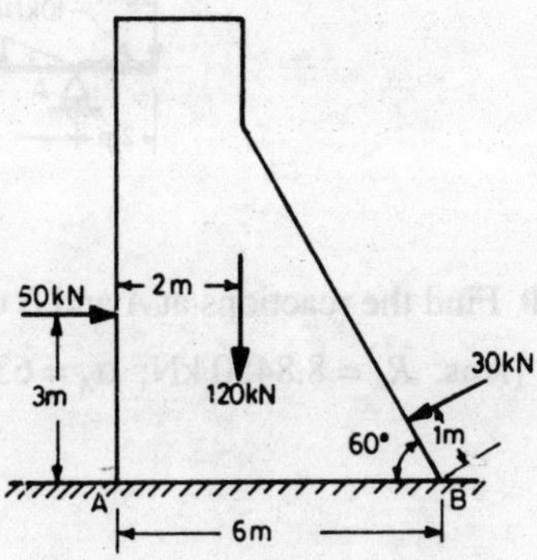

Fig. 3.58

3.7 Beam *AB* 8.5 m long, is hinged at *A* and is supported on roller at *B*. The roller support is inclined at 45° to the horizontal. Find the reactions at *A* and *B*, if the loads on it are as shown in Fig. 3.59.

[*Ans.* R_A = 89.4312 kN; α_A = 54.76°; R_B = 101.2517 kN]

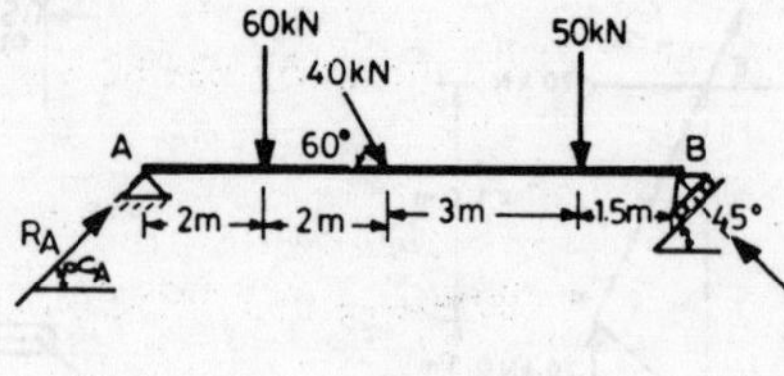

Fig. 3.59

3.8 Determine the reaction at *A* and *B* on the over-hanging beam shown in Fig. 3.60.

[*Ans.* $R_A = 23.3666$ kN; $\alpha = 24.79°$; $R_B = 71.011$ kN]

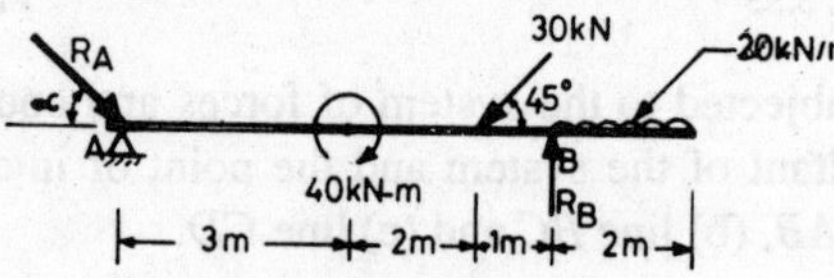

Fig. 3.60

3.9 An over-hanging beam is on rollers at *A* and is hinged at *B* and is loaded as shown in Fig. 3.61. Determine the reactions at *A* and *B*.

[*Ans.* $R_A = 160$ kN; $R_B = 90$ kN]

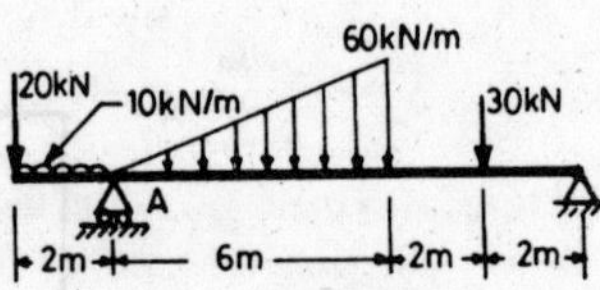

Fig. 3.61

3.10 Find the reactions at *A* and *B* on the structure shown in Fig. 3.62.

[Ans. $R_A = 8.8450$ kN; $\alpha_A = 63.11°$; $R_B = 3.1111$ kN]

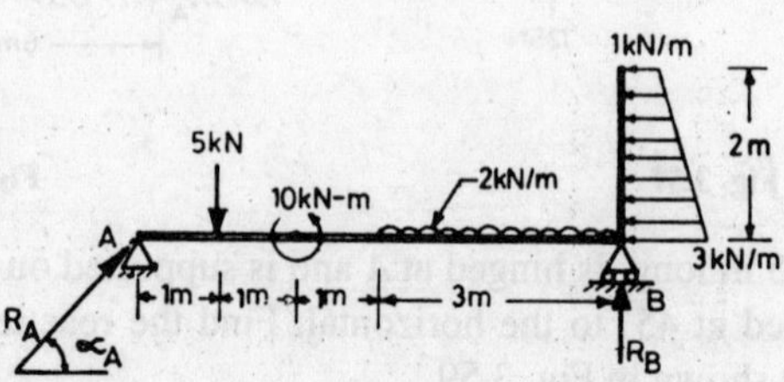

Fig. 3.62

3.11 Determine the reactions at supports *A*, *C* and *D* in the structure shown in Fig. 3.63.

[*Ans.* $R_A = 20$ kN; $R_C = 67.5$ kN ; $R_D = 7.5$ kN downward]

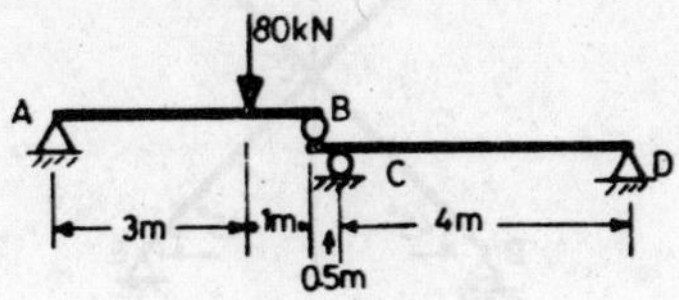

Fig. 3.63

3.12 Determine the reactions at the supports *A*, *B*, *C* and *D* for the arrangement of beams shown in Fig. 3.64. Neglect the self-weight of the beams.

[*Ans.* $R_A = 203.8013$; $R_B = 113.33$ kN; $R_C = 20.5726$ kN; $R_D = 82.2890$ kN]

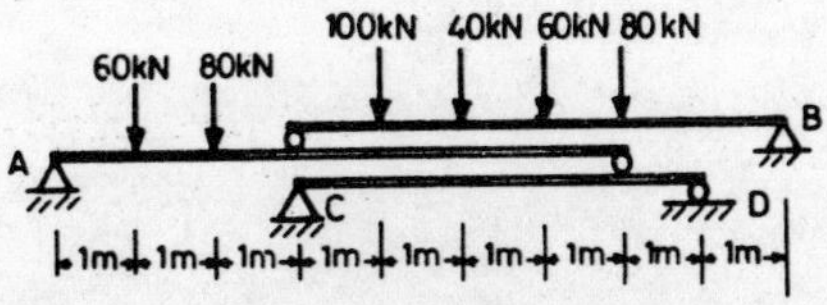

Fig. 3.64

3.13 A 1000-N cylinder is supported by a horizontal rod *AB* and a smooth uniform rod *CD* which weighs 500 N as shown in Fig. 3.65. Assuming the pins at *A*, *B*, *C* and *D*, to be frictionless and weight of *AB* negligible, find the reactions at *C* and *D*.

[*Ans.* $V_C = 937.1$ N (upward); $H_C = 577.3$ N (towards left) $R_D = 562.9$ N (upward)]

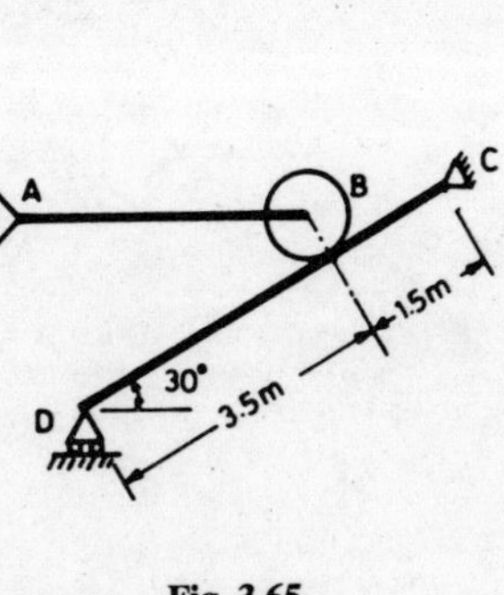

Fig. 3.65

3.14 The frame shown in Fig. 3.66 is supported by a hinge at *E* and by a roller at *D*. Determine the horizontal and vertical components of the reactions at hinge *C* as it acts upon member *BD*.

[*Ans.* $H_C = 140$ N (towards right); $V_C = 35$ N (upward)]

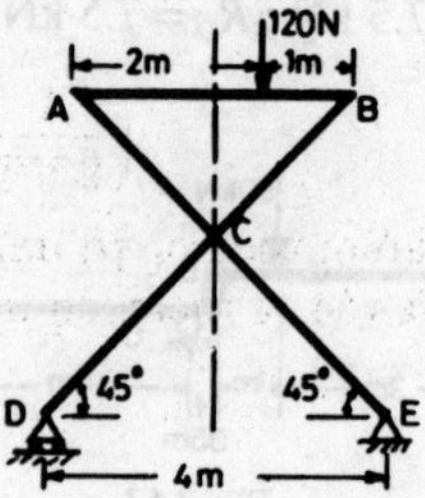

Fig. 3.66

CHAPTER 4

Friction

When a body moves or tends to move over another body, a force opposing the motion develops at the contact surfaces. This force which opposes the movement or the tendency of movement is called **frictional force** or simply **friction**. Friction is due to the resistance offered to motion by minutely projecting particles at the contact surfaces. In this chapter, the concepts related to friction are explained and the laws of friction are presented. Application of these laws to many engineering problems including wedge and rope/belt friction are illustrated.

4.1 FRICTIONAL FORCE

Frictional force has the remarkable property of adjusting itself in magnitude to the force producing or tending to produce the motion so that the motion is prevented. However, there is a limit beyond which the magnitude of this force cannot increase. If the applied force is more than this maximum frictional force, there will be movement of one body over the other. This maximum value of frictional force, which comes into play, when the motion is *impending*, is known as **Limiting Friction**. It may be noted that when the applied force is less than the limiting friction, the body remains at rest and such frictional force is called **Static Friction**, which may have any value between zero and the limiting friction. If the value of the applied force exceeds the limiting friction, the body starts moving over the other body and the frictional resistance experienced by the body while moving is known as **Dynamic Friction**. Dynamic friction is found to be less than limiting friction. Dynamic friction may be grouped into the following two:

(a) **Sliding Friction**: It is the friction experienced by a body when it slides over the other body.

(b) **Rolling Friction**: It is the friction experienced by a body when it rolls over a surface.

It is experimentally found that the magnitude of limiting friction bears a constant ratio to the normal reaction between the two surfaces and this ratio is called **Coefficient of Friction**. Thus in Fig. 4.1,

Coefficient Friction $= \frac{F}{N}$

Where, F = limiting friction

and N = normal reaction between the contact surfaces.

Coefficient of friction is denoted by μ. Thus,

$$\mu = \frac{F}{N}$$

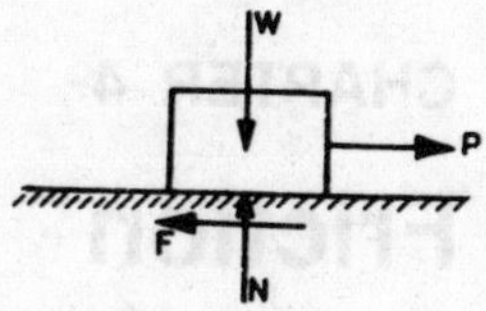

Fig. 4.1

4.2 LAWS OF FRICTION

The principles discussed in **4.1** are mainly due to the experimental studies by Coulomb (1781) and by Morin (1831). These principles constitute the laws of dry friction and may be listed as follows:

(1) The force of friction always acts in a direction opposite to that in which the body tends to move;

(2) Till the limiting value is reached, the magnitude of friction is exactly equal to the force which tends to move the body;

(3) The magnitude of the limiting friction bears a constant ratio to the normal reaction between the two surfaces;

(4) The force of friction depends upon the roughness/smoothness of the surfaces;

(5) The force of friction is independent of the area of contact between the two surfaces.

4.3 ANGLE OF FRICTION, ANGLE OF REPOSE AND CONE OF FRICTION

Angle of Friction

Consider the block shown in Fig. 4.2 subject to pull *P*. Let *F* be the frictional force developed and *N* the normal reaction. Thus, at the contact surface, the reactions are *F* and *N*. They can be combined graphically to get the reaction *R* which acts at angle θ to normal reaction. This angle θ , called the angle of friction, is given by:

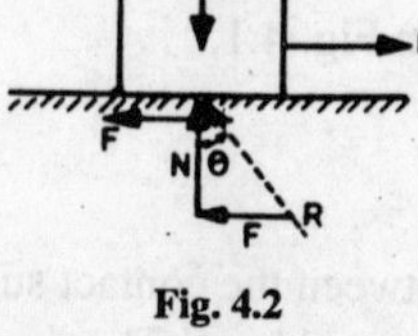

Fig. 4.2

$$\tan\theta = \frac{F}{N}$$

As frictional force increases the angle θ increases and it can reach maximum value α when limiting value of friction is reached. At this stage

$$\tan\alpha = \frac{F}{N} = \mu \qquad \textit{Eqn. } 4.1$$

and this value of α is called **Angle of Limiting Friction**. Hence, the angle of limiting friction can be defined as the angle between the resultant reaction and the normal to the plane on which the motion of the body is impending.

Angle of Repose

It is well known that when grains (food grain, soil, sand, etc.) are heaped, there exists a limit for the inclination of the surface. Beyond this inclination the grains starts rolling down. This limiting angle upto which the grains repose (sleep) is called the angle of repose.

Now, consider a block of weight *W* resting on an inclined plane which makes an angle θ with the horizontal as shown in Fig. 4.3. When θ is small the block will rest on the plane. If θ is increased gradually a stage is reached at which the block starts sliding. This angle between those two contact surfaces is called the angle of repose.

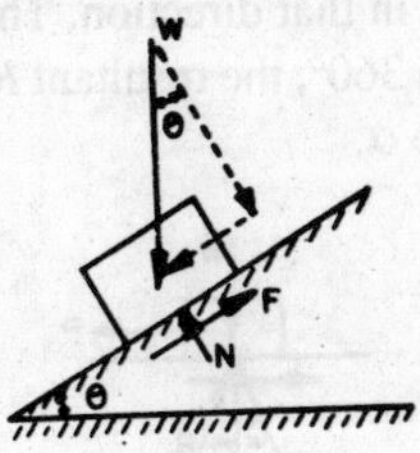

Fig. 4.3

Thus, *the maximum inclination of the plane on which a body, free from external forces, can repose (sleep)* is called **Angle of Repose**.

Now, consider the equilibrium of the block shown in Fig. 4.3. Since the surface of contact is not smooth, not only normal reaction, but frictional force also develops. Since the body tends to slide downward, the frictional resistance will be up the plane.

Σ forces normal to the plane = 0, gives

$$N = W\cos\theta \qquad \textit{Eqn. } 4.2$$

Σ forces parallel to the plane = 0, gives

$$F = W\sin\theta \qquad \textit{Eqn. } 4.3$$

Dividing Eqn. 4.3 by Eqn. 4.2 we get,

$$\tan\theta = \frac{F}{N} \qquad \text{Eqn.4.4}$$

If ϕ is the value of θ when motion is impending, frictional force will be limiting friction and hence

$$\tan\phi = \frac{F}{N}$$

$$= \mu$$

$$= \tan\alpha$$

$$\text{or} \quad \phi = \alpha$$

Thus, *the value of angle of repose is the same as the value of limiting angle of friction.*

Cone of Friction

When a body is having impending motion in the direction of P, the frictional force will be the limiting friction and the resultant reaction R will make limiting friction angle α with the normal as shown in Fig. 4.4. If the body is having impending motion in some other direction, again the resultant reaction makes limiting frictional angle α with the normal in that direction. Thus, when the direction of force P is gradually changed through 360°, the resultant R generates a right circular cone with semicentral angle equal to α.

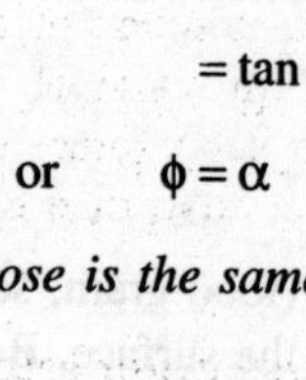

Fig. 4.4

If the resultant reaction is on the surface of this inverted right circular cone whose semi-central angle is limiting frictional angle (α), the motion of the body is impending. If the resultant is within this cone the body is stationary. This inverted cone with semi-central angle, equal to limiting frictional angle α, is called **Cone of Friction.**

Ex. 4.1: Block A weighing 1000 N rests over block B which weighs 2000 N as shown in Fig. 4.5(a). Block A is tied to wall with a horizontal string. If the coefficient of friction between A and B is 1/4 and between B and the floor is 1/3, what should be the value of P to move the block B if (a) P is horizontal? (b) P acts 30° upwards to horizontal?

(a) *When P is horizontal:*

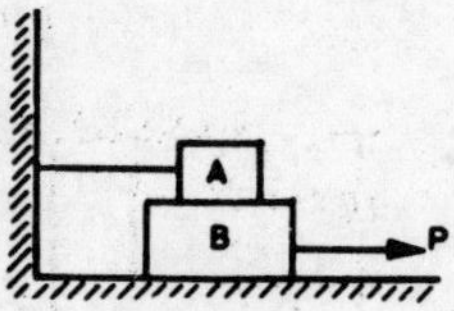

Fig. 4.5(a)

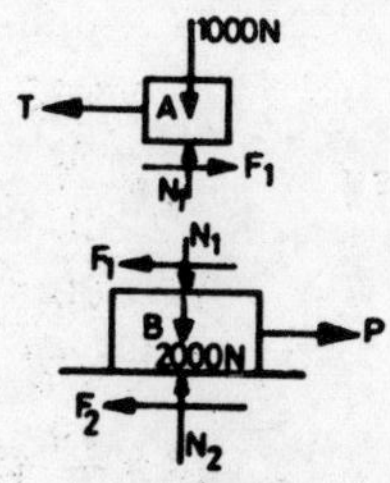

Fig. 4.5(b)

The free body diagrams of the two blocks are shown in Fig. 4.5(b). Note the frictional forces F_1 and F_2 are to be marked in the opposite direction of impending relative motion. Considering block A,

$$\Sigma V = 0$$

$$N_1 = 1000 \text{ N}$$

since F_1 is limiting friction, $\dfrac{F_1}{N_1} = \dfrac{1}{4}$

$$\therefore \quad F_1 = 250 \text{ N}$$

$$\Sigma H = 0$$

$$T = F_1$$

$$= 250 \text{ N}$$

Considering block B,

$$\Sigma V = 0$$

$$N_2 - 2000 - N_1 = 0$$

$$N_2 = 3000 \text{ N} \quad \text{since } N_1 = 1000 \text{ N}$$

Since F_2 is the limiting friction $F_2 = \mu_2 N_2$

$$= \frac{1}{3} \times 3000 = 1000 \text{ N}$$

$$\Sigma H = 0$$

$$P - F_1 - F_2 = 0$$

$$\mathbf{P} = F_1 + F_2 = 250 + 1000$$

$$= 1250 \text{ N}$$ *Ans.*

(b) *When P is enclined*

Free body diagram for this case is shown in Fig. 4.5(c).

As in the previous case, here also N_1 = 1000 N and F_1 = 250 N. Consider the equilibrium of block B.

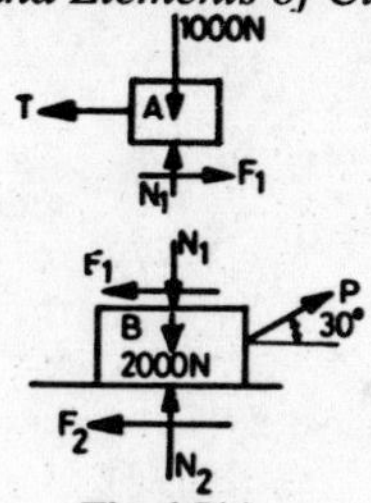

Fig. 4.5(c)

$\Sigma V = 0$

$$N_2 - 2000 - N_1 + P \sin 30° = 0$$

$$N_2 + 0.5\,P = 3000 \text{ N} \qquad \text{since } N_1 = 1000 \text{ N}$$

From law of friction,

$$F_2 = \frac{1}{3} N_2$$

$$= \frac{1}{3}(3000 - 0.5\,P)$$

$$= 1000 - \frac{0.5}{3} P$$

$$\Sigma H = 0$$

$$P \cos 30° - F_1 - F_2 = 0$$

$$P \cos 30^0 - 250 - (1000 - \frac{0.5}{3} P) = 0$$

$\therefore$ $\qquad$ **P = 1210.43 N** $\qquad$ *Ans.*

Ex. 4.2: What should be the value of θ in Fig. 4.6(a) which will make the motion of 900 N block down the plane to impend? The coefficient of friction for all contact surfaces is $\frac{1}{3}$.

Fig. 4.6(a)

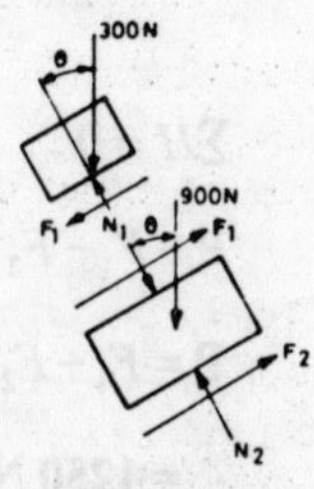

Fig. 4.6(b)

900 N block is on the verge of moving downward. Hence frictional forces F_1 and F_2 [Fig. 4.6(b)] act up the plane on 900 N block. Free body diagram of the blocks is as shown in Fig. 4.6(b).

For 300 N block:

Σ forces normal to plane = 0

$N_1 - 300 \cos\theta = 0$

or $N_1 = 300 \cos\theta$ ...(1)

From law of friction $F_1 = \frac{1}{3} N_1 = 100 \cos\theta$ ...(2)

For 900 N block:

Σ forces normal to the plane = 0

$N_2 - N_1 - 900 \cos\theta = 0$

or $N_2 = N_1 + 900 \cos\theta$

Substituting the value of N_1 from (1) we get

$N_2 = 1200 \cos\theta$ ...(3)

From law of friction

$$F_2 = \frac{1}{3} N_2 = 400 \cos\theta \qquad \text{...(4)}$$

Σ forces parallel to the plane = 0

$$F_1 + F_2 - 900 \sin\theta = 0$$

i.e. $100 \cos\theta + 400 \cos\theta = 900 \sin\theta$

$$\tan\theta = \frac{5}{9}$$

$\therefore$ $\theta = \mathbf{29.05°}$ *Ans.*

Ex. 4.3: A weight 500 N just starts moving down a rough inclined plane supported by a force of 200 N acting parallel to the plane and it is at the point of moving up the plane when pulled by a force of 300 N parallel to the plane. Find the inclination of the plane and the coefficient of friction between the inclined plane and the weight.

Free body diagram of the block when it just starts moving down is shown in Fig. 4.7(a) and when just starts moving up is shown in the Fig. 4.7(b). Now, frictional forces oppose the direction of the movement of the block and since it is limiting case $\frac{F}{N} = \mu$.

When the block just starts moving down [Fig. 4.7 (a)]:

Σ forces perpendicular to the plane = 0 ...(1)

$$N = 500 \cos\theta$$

From law of friction, $F_1 = \mu N$

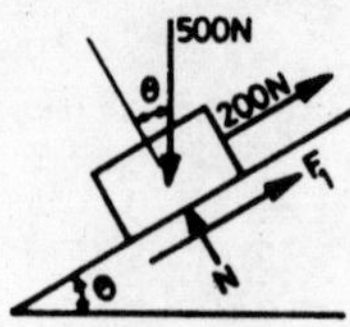

Fig. 4.7(a)

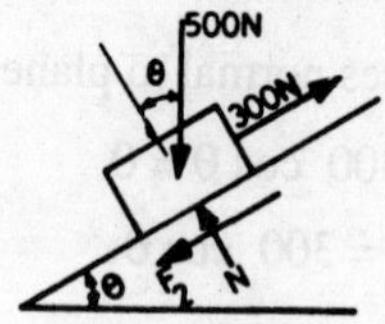

Fig. 4.7(b)

i.e. $F_1 = 500\,\mu\cos\theta$ ...(2)

Σ forces parallel to the plane = 0

$$500\sin\theta - F_1 - 200 = 0$$

i.e $500\sin\theta - 500\ \mu\ \cos\theta = 200$ ...(3)

When the block just starts moving up the plane [Fig. 4.7(b)]:

Σ forces perpendicular to the plane = 0

$N = 500\cos\theta$ ...(4)

From the law of friction, $F_2 = 500\,\mu\cos\theta$ ...(5)

Σ forces parallel to the plane = 0

$$500\sin\theta + F_2 - 300 = 0$$

i.e $500\sin\theta + 500\,\mu\cos\theta = 300$ ...(6)

Adding (3) and (6) we get

$$1000\sin\theta = 500$$

$$\sin\theta = \frac{1}{2}$$

or $\theta = 30°$ ***Ans.***

Substituting it in Eqn. (6) we get:

$$500\,\mu\cos 30° = 300 - 500\sin 30°$$

$$= 50$$

$\therefore$ **$\mu = 0.11547$** ***Ans.***

Ex. 4.4: What is the value of P in the system shown in Fig. 4.8(a) to cause the motion to impend? Assume the pulley is smooth and coefficient of friction between the other contact surfaces is 0.20.

Free body diagrams of the blocks are as shown in Fig. 4.8(b). Considering 750 N block:

Σ forces normal to the plane = 0

$$N_1 - 750\ \cos\ 60° = 0$$

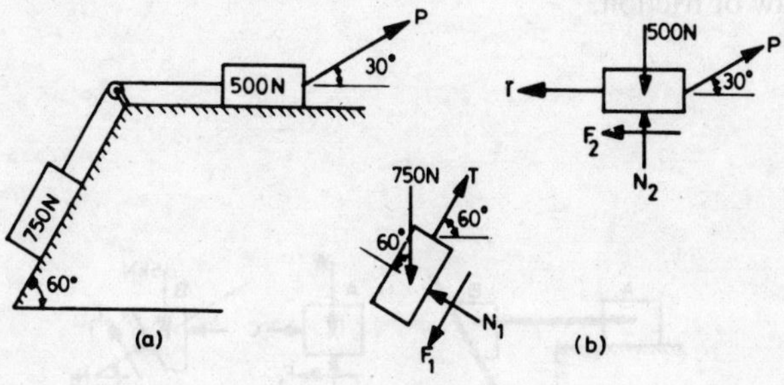

Fig. 4.8

$N_1 = 375$ N

Since the motion is impending, from law of friction,

$$F_1 = \mu N_1 = 0.2 \times 375$$

$$= 75 \text{ N}$$

Σ forces parallel to the plane = 0

$$T - F_1 - 750 \sin 60° = 0$$

$$T = 75 + 750 \sin 60°$$

$$= 724.52 \text{ N}$$

Considering 500 N body:

$$\Sigma V = 0$$

$$N_2 - 500 + P \sin 30° = 0$$

$$N_2 + 0.5 P = 500$$

From law of friction,

$$F_2 = 0.2 N_2$$

$$= 0.2 (500 - 0.5 P)$$

$$= 100 - 0.1 P$$

$$\Sigma H = 0$$

$$P \cos 30° - T - F_2 = 0$$

$$P \cos 30° - 724.52 - 100 + 0.1 P = 0$$

P = 853.52 N *Ans.*

Ex. 4.5: Two blocks connected by a horizontal link *AB* are supported on two rough planes as shown in Fig. 4.9(a). The coefficient of friction for the block on the horizontal plane is 0.4. The limiting angle of friction for block *B* on the inclined plane is 20°. What is the smallest weight *W* of the block *A* for which equilibrium of the system can exist if weight of block *B* is 5 kN?

Free body diagrams for block A and B are as shown in Fig. 4.9(b).
Consider block B.
From law of friction,

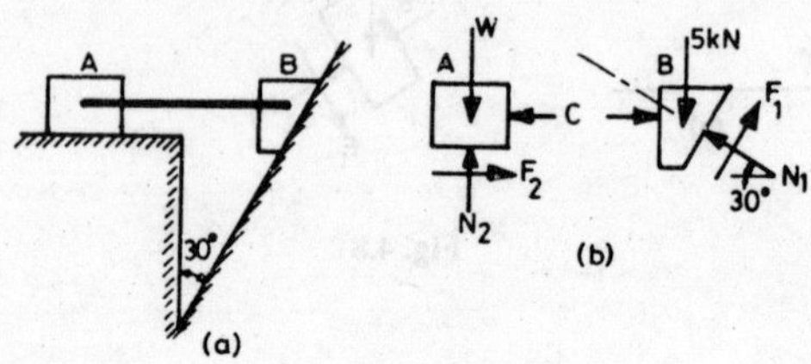

Fig. 4.9

$F_1 = N_1 \tan 20°$ [Since $\mu = \tan 20°$]

$\Sigma V = 0$

$N_1 \sin 30° + F_1 \sin 60° - 5 = 0$

$0.5N_1 + N_1 \tan 20° \sin 60° = 5$

$\therefore \quad N_1 = 6.133 \text{ kN}$

$\therefore \quad F_1 = 6.133 \tan 20° = 2.232 \text{ kN}$

$\Sigma H = 0$

$C + F_1 \cos 60° - N_1 \cos 30° = 0$

$C = 6.133 \cos 30° - 2.232 \cos 60°$

$= 4.196 \text{ kN}$

Now consider the equilibrium of block A.

$\Sigma H = 0$

$F_2 = C = 4.196 \text{ kN}$

From law of friction $F_2 = \mu N_2$

i.e $N_2 = \dfrac{4.196}{0.4} = 10.49 \text{ kN}$

$\Sigma V = 0$

W = N2 = 10.49 kN *Ans.*

Ex. 4.6: Two blocks A and B weighing 2000 N each are to be held from slipping by the thrust of two weightless link rods each of which is connected by pin joints at one end to the blocks and interconnected by a pin joint at other end O, and subjected to a horizontal force P needed to keep the blocks from slipping as shown in Fig. 4.10. Coefficient of friction is 0.25 for all contact surfaces.

Let C_1 be the force in link *AO* and C_2 be the force in link *OB*. The Free body diagrams of blocks *A*, *B* and the hinge *O* are shown in Fig. 4.10(b). Block *A* is on the verge of slipping.

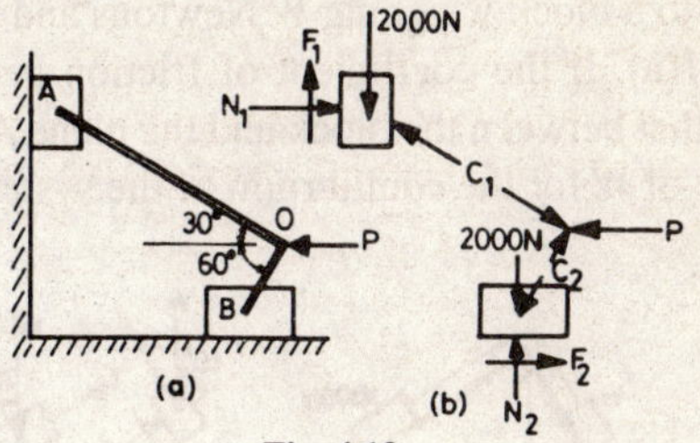

Fig. 4.10

Hence from law of friction,

$$F_1 = \mu N_1 = 0.25\ N_1$$

Considering block *A*,

$$\Sigma H = 0$$

$$N_1 = C_1 \cos 30°$$

$$\Sigma V = 0$$

$$F_1 + C_1 \sin 30° - 2000 = 0$$

$$0.25\ N_1 + C_1 \times 0.5 = 2000$$

$$(0.25 \cos 30° + 0.5)\ C_1 = 2000$$

$$C_1 = 2791.32 \text{ N}$$

Applying Lami's theorem to the equilibrium of the joint,

$$\frac{P}{\sin 90°} = \frac{C_2}{\sin 150°} = \frac{C_1}{\sin 120°}$$

$\therefore$ $\quad\mathbf{P} = \mathbf{3223.14\ N}$ *Ans.*

$\quad\mathbf{C_2} = \mathbf{1611.57\ N}$ *Ans.*

The above solution holds good provided block *B* is not slipping. To verify this fact consider the block *B*.

$$\Sigma H = 0$$

$$F_2 = C_2 \cos 60° = 805.79 \text{ N}$$

$$\Sigma V = 0$$

$$N_2 = 2000 + C_2 \sin 60°$$

$$= 3395.60 \text{ N}$$

Limiting friction $= \mu\ N_2 = 0.25\ N_2 = 848.92$ N

Since the actual frictional force F_2 developed is less than the limiting frictional force, block *B* is stationary and hence **P = 3223.14 N is correct answer**. Note that if F_2 calculated is more than the limiting friction μN_2, there is no possibility of maintaining equilibrium condition in the position shown.

Ex. 4.7: Two identical planes *AC* and *BC* inclined at 60° and 30° to the horizontal meet at *C*. A load of 1000 N rests on the inclined plane *BC* and is tied by a rope passing over a pulley to a block weighing *W* Newtons and resting on the plane *AC* as shown in Fig. 4.11(a). If the coefficient of friction between the load and the plane *BC* is 0.28 and that between the block and the plane *AC* is 0.20, find the least and the greatest value of *W* for the equilibrium of the system.

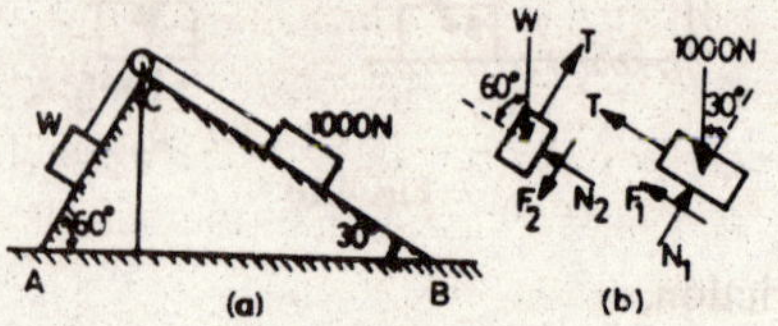

Fig. 4.11(a),(b)

For the least value of *W* for equilibrium, the motion of 1000 N block is impending downward. For such a case the free body diagram of blocks are shown in Fig. 4.11(b). Considering the 1000 N block:

Σ forces normal to plane = 0

$N_1 = 1000 \cos 30° = 866.03$ N

From the law of friction $F_1 = 0.28\, N_1$

$$= 242.49 \text{ N}$$

Σ forces parallel to the plane = 0

$$T = -F_1 + 1000 \sin 30°$$

$$= 257.51 \text{ N}$$

Now consider the equilibrium of block of weight *W*:

Σ Forces normal to the plane = 0

$$N_2 = W \cos 60° = 0.5\, W$$

$$\therefore \quad F_2 = 0.2\, N_2 = 0.1\, W$$

Σ forces parallel to the plane = 0

$$F_2 + W \sin 60° = T$$

$$0.1\, W + W \sin 60° = 257.51$$

$\therefore$ **W = 266.57 N** *Ans.*

For the greatest value of *W*, the 1000 N block is on the verge of moving up the plane. For such a case, the free body diagrams of the blocks are as shown in Fig. 4.11(c).

Considering block of 1000 N,

$N_1 = 866.03$ N

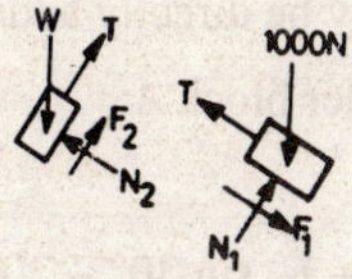

Fig. 4.11(c)

$F_1 = 242.49$ N

$T = 1000 \sin 30° + F_1 = 742.49$ N

Considering block of weight W,

$N_2 = W \cos 60° = 0.5\ W$

$F_2 = 0.2\ N_2 = 0.1\ W$

and $W \sin 60° - F_2 = T$

$W (\sin 60° - 0.1) = 742.49$

W = 969.28 N *Ans.*

Ex. 4.8: Two blocks A and B each weighing 1500 N are connected by a uniform horizontal bar which weighs 1000 N. If the angle of limiting friction under each block is 15°, find the force P directed parallel to the 60° inclined plane that will cause motion impending to the right [ref. Fig. 4.12(a)].

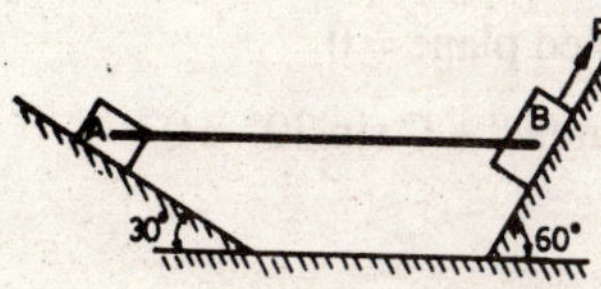

Fig. 4.12(a)

Free body diagrams of blocks A, beam AB and block B are as shown in Fig. 4.12(b), (c) and (d), respectively.

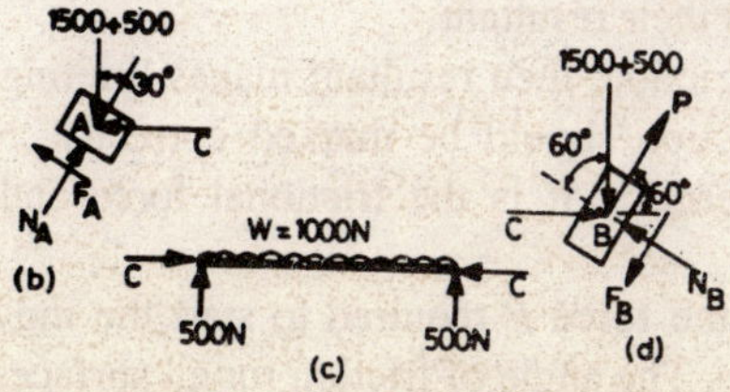

Fig. 4.12(b), (c), (d)

From vertical equilibrium of the beam *AB*, it may be found that 500 N force is transferred at *A* and *B* which may be directly added to self-weights of the blocks.

Now consider block *A*:

$\Sigma V = 0$

$N_A \cos 30° + F_A \sin 30° - 2000 = 0$

But from the law of friction $F_A = N_A \tan 15°$

$\therefore\ N_A (\cos 30° + \sin 30° \tan 15°) = 2000$

$N_A = 2000$ N

$\therefore\ F_A = 2000 \tan 15° = 535.90$ N

$\Sigma H = 0$

$C - N_A \sin 30° + F_A \cos 30° = 0$

$C = 2000 \times \sin 30° - 535.9 \cos 30°$

$= 535.90$ N

Consider block *B*:

Σ forces normal to the inclined plane = 0

$N_B - 2000 \cos 60° - C \cos 30° = 0$

$N_B = 1464.10$ N

$\therefore\quad F_B = N_B \tan 15° = 392.30$ N

Σ forces parallel to the inclined plane = 0

$P - F_B - 2000 \sin 60° + C \sin 30° = 0$

P = 1856.40 N *Ans.*

4.4 WEDGES

Wedges are small pieces of materials with two of their opposite surfaces not parallel. They are used to lift, slightly, heavy blocks, machinery, precast beams etc, required for final alignment or to make place for inserting lifting devices. The weight of the wedge is very small compared to the weight lifted. Hence, in all problems, the weight of wedges will be neglected. In the analysis of many problems, instead of treating normal reaction and frictional force independently, it is advantageous to treat their resultant.

If *F* is limiting friction, then resultant makes limiting frictional angle α with the normal. Its direction should be marked correctly. Note that the tangential component of the reaction *R* is the frictional force and it will always oppose impending motion.

Ex. 4.9: Determine the force *P* required to start the movement of the wedge as shown in Fig. 4.13(a). The angle of friction for all surfaces of contact is 15°.

As wedge is driven, it moves towards left and the block upwards. When motion is impending limiting friction develops. Hence resultant force makes limiting angle 15° with normal. The free body diagrams of the block and wedge are shown in Fig. 4.13(b). The forces on block and wedge are redrawn in Fig. 13(c) and (d) so that Lami's theorem can be applied conveniently. Applying Lami's theorem to the system of forces on block:

$$\frac{R_1}{\sin 145°} = \frac{R_2}{\sin 75°} = \frac{20}{\sin 140°}$$

$$R_1 = 17.847 \text{ kN}$$

$$R_2 = 30.044 \text{ kN}$$

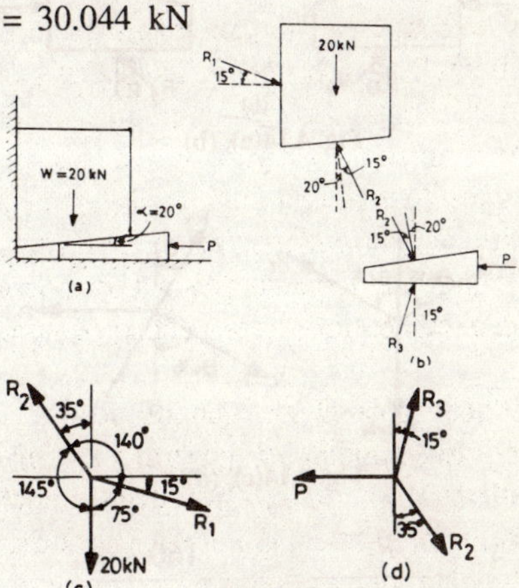

Fig. 4.13

Applying Lami's theorem to system of forces on the wedge, we get,

$$\frac{P}{\sin 130°} = \frac{R_2}{\sin 105°}$$

$$\mathbf{P = 23.835 \text{ kN}} \qquad \textit{Ans.}$$

Ex. 4.10: A weight of 160 kN is to be raised by means of the widges *A* and *B* as shown in Fig. 4.14(a). Find the value of force *P* for impending motion of block *C* upwards, if coefficient of friction is 0.25 for all surfaces. Weights of the block *C* and the wedges may be neglected.

Let α be angle of limiting friction. Then,

$$\theta = \tan^{-1}(0.25) = 14.036°$$

The free body diagrams of *A*, *B* and *C* are as shown in Fig. 4.14(b). The problem being symmetric, the forces R_1 and R_2 on wedges *A* and *B* are the same. The system of forces on block *C* and on wedge *A* are shown in the form convenient for applying Lami's theorem [ref. Fig. 4.14(c) and (d)].

Consider the equilibrium of block C.

$$\frac{R_1}{\sin(180 - 16 - \theta)} = \frac{160}{\sin 2(\theta + 16)}$$

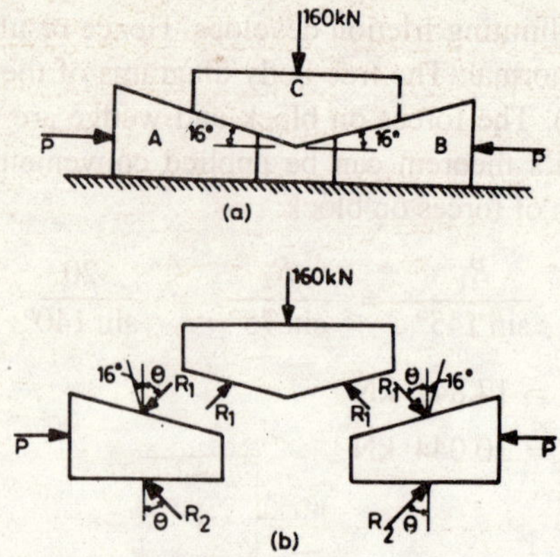

Fig. 4.14(a), (b)

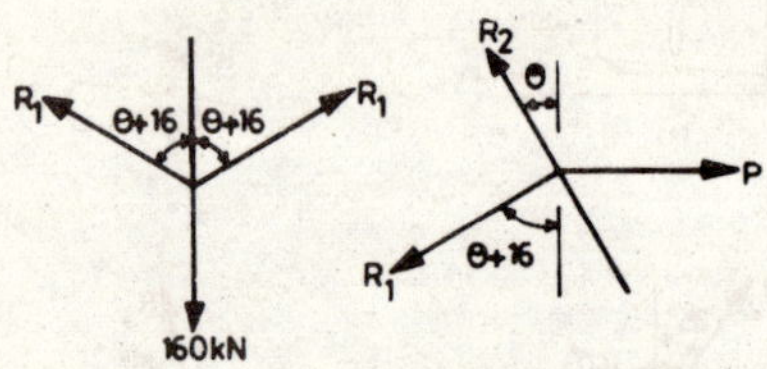

Fig. 4.14(c), (d)

i.e.

$$\frac{R_1}{\sin 149.96} = \frac{160}{\sin 60.072°}$$

$$R_1 = 92.41 \text{ kN}$$

Consider the equilibrium of wedge A:

$$\frac{P}{\sin(180 - \theta - \theta - 16)} = \frac{R_1}{\sin(90 + \theta)}$$

$$P = 66.256 \text{ kN} \qquad \textit{Ans.}$$

4.5 PROBLEMS INVOLVING NONCONCURRENT FORCE SYSTEM

In the problems which follow, apart from law of friction, the three equations of equilibrium listed in 3.11 will be used in the analysis.

Ex. 4.11: A ladder of length 4 m weighing 200 N is placed against a vertical wall as shown in Fig. 4.15(a). The coefficient of friction between the wall and the ladder is 0.2 and that between the floor and the ladder is 0.3. The ladder in addition to its own weight has to support a man weighing 600 N at a distance of 3 m from *A*. Calculate the minimum horizontal force to be applied at *A* to prevent slipping.

The free body diagram of the ladder is as shown in Fig. 4.15(b).

$$\Sigma M_A = 0$$

$$N_B\, 4 \sin 60° + F_B\, 4 \cos 60° - 600 \times 3 \cos 60° - 200 \times 2 \cos 60° = 0$$

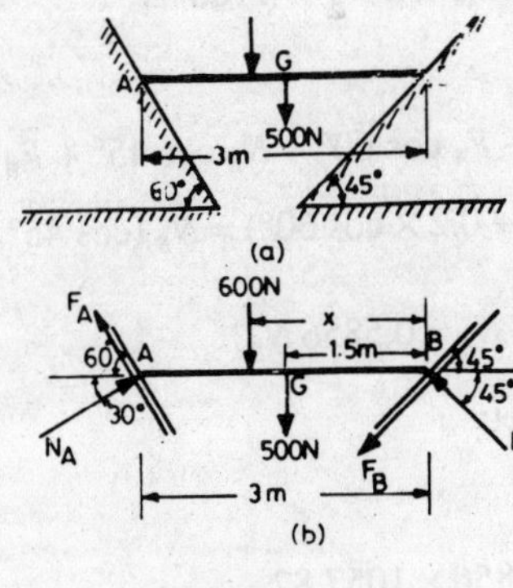

Fig. 4.17(a), (b)

$$0.6732\ N_A + 0.5657\ N_B = 1100 \qquad ...(3)$$

$\Sigma H = 0$

$$N_A \cos 30° - F_A \cos 60° - N_B \cos 45° - F_B \cos 45° = 0$$

$$N_A (\cos 30° - 0.2 \times \cos 60°) = N_B (\cos 45° + 0.2 \times \cos 45°)$$

$$N_A = \frac{0.8485}{0.7660} N_B = 1.1077\ N_B \qquad ...(4)$$

Substituting this value of N_A in Eqn. (3) we get;

$$N_B = 838.79 \text{ N}$$

$$\therefore \text{N}_A = 929.13 \text{ N}$$

$$\Sigma M_B = 0,$$

$$600\,x + 500 \times 1.5 - F_A \sin 60° \times 3 - N_A \sin 30^0 \times 3 = 0$$

$\therefore$ **x = 1.877 m from B** *Ans.*

When the load is close to end B, the end B may slide down and end A may slide up. Let x be the distance of load from the end B. For such case the free body diagram is shown in Fig. 4.17(c).

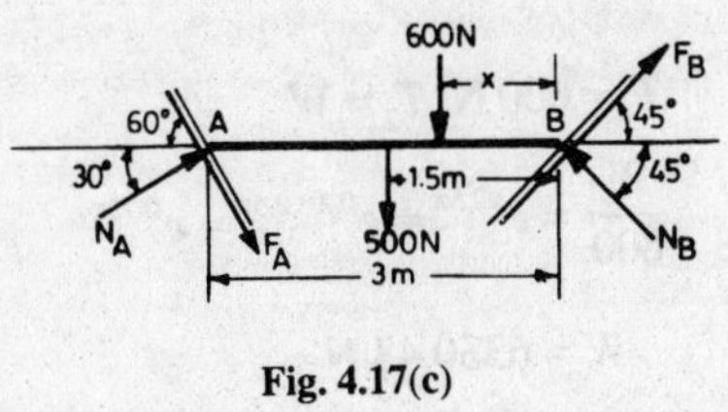

Fig. 4.17(c)

$\Sigma V = 0$

$$N_A \sin 30° - F_A \sin 60° + N_B \sin 45° + F_B \sin 45° = 1100$$

$$N_A (\sin 30° - 0.2 \times \sin 60°) + N_B$$
$$(\sin 45° + 0.2 \times \sin 45°) = 1100$$

$$0.3268\, N_A + 0.8485\, N_B = 1100 \quad \text{...(5)}$$

$$\Sigma H = 0$$

$$N_A \cos 30° + F_A \cos 60° - N_B \cos 45° + F_B \cos 45° = 0$$

$$N_A (\cos 30° + 0.2 \times \cos 60°) = N_B (\cos 45° - 0.2 \times \cos 45°)$$

$$N_A = \frac{0.56.57}{0.9660} N_B = 0.5856\, N_B \quad \text{...(6)}$$

Substituting it in (5), we get:

$$N_B = 1057.82 \text{ N}$$

and hence $N_A = 0.5856 \times 1057.82$

$$= 619.67 \text{ N}$$

$$\Sigma M_B = 0$$

$$600\, x + 500 \times 1.5 - N_A \sin 30° \times 3 + F_A \sin 60° \times 3 = 0$$

$\therefore \quad x = -\, 0.237$ m

It means that the motion will be impending when the load is at 0.237 m to the right of B, which is not a possible case of loading. Hence load can be placed even on point B safely. Thus, 600 N load can be placed anywhere between point B and a point 1.85 m from B.

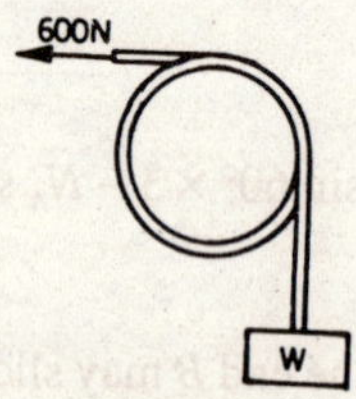

Fig. 4.18

(1) Let the impending motion of the weight be downward. Then,

$$T_1 = 600 \text{ N};\ T_2 = W$$

$$\frac{W}{600} = e^{\mu 2.5\pi} = e^{0.3 \times 2.5\pi} = e^{0.75\pi}$$

$$W = 6330.43 \text{ N}$$

(2) Let the impending motion of weight be upwards. Then,

$$T_1 = W; \quad T_2 = 600 \text{ N}$$

$$T_2 = T_1\, e^{\mu\theta}$$

$$600 = W e^{0.75\pi}$$

$$W = 56.87 \text{ N}$$

Thus, a **600 N force can support a range of loads beween 56.87 N to 6330.43 N weight on the other side of drum.** *Ans.*

Ex. 4.15: In Fig. 4.19(a) the coefficient of friction is 0.20 between the rope and the fixed drum, and between other surface of contact $\mu = 0.3$. Determine the minimum weight W to prevent downward motion of the 1000 N body.

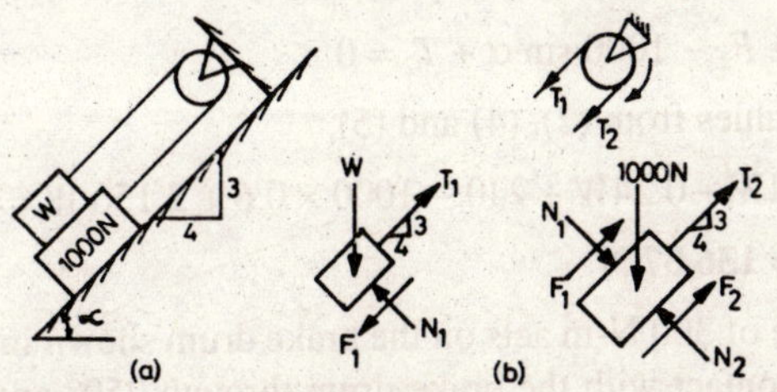

Fig. 4.19

Since 1000 N weight is on the verge of sliding downwards the rope connecting it is the tight side and the rope connecting W is the slack side. Free body diagrams for W and 1000 N body are shown in Fig. 4.19(b).

Now, $\cos\alpha = 0.8$

$\sin\alpha = 0.6$

Consider the equilibrium of weight W,

Σ Forces perpendicular to the plane = 0

$$N_1 = W\cos\alpha$$

$$N_1 = 0.8\,W \quad \text{...(1)}$$

$$\therefore F_1 = \mu N_1 = 0.3 \times 08\,W$$

$$F_1 = 0.24\,W \quad \text{...(2)}$$

Σ Forces parallel to the plane = 0

$$T_1 = F_1 + W\sin\alpha = 0.24\,W + 0.6\,W$$

$$= 0.84\,W \quad \text{...(3)}$$

Angle of contact of rope with the pulley = 180° = π radians

Applying friction equation, we get:

$$T_2 = T_1 e^{\mu} = T_1 e^{0.3\pi}$$

$$T_2 = 2.566\ T_1$$

Substituting the value of T_1 from (3)

$$T_2 = 2.156\ W \quad \text{...(4)}$$

Now, consider 1000 N body,

$$\Sigma \text{ forces perpendicular to the plane} = 0$$

$$N_2 - N_1 - 1000 \cos \alpha = 0$$

Substituting the value of N_1 from (1),

$$N_2 = 0.8\ W + 1000 \times 0.8 = 0.8W + 800$$

$$\therefore\ F_2 = 0.3\ N_2 = 0.24W + 240 \quad \text{...(5)}$$

$$\Sigma \text{ forces parallel to the plane} = 0$$

$$F_1 + F_2 - 1000 \sin \alpha + T_2 = 0$$

Substituting the values from (2), (4) and (5),

$$0.24W + 0.24W + 240 - 1000 \times 0.6 + 2.156\ W = 0$$

$$\mathbf{W = 136.57\ N} \quad \textit{Ans.}$$

Ex. 4.16: A torque of 300 N-m acts on the brake drum shown in Fig. 4.20(a). If the brake band is in contact with the brake drum through 250° and the coefficient of friction is 0.3, determine the force P applied at the end of the brake lever for the position shown in the figure.

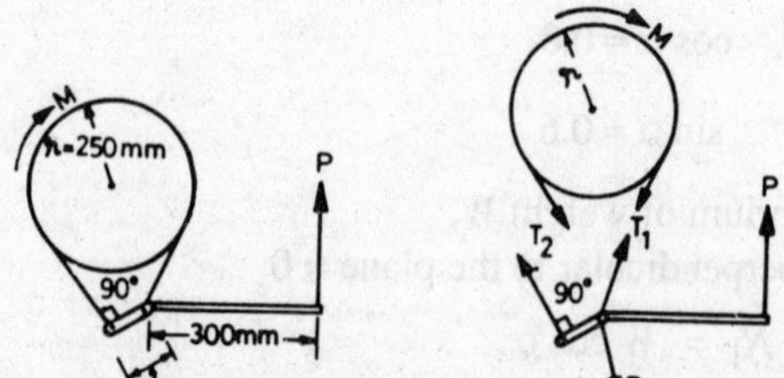

Fig. 4.20

Fig. 4.20(b) shows free body diagrams of brake drum and the lever arm.

Now $\quad T_2 = T_1\ e^{\mu\theta}$

$\theta = \frac{250\pi}{180}$ radians and $\mu = 0.3$

$\therefore\ \mu\,\Theta = 1.309$

$\therefore\ T_2 = T_1\ e^{1.309} = 3.7025\ T_1$

Now, $(T_2 - T_1)\ r = M$

$$(3.7025 - 1)\, T_1 \times 250 = 300 \times 10^3$$

$$\therefore \quad T_1 = 444.04 \text{ N}$$

$$\therefore \quad T_2 = 1644.058 \text{ N}$$

Consider the lever arm. Taking moment about the hinge, we get

$$T_2 \times 50 = P \times 300$$

$$P = 274.0 \text{ N} \qquad \textit{Ans.}$$

EXERCISE

4.1: A pull of 180 N applied upward at 30° to a rough horizontal plane was required to just move a body resting on the plane while a push of 220 N applied along the same line of action was required to just move the same body downwards. Determine the weight of the body and the coefficient of friction.

[*Ans* $W = 990$ N; $\mu = 0.1722$]

4.2: The block *A* shown in Fig. 4.21 weighs 2000 N. The cord attached to *A* passes over a frictionless pulley and supports a weight equal to 800 N. The value of coefficient of friction between *A* and the horizontal plane is 0.35. Solve for horizontal force *P*: (1) if motion is impending towards the left, and (2) if the motion is impending towards the right.

[*Ans.* (1) 1252.82 N; (2) 132.82 N]

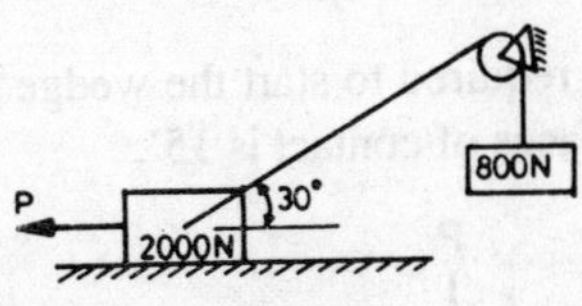

Fig. 4.21

4.3: A 3000 N block is placed on an inclined plane as shown in Fig. 4.22. Find the maximum value of *W* for equilibrium if tipping does not occur. Assume coefficient of friction as 0.2

[*Ans.* 1014.96 N]

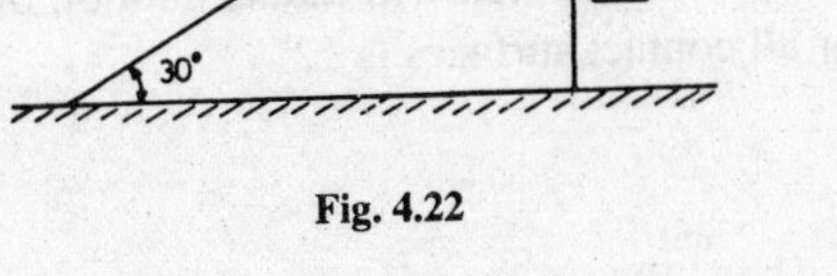

Fig. 4.22

4.4: Find whether block *A* is moving up or down the plane in Fig. 4.23 for the data given below. Weight of block *A* = 300N. Weight of block *B* = 600 N. Coefficient of limiting friction between plane *AB* and block *A* is 0.2. Coefficient of limiting friction between plane *BC* and block *B* is 0.25. Assume pulley as smooth.

[*Ans*. The block *A* moves up]

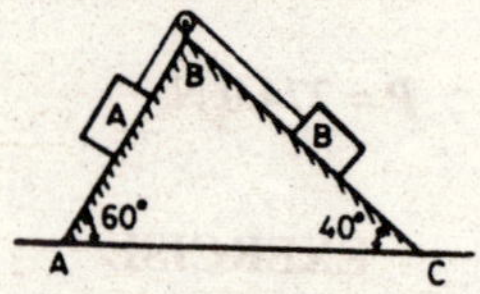

Fig. 4.23

4.5: Two identical blocks *A* and *B* are connected by a rod and they rest against vertical and horizontal planes respectively as shown in Fig. 4.24. If sliding impends when θ = 45°, determine the coefficient of friction, assuming it to be the same for both floor and wall.

[*Ans*. 0.414]

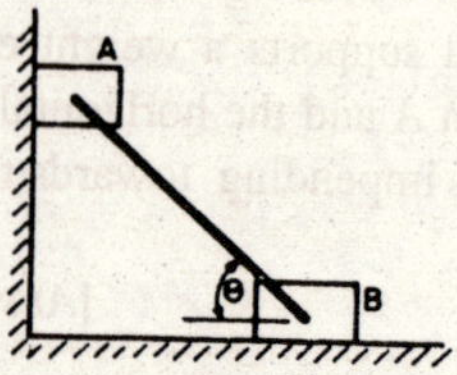

Fig. 4.24

4.6: Determine the force *P* required to start the wedge as shown in Fig. 4.25. The angle of friction for all surfaces of contact is 15°.

[*Ans*. 26.6784 kN]

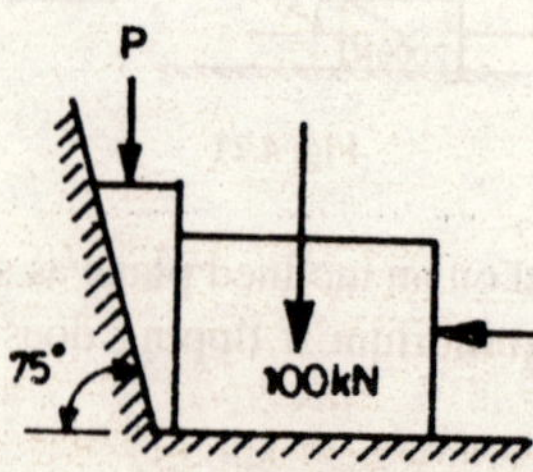

Fig. 4.25

4.7: Two blocks *A* and *B* weighing 3 kN and 15 kN, respectively, are held in position against an inclined plane by applying a horizontal force *P* as shown in Fig. 4.26. Find the least value of *P* which will induce motion of the block *A* upwards. Angle of friction for all contact surfaces is 12°.

[*Ans*.173.575 kN]

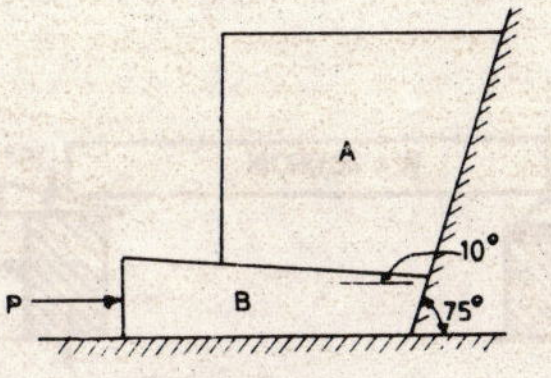

Fig. 4.26

4.8: In Fig. 4.27, *C* is a stone block weighing 6 kN. It is being raised slightly by means of two wooden wedges *A* and *B* with a force *P* on wedge *B*. The angle between the contacting surfaces of the wedge is 5°. If coefficient of friction is 0.3 for all contacting surfaces, compute the value of *P* required to impend upward motion of the block *C*. Neglect weight of the wedges.

[*Ans*. 2.344 kN]

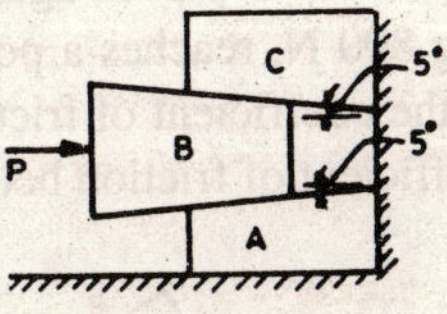

Fig. 4.27

4.9: Find the horizontal force *P* required to pull the block *A* of weight 150 N which carries block *B* of weight 1280 N as shown in Fig. 4.28. Take angle of limiting friction between floor and block *A* as 14° and that between vertical wall and block *B* as 13°, and coefficient of limiting friction between the blocks as 0.3.

[*Ans*. *P* = 318.06 N]

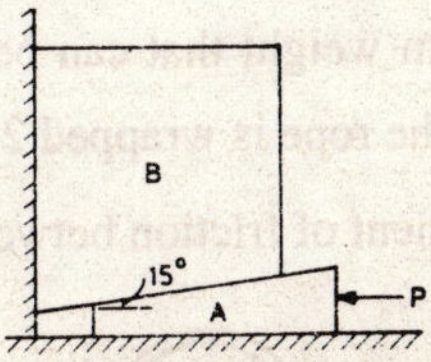

Fig. 4.28

4.10: The level of a precast beam weighing 20,000 N is to be adjusted by driving a wedge as shown in Fig. 4.29. If coefficient of friction between the wedge and pier is 0.35 and that between beam and the wedge is 0.25, determine the minimum force *P* required on the wedge to make adjustment of the beam. Angle of the wedge is 15°.

(*Hint*: Vertical component of reaction on wedge at contact with beam = 1/2 vertical load on beam = 10,000 kN)

[*Ans*. 9057.4 N]

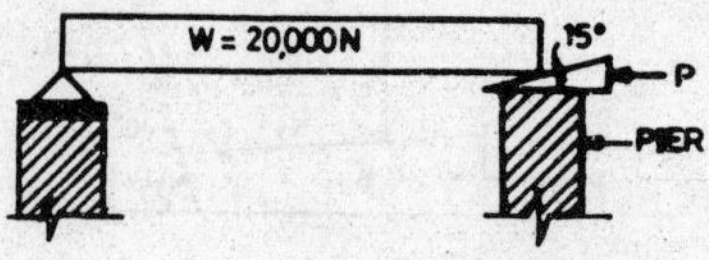

Fig. 4.29

4.11: A ladder 5 m long rests on a horizontal ground and leans against a smooth vertical wall at an angle of 70° with the horizontal. The weight of the ladder is 300 N. The ladder is on the verge of sliding when a man weighing 750 N stands on a rung 1.5 m high. Calculate the coefficient of friction between the ladder and the floor.

[*Ans.* $\mu = 0.1837$]

4.12: A 4 m ladder weighing 200 N is placed against a vertical wall as shown in Fig. 4.31. As a man weighing 800 N, reaches a point 2.7 m from *A*, the ladder is about to slip. Assuming that the coefficient of friction between the ladder and the wall is 0.2, determine the coefficient of friction between the ladder and the floor.

[*Ans.* 0.3548]

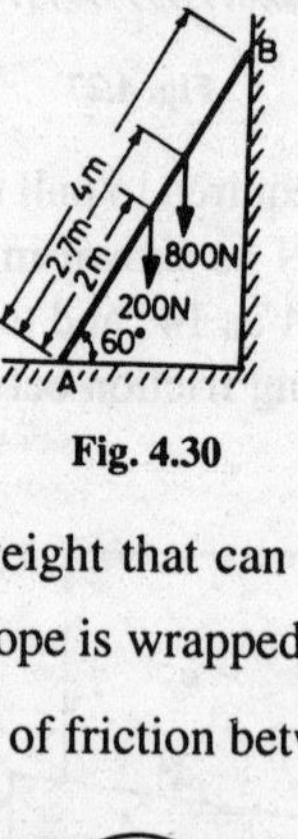

Fig. 4.30

4.13: Determine the maximum weight that can be lowered by a person who can exert a 300 N pull on rope if the rope is wrapped $2\frac{1}{2}$ turns round a horizontal spur as shown in Fig. 4.31. Coefficient of friction between spur and the rope is 0.3.

[*Ans.* 33395.33 N]

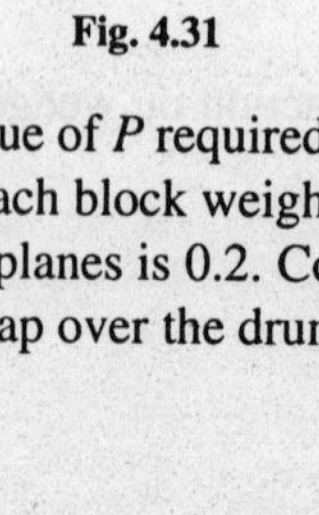

Fig. 4.31

4.14: Determine the minimum value of *P* required to cause motion of blocks *A* and *B* towards right (ref. Fig. 4.32). Each block weighs 3000 N and coefficient of friction between blocks and inclined planes is 0.2. Coefficient of friction between the drum and rope is 0.1. Angle of wrap over the drum is 90°.

[*Ans.* 3065.18 N]

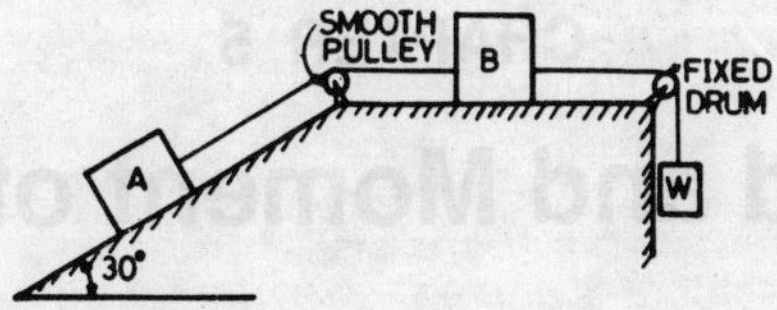

Fig. 4.32

4.15: Block *A* shown in Fig. 4.33 weighs 2000 N. The cord attached to *A* passes over a fixed drum and supports a weight equal to 800 N. The value of coefficient of friction between *A* and the horizontal plane is 0.25 and between the rope and the fixed drum is 0.1. Solve for *P*: (1) if motion is impending towards the left, (2) if the motion is impending towards the right.

[*Ans*. (1) 1230.94 N; (2) 143.0 N]

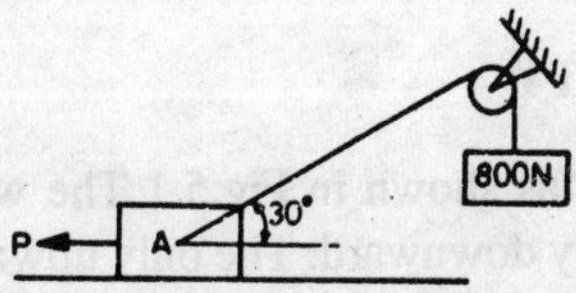

Fig. 4.33

4.16: The dimension of a brake drum is as shown in Fig. 4.34. Determine the torque *M* exerted on the drum if the load *P* = 50 N. Assume coefficient of kinetic friction between rope and drum to be 0.15.

[*Ans*. 747.685 N-m]

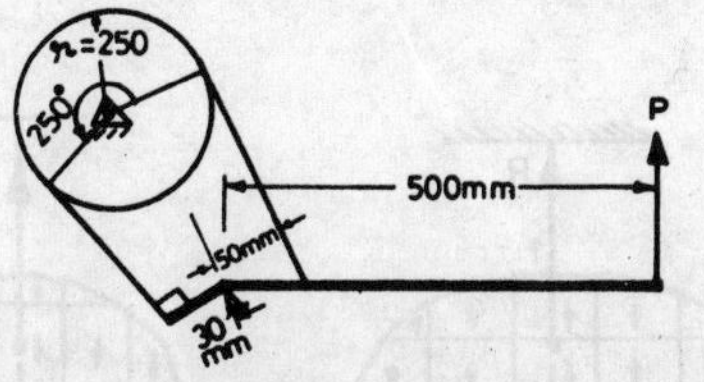

Fig. 4.34

CHAPTER 5

Centroid and Moment of Inertia

Centroid and moment of inertia are two important properties of a section, which are required frequently in the analysis of many engineering problems. In this chapter the meaning of centre of gravity and centroid is explained. Procedure for locating centroid of plane figures is illustrated. The term moment of inertia is defined and the method of finding moment of inertia of a given sectional area is explained. Determination of the centroid and the moment of inertia of sections will be of great importance in the study of subjects like strength of materials, structural design and machine design.

5.1 CENTRE OF GRAVITY

Consider a suspended body as shown in Fig.5.1 The weights of various parts of this body are acting vertically downward. The only upward force is the force in the string. To satisfy the equilibrium condition the resultant weight of the body W must act along the line of the string (1)–(1). Now, if the position is changed and the body is suspended again, it will reach equilibrium in a particular position. Let the line of action of resultant weight be (2)–(2) intersecting line (1)–(1) at G. It is obvious that if the body is suspended in any other position, the line of action of resultant weight W passes through G. This point is called the centre of gravity. Thus, *centre of gravity can be defined as the point through which resultant of force of gravity (weight) of the body acts.*

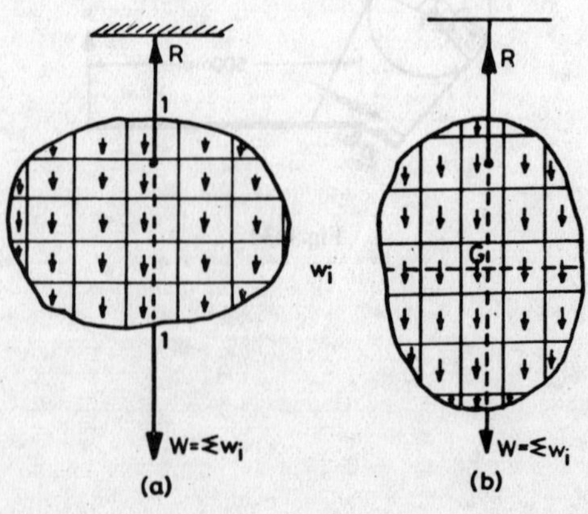

Fig. 5.1

5.2 CENTRE OF GRAVITY OF A FLAT PLATE

Consider a flat plate of thickness t as shown in Fig. 5.2. Let W_i be the weight of any elemental portion acting at a point (x_i, y_i). Let W be the total weight of the plate acting at the point $(\bar{x}, \bar{y})$. According to definition of centre of gravity, the point $(\bar{x}, \bar{y})$ is the centre of gravity. Now,

$$\text{Total weight } W = \Sigma W_i \qquad \textit{Eqn. } 5.1$$

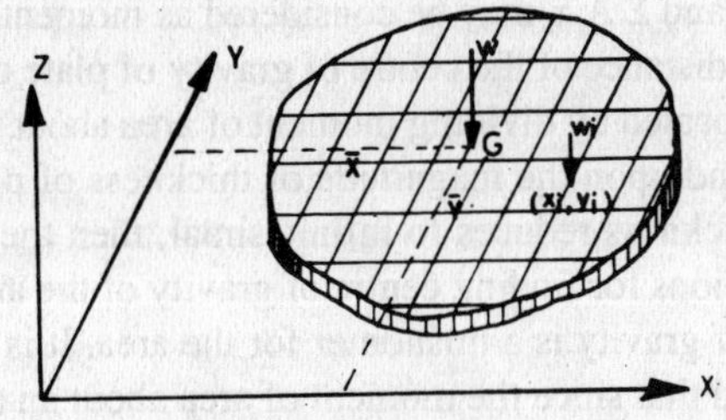

Fig. 5.2

Taking moment about x axis and equating moment of resultant to moment of component forces, we get

$$W\bar{y} = W_1y_1 + W_2y_2 + W_3y_3 + \cdots$$

$$= \sum W_iy_i$$

$$\therefore \quad \bar{y} = \frac{\sum W_iy_i}{W} \qquad \textit{Eqn.}5.2$$

Similarly, taking moment about y axis we get,

$$W\bar{x} = W_1x_1 + W_2x_2 + W_3x_3 + \cdots$$

$$= \sum W_ix_i$$

$$\bar{x} = \frac{\sum W_ix_i}{W} \qquad \textit{Eqn.}5.3$$

5.3 CENTROID

Let A_i be the area of the ith element of plate of uniform thickness in the plate. If γ is the unit weight of the material of plate and t its uniform thickness, then

$$W_i = \gamma A_i t$$

$$\therefore \quad \text{Total weight,} \quad W = \sum \gamma A_i t = \gamma t \sum A_i$$

$$= \gamma t A \qquad \textit{Eqn.}5.4$$

where, $A = \Sigma\ A_i$ is total area.

From equations 5.2 and 5.3 we get

$$\bar{y} = \frac{\Sigma A_i \gamma t\, y_i}{\gamma t A} = \frac{\Sigma A_i y_i}{A} \qquad \textit{Eqn.5.5}$$

and

$$\bar{x} = \frac{\Sigma A_i \gamma t\, x_i}{\gamma t A} = \frac{\Sigma A_i x_i}{A} \qquad \textit{Eqn.5.6}$$

since γ and t are constants.

The terms $\Sigma\, A_i y_i$ and $\Sigma\, A_i x_i$ may be considered as moment of area about x axis and y axis. Thus, the distance of the centre of gravity of plate of uniform thickness from an axis can be located by dividing moment of area about that axis by the total area. It will not depend upon the magnitude of thickness of plate and unit weight of material. If the thickness reduces to infinitesimal, then the plate reduces to an area. Still the expressions for finding centre of gravity of the area are same as Eqn. 5.5 and 5.6. Centre of gravity is a misnomer for the area. It is to be called as **centroid**. It may be noted that since the moment of area about an axis divided by total area gives the distance of centroid from that axis, the moment of area is zero about any centroidal axis.

5.4 DIFFERENCE BETWEEN CENTRE OF GRAVITY AND CENTROID

From the above discussion we can draw the following differences between centre of gravity and centroid:

(1) The term centre of gravity applies to bodies with mass and weight, and centroid applies to plane areas.

(2) Centre of gravity of a body is a point through which the resultant gravitational force (weight) acts for any orientation of the body whereas centroid is a point in a plane area such that the moment of area about any axis through that point is zero.

5.5 USE OF AXIS OF SYMMETRY

Centroid of an area lies on the axis of symmetry if it exists. This is useful theorem to locate the centroid of an area. This theorem can be proved as follows:

Consider the area shown in Fig. 5.3. In this figure $y - y$ is the axis of symmetry. From Eqn. 5.6, the distance of centroid from this axis is given by:

$$\frac{\Sigma A_i x_i}{A}$$

Consider the two elemental areas shown in Fig. 5.3, which are equal in size and are equidistant from the axis, but on either side. Now the sum of moments of these areas cancel each other since the areas and distances are the same, but signs of distances are opposite. Similarly, we can go on considering on area on one side of symmetric axis and corresponding image area on the other side, and prove that total moments of area ($\Sigma A_i x_i$) about the symmetric axis is zero. Hence the distance of centroid from the symmetric axis is zero, i.e. centroid always lies on symmetric axis.

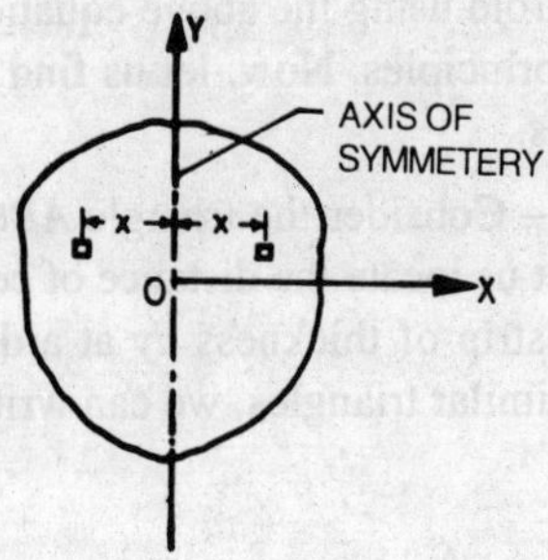

Fig. 5.3

Making use of the symmetry we can conclude that:

(1) Centroid of a circle is its centre (Fig. 5.4);

(2) Centroid of a rectangle of sides b and d is at a distance $\frac{b}{2}$ and $\frac{d}{2}$ from the Fig. 5.5

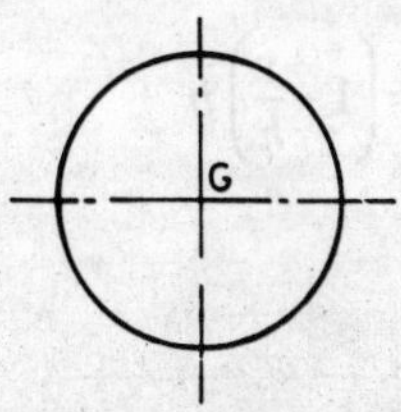

Fig. 5.4

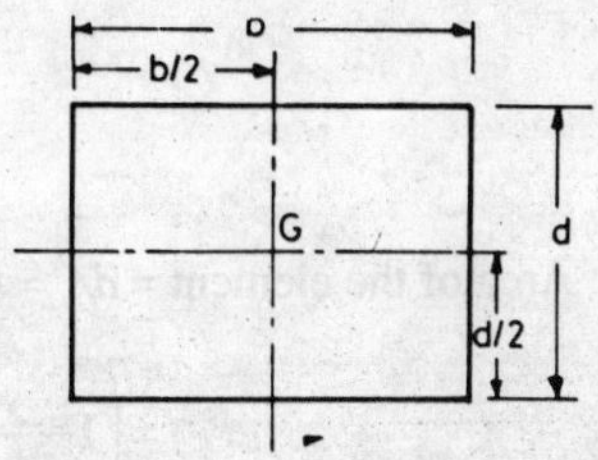

Fig. 5.5

5.6 DETERMINATION OF CENTROID OF SIMPLE FIGURES FROM FIRST PRINCIPLE

For simple figures like triangle and semicircle, we can write general expression for the elemental area and its distance from an axis. Then equations 5.5 and 5.6 reduce to:

$$\bar{y} = \frac{\int y dA}{A} \qquad \text{Eqn.5.7}$$

$$\bar{x} = \frac{\int x dA}{A} \qquad \textit{Eqn.5.8}$$

The location of the centroid using the above equations may be considered as finding centroid from first principles. Now, let us find centroid of some standard figures from first principles.

Centroid of a Triangle – Consider the triangle ABC of base width b and height h as shown in Fig. 5.6. Let us locate the distance of centroid from the base. Let b_1 be the width of elemental strip of thickness dy at a distance y from the base. Since ΔAEF and ΔABC are similar triangles, we can write:

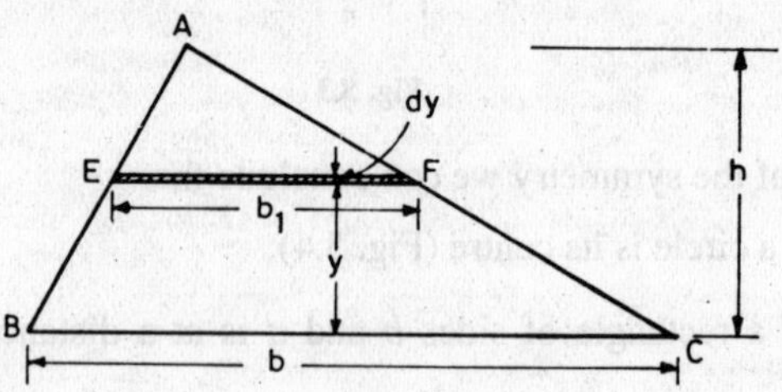

Fig. 5.6

$$\frac{b_1}{b} = \frac{h-y}{h}$$

$$b_1 = \left(\frac{h-y}{h}\right) b = \left(1 - \frac{y}{h}\right) b$$

$\therefore$ Area of the element $= dA = b_1 dy$

$$= \left(1 - \frac{y}{h}\right) b\, dy$$

Area of the triangle $A = \frac{1}{2} bh$

$\therefore$ From *Eqn.* 5.7

$$\bar{y} = \frac{\text{Moment of area}}{\text{Total area}} = \frac{\int y dA}{A}$$

Now,

$$\int y dA = \int_0^h y\left(1 - \frac{y}{h}\right) b\, dy$$

$$= \int_0^h \left(y - \frac{y^2}{h} \right) b\, dy$$

$$= b \left[\frac{y^2}{2} - \frac{y^3}{3h} \right]_0^h$$

$$= \frac{bh^2}{6}$$

$$\therefore \qquad \bar{y} = \int \frac{y dA}{A} = \frac{bh^2}{6} \times \frac{1}{\frac{1}{2} bh}$$

$$\therefore \qquad \bar{y} = \frac{h}{3}$$

Thus the centroid of a triangle is at a distance $\frac{h}{3}$ from the base (or $\frac{2h}{3}$ from the apex) of the triangle where h is the height of the triangle.

Centroid of a Semicircle – Consider the semicircle of radius R as shown in Fig. 5.6. Due to symmetry centroid must lie on y axis. Let its distance from diametral axis be $\bar{y}$. To find $\bar{y}$, consider an element at a distance r from the centre O of the semicircle, radial width being dr and bound by radii at θ and $\theta + d\theta$.
Area of element = $r\, d\theta\, dr$.

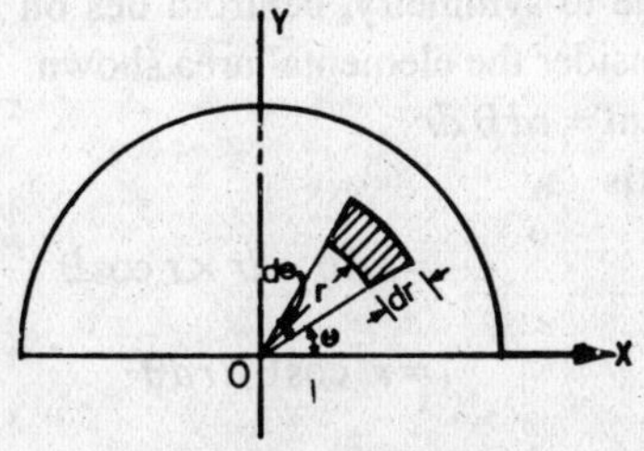

Fig. 5.7

Its moment about diametral axis x is given by:

$$r d\theta \times dr \times r \sin\theta$$

$$= r^2 \sin\theta\, dr d\theta$$

$\therefore$ Total moment of area about diametral axis,

$$\int_0^{\pi}\int_0^{R} r^2 \sin\theta \, dr \, d\theta$$

$$= \int_0^{\pi} \left[\frac{r^3}{3}\right]_0^R \sin\theta \, d\theta$$

$$= \frac{R^3}{3}[-\cos]_0^{\pi}$$

$$= \frac{R^3}{3}[1+1] = \frac{2R^3}{3}$$

Area of semi circle A $= \frac{1}{2}\pi R^2$

$$\therefore \quad \bar{y} = \frac{\text{Moment of area}}{\text{Total area}} = \frac{\frac{2R^3}{3}}{\frac{1}{2}\pi R^2}$$

$$= \frac{4R}{3\pi}$$

Thus, the centroid of the circle is at a distance $\frac{4R}{3\pi}$ from the diametral axis.

Centroid of Sectoral a Circle – Consider the sector of a circle of angle 2 α as shown in Fig. 5.8. Due to symmetry, centroid lies on x axis. To find its distance from the centre O, consider the elemental area shown.

Area of the element = $rd\,\theta\,dr$

Its moment about y axis

$$= rd\theta \times dr \times r\cos\theta$$

$$= r^2 \cos\theta \, dr d\theta$$

$\therefore$ Total moment of area about y axis

$$= \int_{-\alpha}^{\alpha}\int_0^{R} r^2 \cos\theta \, dr d\theta$$

$$= \left[\frac{r^3}{3}\right]_0^R [\sin\theta]_{-\alpha}^{\alpha}$$

$$= \frac{R^3}{3} 2\sin\alpha$$

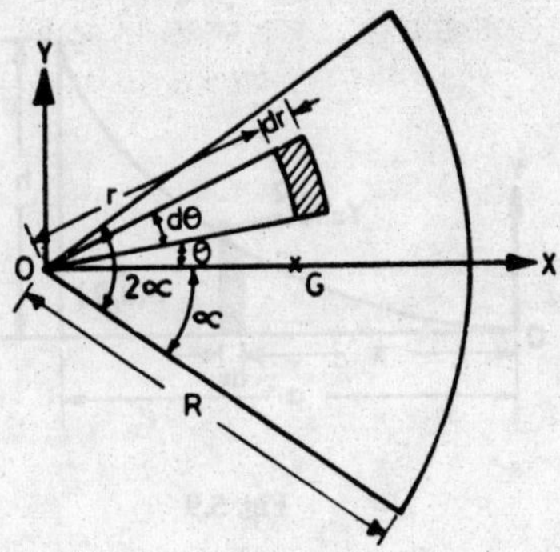

Fig. 5.8

Total area of the sector

$$= \int_{-\alpha}^{\alpha} \int_{0}^{R} r\,dr\,d\theta$$

$$= \int_{-\alpha}^{\alpha} \left[\frac{r^2}{2}\right]_0^R d\theta$$

$$= \frac{R^2}{2}[\theta]_{-\alpha}^{\alpha}$$

$$= R^2\alpha$$

∴ The distance of centroid from centre O

$$= \frac{\text{Moment of area about } y \text{ axis}}{\text{Area of the figure}}$$

$$= \frac{\frac{2R^3}{3}\sin\alpha}{R^2\alpha} = \frac{2R}{3A}\sin\alpha$$

Centroid of Parabolic Spandrel – Consider the parabolic spandrel shown in Fig. 5.9. Height of the element at a distance x from O is $y = kx^2$

Width of element $= dx$

∴ Area of the element $= kx^2 dx$

∴ Total area of spandrel $= \int_0^a kx^2\,dx$

$$\left[\frac{kx^3}{3}\right]_0^a = \frac{ka^3}{3}$$

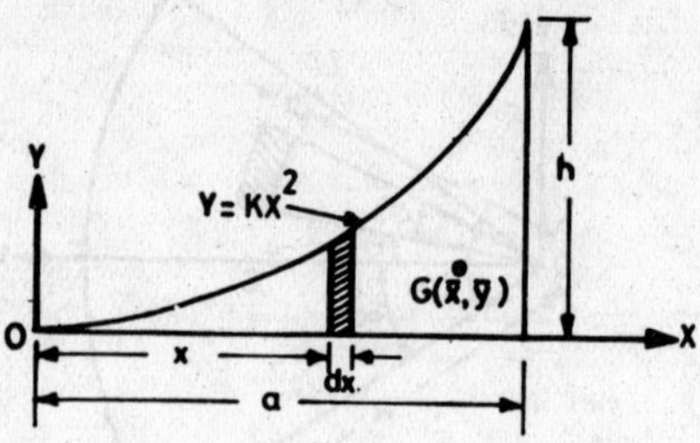

Fig. 5.9

Moment of area about y axis

$$= \int_0^a kx^2\,dx \times x = \int_0^a kx^3\,dx$$

$$= \frac{ka^4}{4}$$

Moment of area about x axis

$$= \int_0^a kx^2\,dx\,\frac{kx^2}{2} = \int_0^a \frac{k^2x^4}{2}\,dx$$

$$= \frac{k^2a^5}{10}$$

$$\therefore \qquad \bar{x} = \frac{ka^4}{4} \div \frac{ka^3}{3} = \frac{3a}{4}$$

$$\bar{y} = \frac{k^2a^5}{10} \div \frac{ka^3}{3} = \frac{3}{10}ka^2$$

From the Fig. 5.9, at $x = a$, $y = h$

$$\therefore \quad h = ka^2 \text{ or } k = \frac{h}{a^2}$$

$$\therefore \quad \bar{y} = \frac{3}{10} \times \frac{h}{a^2}a^2 = \frac{3h}{10}$$

Thus, centroid of spandrel is $\left(\frac{3a}{4}, \frac{3h}{10}\right)$

Centroids of some common figures are shown in Table 5.1.

Table 5.1 Centroid of Some Common Figures

Shape	*Figure*	$\bar{x}$	$\bar{y}$	*Area*
Triangle	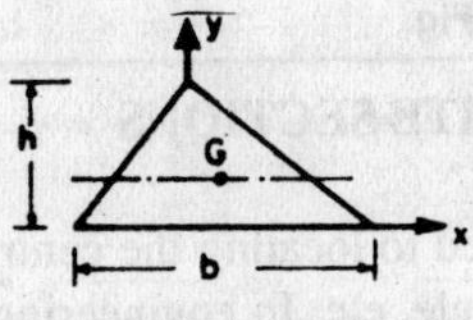 Fig.	–	$\frac{h}{3}$	$\frac{bh}{2}$
Semicircle	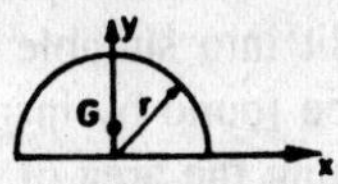 Fig.	0	$\frac{4R}{3\pi}$	$\frac{\pi R^2}{2}$
Quarter circle	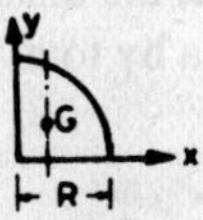 Fig.	$\frac{4R}{3\pi}$	$\frac{4R}{3\pi}$	$\frac{\pi R^2}{4}$
Sector of a circle	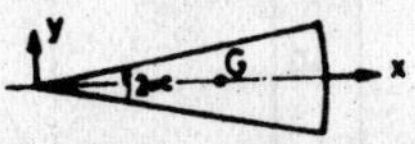 Fig.	$\frac{2R}{3\alpha}\sin\alpha$	0	αR^2
Parabola	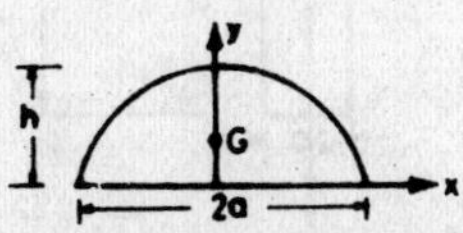 Fig.	0	$\frac{3h}{5}$	$\frac{4ah}{3}$
Semi Parabola	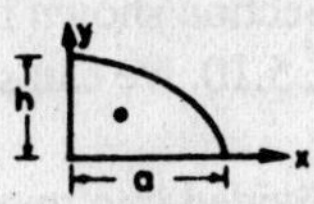 Fig.	$\frac{3}{8}a$	$\frac{3h}{5}$	$\frac{2ah}{3}$

(Contd...)

Parabolic Spandrel	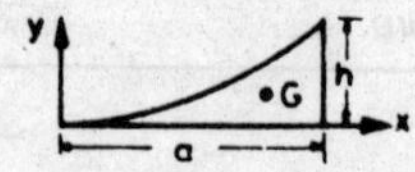Fig.	$\frac{3a}{4}$	$\frac{3h}{10}$	$\frac{ah}{3}$

5.7 CENTROID OF COMPOSITE SECTIONS

So far, the discussion was confined to locating the centroid of simple figures like rectangle, triangle, circle, semicircle, etc. In engineering practice, use of sections which are built up of many simple sections is very common. Such sections may be called as built-up sections or composite sections. To locate the centroid of composite sections, one need not go for the first principle (method of integration). The given composite section can be split into suitable simple figures and then the centroid of each simple figure can be found by inspection or using the standard formulae listed in Table 5.1. Assuming the area of the simple figure as concentrated at its centroid, its moment about an axis can be found by multiplying the area with distance of its centroid from the reference axis. After determining moment of each area about reference axis, the distance of centroid from the axis is obtained by dividing total moment of area by total area of the composite section.

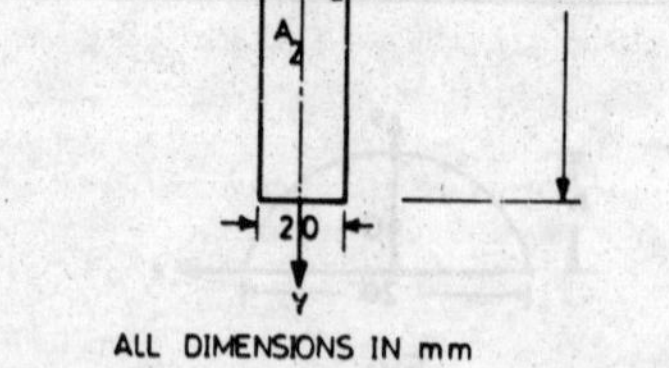

Fig. 5.10

Ex. 5.1: Locate the centroid of the T-Section shown in the Fig. 5.10.

Selecting the axis as shown in Fig. 5.10, we can say due to symmetry centroid lies on y axis, i.e. $\bar{x} = 0$.

Now the given T-section may be divided into two rectangles A_1 and A_2 each of size 100×20 and 20×100. The centroid of A_1 and A_2 are g_1 (0, 10) and g_2 (0, 70), respectively.

$\therefore$ The distance of centroid from top is given by:

$$\bar{y} = \frac{100 \times 20 \times 10 + 20 \times 100 \times 70}{100 \times 20 + 20 \times 100}$$

$$= 40 \text{ mm}$$

Hence, **centroid of T-section is on the symmetric axis at a distance 40 mm from the top.** *Ans.*

Ex. 5.2: Find the centroid of the unequal angle 200 × 150 × 12 mm, shown in Fig. 5.11.

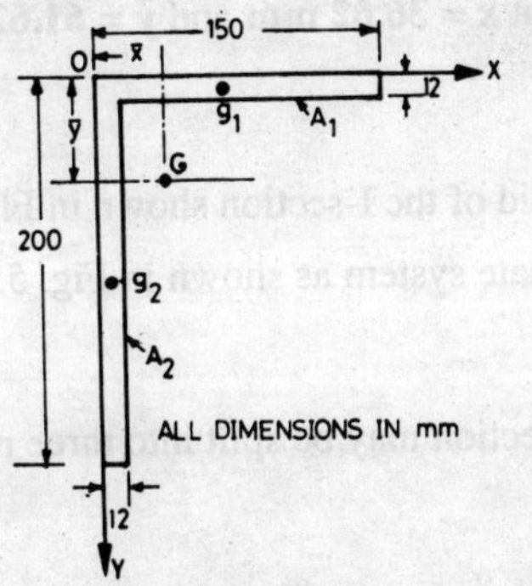

Fig. 5.11

The given composite figure can be divided into two rectangles:

$$A_1 = 150 \times 12 = 1800 \text{ mm}^2$$

$$A_2 = (200 - 12) \times 12 = 2256 \text{ mm}^2$$

Total area $A = A_1 + A_2 = 4056 \text{ mm}^2$

Selecting the reference axis x and y as shown in Fig. 5.11. The centroid of A_1 is g_1 (75, 6) and that of A_2 is:

$$g_2\left[6,\ 12 + \frac{1}{2}(200 - 12)\right]$$

i.e., g_2 (6, 106).

$$\therefore \quad \bar{x} = \frac{\text{Moment about } y \text{ axis}}{\text{Total area}}$$

$$= \frac{A_1 x_1 + A_2 x_2}{A}$$

$$= \frac{1800 \times 75 + 2256 \times 6}{4056}$$

$$= 36.62 \text{ mm}$$

$$\bar{y} = \frac{\text{Moment about } x \text{ axis}}{\text{Total area}}$$

$$= \frac{A_1 y_1 + A_2 y_2}{A}$$

$$= \frac{1800 \times 6 + 2256 \times 106}{4056}$$

$$= 61.62 \text{ mm}$$

Thus, the centroid is at $\bar{x}$ = **36.62 mm** and $\bar{y}$ = **61.62 mm** as shown in the figure.

Ans.

Ex. 5.3: Locate the centroid of the I-section shown in Fig. 5.12.

Selecting the co-ordinate system as shown in Fig. 5.12, due to symmetry centroid must lie on y axis,
i.e. $\bar{x} = 0$.

Now, the composite section may be split into three rectangles.

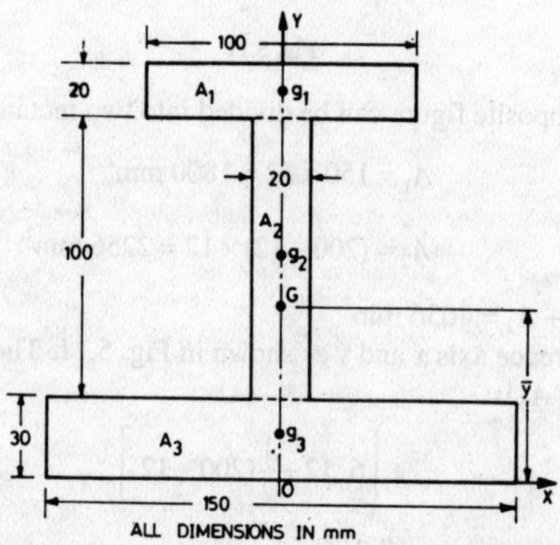

Fig. 5.12

$A_1 = 100 \times 20 = 2000 \text{ mm}^2$.
Centroid of A_1 from the origin is:

$$y_1 = 30 + 100 + \frac{20}{2} = 140 \text{ mm}$$

Similarly $$A_2 = 100 \times 20 = 2000 \text{ mm}^2$$

$$y_2 = 30 + \frac{100}{2} = 80 \text{ mm}$$

$$A_3 = 150 \times 30 = 4500 \text{ mm}^2, \text{ and}$$

$$y_3 = \frac{30}{2} = 15 \text{ mm}$$

$$\therefore \quad \overline{Y} = \frac{A_1Y_1 + A_2Y_2 + A_3Y_3}{A}$$

$$= \frac{2000 \times 140 + 2000 \times 80 + 4500 \times 15}{2000 + 2000 + 4500}$$

$$= 59.71 \text{ mm}$$

Thus, **the centroid is on the symmetric axis at a distance 59.71 mm from the bottom** as shown in Fig. 5.12. *Ans.*

Ex. 5.4: Determine the centroid of the section of the concrete dam shown in Fig. 5.13.

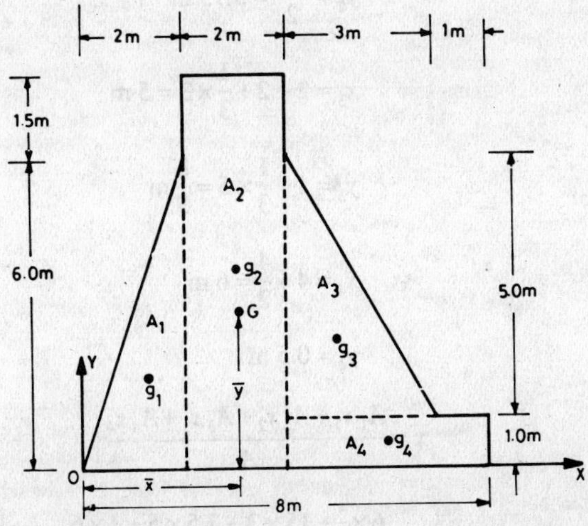

Fig. 5.13

Let the axis be selected as shown in Fig. 5.13. Note that it is convenient to take axes in such a way that the centroids of all simple figures are having positive coordinates. If coordinate of any simple figure comes out to be negative, one should be careful in assigning the sign of moment of area of that figure.

The composite figure can be conveniently divided into two triangles and two rectangles, as shown in Fig. 5.13.

Now, $$A_1 = \frac{1}{2} \times 2 \times 6 = 6 \text{ m}^2$$

$$A_2 = 2 \times 7.5 = 15 \text{ m}^2$$

$$A_3 = \frac{1}{2} \times 3 \times 5 = 7.5 \text{ m}^2$$

$$A_4 = 1 \times 4 = 4 \text{ m}^2$$

$$A = \text{total area} = 32.5 \text{ m}^2$$

Centroides of simple figures are:

$$x_1 = \frac{2}{3} \times 2 = \frac{4}{3} \text{ m}$$

$$y_1 = \frac{1}{3} \times 6 = 2 \text{ m}$$

$$x_2 = 2 + 1 = 3 \text{ m}$$

$$y_2 = \frac{7.5}{2} = 3.75 \text{ m}$$

$$x_3 = 2 + 2 + \frac{1}{3} \times 3 = 5 \text{ m}$$

$$y_3 = 1 + \frac{1}{3} \times 5 = \frac{8}{3} \text{ m}$$

$$x_4 = 4 + \frac{4}{2} = 6 \text{ m}$$

$$y_4 = 0.5 \text{ m}$$

$$\bar{x} = \frac{A_1 x_1 + A_2 x_2 + A_3 x_3 + A_4 x_4}{A}$$

$$= \frac{6 \times \frac{4}{3} + 15 \times 3 + 7.5 \times 5 + 4 \times 6}{32.5}$$

$$= 3.523 \text{ m}$$

$$\bar{y} = \frac{A_1 y_1 + A_2 y_2 + A_3 y_3 + A_4 y_4}{A}$$

$$= \frac{6 \times 2 + 15 \times 3.75 + 7.5 \times \frac{8}{3} + 4 \times 0.5}{32.5}$$

$= 2.777$ m

The centroid is at $\bar{x}$ = **3.523 m** *Ans.*

and $\bar{y}$ = **2.777 m** *Ans.*

Ex. 5.5: Determine the centroid of the area shown in Fig. 5.14 with respect to the axes shown.

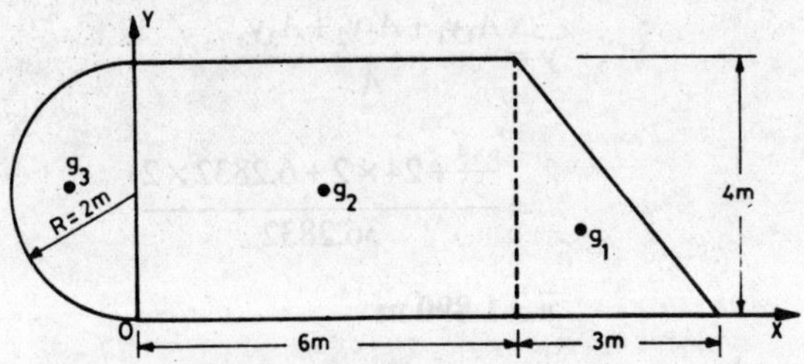

Fig. 5.14

The composite section is divided into three simple figures, a triangle, a rectangle and a semicircle.

Now, area of triangle $A_1 = \dfrac{1}{2} \times 3 \times 4 = 6 \text{ m}^2$

Area of rectangle $A_2 = 6 \times 4 = 24 \text{ m}^2$

Area of semicircle $A_3 = \dfrac{1}{2} \times \pi \times 2^2 = 6.2832 \text{ m}^2$

$\therefore$ Total area $A = 36.2832 \text{ m}^2$

The coordinates of centroids of these three simple figures are:

$$x_1 = 6 + \frac{1}{3} \times 3 = 7 \text{ m}$$

$$y_1 = \frac{4}{3} \text{ m}$$

$$x_2 = 3 \text{ m}$$

$$y_2 = 2 \text{ m}$$

$$x_3 = \frac{-4R}{3\pi} = -\frac{4 \times 2}{3\pi} = -0.8488 \text{ m}$$

$$y_3 = 2 \text{ m}$$

(Note carefully the sign of x_3).

$$\bar{x} = \frac{A_1 x_1 + A_2 x_2 + A_3 x_3}{A}$$

$$= \frac{6 \times 7 + 24 \times 3 + 6.2832 \times (-0.8488)}{36.2832}$$

i.e. $\bar{x} = \mathbf{2.995\ m}$ *Ans.*

$$\bar{y} = \frac{A_1 y_1 + A_2 y_2 + A_3 y_3}{A}$$

$$= \frac{\frac{6 \times 4}{3} + 24 \times 2 + 6.2832 \times 2}{36.2832}$$

i.e. $\bar{y} = \mathbf{1.890\ m}$ *Ans.*

Ex. 5.6: In a gusset plate, there are six rivet holes of 21.5 mm diameter as shown in Fig. 5.15. Find the position of the centroid of the gusset plate.

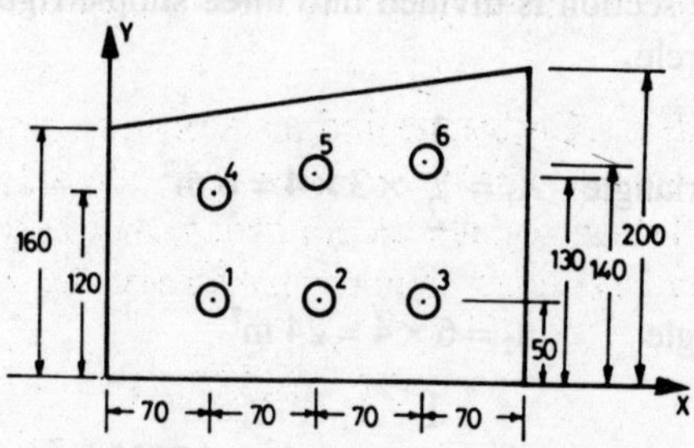

Fig. 5.15

The composite area is equal to a rectangle of size 160 × 280 mm plus a triangle of size 280 mm base width and 40 mm height and minus areas of six holes. In this case also the Eqns. 5.5 and 5.6 can be used for locating centroid by treating area of holes as negative. The area of simple figures and their centroids are as shown in Table given below:

Figure	*Area in mm²*	x_i *in mm*	y_i *in mm*
Rectangle	$160 \times 280 = 44{,}800$	140	80
Triangle	$\frac{1}{2} \times 280 \times 40 = 5600$	$\frac{560}{3}$	$160 + \frac{40}{3} = 173.33$
1st hole	$\frac{-\pi \times 21.5^2}{4} = -363.05$	70	50
2nd hole	−363.05	140	50

(Contd...)

Figure	Area in mm^2	x_i in mm	y_i in mm
3rd hole	−363.05	210	50
4th hole	−363.05	70	120
5th hole	−363.05	140	130
6th hole	−363.05	210	140

$$\therefore \qquad A = \sum A_i = 48,221.70$$

$$\therefore \ \sum A_i x_i = 44,800 \times 140 + 5600 \times \frac{560}{3} - 363.05(70 + 140 + 210 + 70 + 140 + 210)$$

$$= 70,12,371.3 \text{ mm}^3$$

$$\bar{x} = \frac{\sum A_i x_i}{A} = 145.42 \text{ mm}$$

$$\sum A_i y_i = 44,800 \times 80 + 5600 \times 173.33 - 363.05(50 \times 3 + 120 + 130 + 140)$$

$$= 43,58,601 \text{ mm}^3$$

$$\bar{y} = \frac{\sum A_i y_i}{A} = \frac{43,58,601}{48221.70} = 90.39 \text{ mm}$$

Thus, the coordinates of centroid of composite figure is given by:

$$\mathbf{\bar{x} = 145.42 \text{ mm}} \qquad \textit{Ans.}$$

$$\mathbf{\bar{y} = 90.39 \text{ mm}} \qquad \textit{Ans.}$$

Ex. 5.7: Determine the coordinates x_c and y_c of the centre of a 100 mm diameter circular hole cut in a thin plate so that this point will be the centroid of the remaining shaded area shown in Fig. 5.16 (All dimensions are in mm).

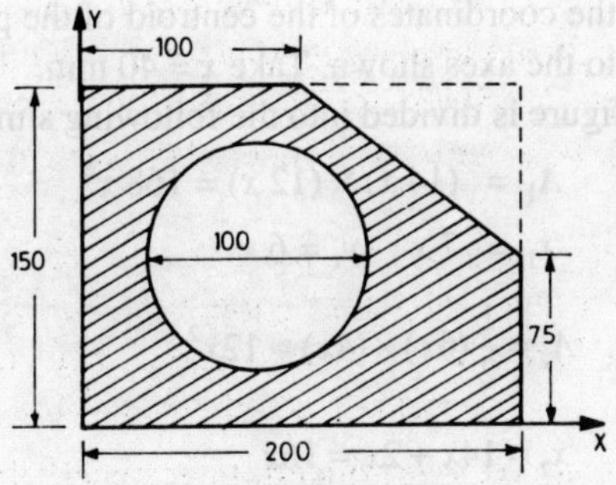

Fig. 5.16

If x_c and y_c are the coordinates of the centre of the circle, centroid also must have the coordinates x_c and y_c as per the condition laid down in the problem. The shaded area may be considered as a rectangle of size 200 mm × 150 mm *minus* a triangle of sides 100 mm × 75 mm and a circle of diameter 100 mm.

$$\therefore \quad \text{Total area} = 200 \times 150 - \frac{1}{2} \times 100 \times 75 - \left(\frac{\pi}{4}\right) 100^2$$

$$= 18,396 \text{ mm}^2$$

$$\bar{x} = x_c = \frac{200 \times 150 \times 100 - \frac{1}{2} \times 100 \times 75 \times \left[200 - \left(\frac{100}{3}\right)\right] - \frac{\pi}{4} \times 100^2 \times x_c}{18,396}$$

$$\therefore \quad x_c(18,396) = 200 \times 150 \times 100 - \frac{1}{2} \times 100 \times 75 \times 166.67 - \frac{\pi}{4} \times 100^2 \, x_c$$

$$x_c = \frac{23,75000}{26,250} = \mathbf{90.48 \text{ mm}} \qquad \textit{Ans.}$$

Similarly,

$$18,396 \, y_c = 200 \times 150 \times 75 - \frac{1}{2} \times 100 \times 75 \times (150 - 25) - \frac{\pi}{4} \times 100^2 y_c$$

$$\therefore \quad y_c = \frac{17,81250.0}{26,250} = \mathbf{67.86 \text{ mm}} \qquad \textit{Ans.}$$

Centre of the circle should be located at (90.48, 67.86) so that this point will be the centroid of the remaining shaded area shown in Fig. 5.16.

Note:
The centroid of the given figure will coincide with the centroid of the figure without circular hole. Hence, the centroid of the given figure can be obtained by determining the centroid of the figure without the circular hole.

Ex. 5.8: Determine the coordinates of the centroid of the plane area shown in Fig. 5.17 with reference to the axes shown. Take $x = 40$ mm.

The composite figure is divided into the following simple figures:

(1) A rectangle $\quad A_1 = (14\,x) \times (12\,x) = 168\,x^2$

$$x_1 = 7\,x; \quad y_1 = 6\,x$$

(2) A triangle $\quad A_2 = \frac{1}{2}(6x) \times (4x) = 12x^2$

$$x_2 = 14x + 2x = 16x$$

$$y_2 = \frac{4x}{3}$$

(3) A rectangle to be subtracted

$$A_3 = (-4\,x) \times (4\,x) = -16\,x^2$$

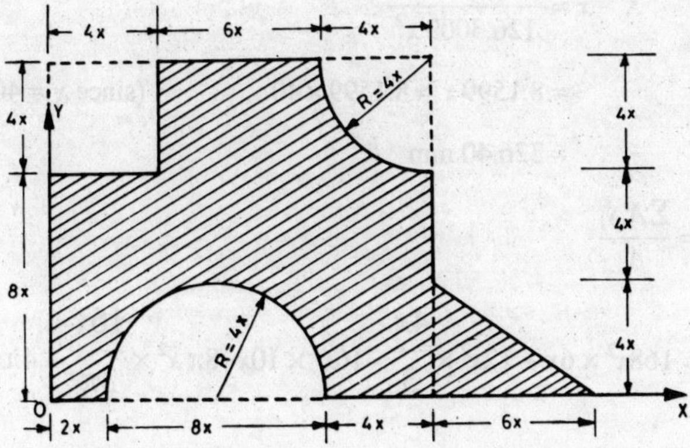

Fig. 5.17

$$x_3 = 2x;\ y_3 = 8x + 2x = 10x$$

(4) A semicircle to be subtracted

$$A_4 = -\frac{1}{2}\pi(4x)^2 = -8\pi x^2$$

$$x_4 = 6x$$

$$y_4 = \frac{4R}{3\pi} = 4\times\frac{(4x)}{3\pi} = \frac{16x}{3\pi}$$

(5) A quarter of a circle to be subtracted

$$A_5 = -\frac{1}{4}\times\pi(4x)^2 = -4\pi x^2$$

$$x_5 = 14x - \frac{4R}{3\pi} = 14x - (4)\left(\frac{4x}{3\pi}\right) = 12.3023\,x$$

$$y_5 = 12x - 4\times\left(\frac{4x}{3\pi}\right) = 10.3023\,x$$

Total area $A = 168x^2 + 12x^2 - 16x^2 - 8\pi x^2 = 4\pi x^2$

$$= 126.3009\,x^2$$

$$\bar{x} = \frac{\sum A_i x_i}{A}$$

$\Sigma\ A_i x_i = 168x^2 \times 7x + 12x^2 \times 16x - 16x^2 \times 2x - 8\pi x^2 \times 6x - 4\pi x^2 \times 12.3023x =$ $1030.6083x^3$

$$\therefore \quad \bar{x} = \frac{1030.6083\,x^3}{126.3009\,x^2}$$

$$= 8.1599\,x = 8.1599 \times 40 \qquad \text{(since } x = 40 \text{ mm)}$$

$$= 326.40 \text{ mm}$$

$$\bar{y} = \frac{\Sigma A_i y_i}{A}$$

$$\Sigma\ A_i x_i = 168x^2 \times 6x + 12x^2 \times \frac{4x}{3} - 16x^2 \times 10x - 8\pi x^2 \times \frac{16x}{3\pi} - 4\pi x^2 \times 10.3023x$$

$$= 691.8708x^3$$

$$\therefore \quad \bar{y} = \frac{691.8708}{126.3009\,x^2}$$

$$= 5.4780\,x$$

$$= 219.12 \text{ mm} \quad \text{(since } x = 40 \text{ mm)}$$

Centroid is at **(326.40, 219.12)** *Ans.*

5.8 MOMENT OF INERTIA

Consider the area shown in Fig. 5.18(a). *dA* is an elemental area with coordinates as *x* and *y*. The term $\Sigma y^2\, dA$ is called **moment of inertia** of the area about *x* axis and is denoted as I_{xx}. Similarly, the moment of inertia about *y* axis is

$$I_{yy} = \Sigma\, x^2\, dA$$

In general, if *r* is the distance of elemental area *dA* from the axis *AB* [Fig. 5.18(b)], the sum of the terms $\Sigma r^2\, dA$ to cover the entire area is called moment of inertia of the area about the axis *AB*. If *r* and *dA* can be expressed in general term, for any element, then the sum becomes an integral. Thus,

$$I_{AB} = \Sigma\, r^2\, dA = \int r^2\, dA \qquad \textit{Eqn. 5.9}$$

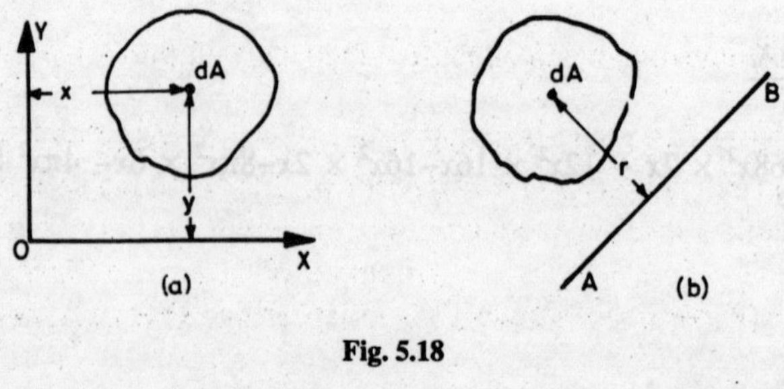

Fig. 5.18

The term *rdA* may be called as moment of area, similar to moment of a force, and hence $r^2\ dA$ may be called as moment of moment of area or the second moment of area. Thus, the moment of inertia is nothing but second moment of area. In fact, the term '**second moment of area**' appears to correctly signify the meaning of the expression $\Sigma\ r^2\ dA$. The term 'moment of inertia' is rather a misnomer. However, the term moment of inertia has come to stay for long time and hence it will be used in this book also.

Though moment of inertia is a purely mathematical term, it is one of the important properties of areas. The strength of members subject to bending depends on the moment of inertia of its cross-sectional area. Students will find this property of area very useful when they study subjects like strength of materials, structural design and machine design.

The moment of inertia is a fourth dimensional term since it is a term obtained by multiplying area by the square of the distance. Hence, in SI units, if metre (m) is the unit for linear measurements used then m^4 is the unit of moment of inertia. If millimetre (mm) is the unit used for linear measurements, then mm^4 is the unit of moment of inertia. In MKS system m^4 or cm^4 and in FPS system ft^4 or in^4 are commonly used as units for moment of inertia.

5.9 POLAR MOMENT OF INERTIA

Moment of inertia about an axis perpendicular to the plane of an area is known as **polar moment of inertia**. It may be denoted as J or I_{zz}. Thus, the moment of inertia about an axis perpendicular to the plane of the area at O in Fig. 5.19 is called polar moment of inertia at point O, and is given by

$$I_{zz} = \Sigma\ r^2\ dA \qquad \textit{Eqn. } 5.10$$

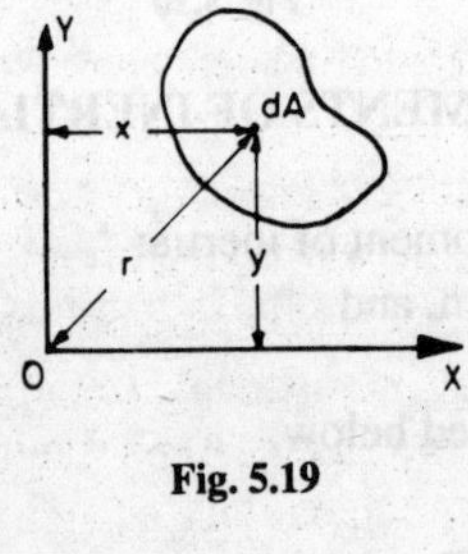

Fig. 5.19

5.10 RADIUS OF GYRATION

Radius of gyration is a mathematical term defined by the relation

$$k = \sqrt{\frac{I}{A}} \qquad \textit{Eqn.} 5.11$$

where K = radius of gyration

I = moment of inertia

and A = the cross-sectional area.

Suffixes with moment of inertia I also accompany the term radius of gyration k. Thus, we can have,

$$k_{xx} = \sqrt{\frac{I_{xx}}{A}}$$

$$k_{yy} = \sqrt{\frac{I_{yy}}{A}}$$

$$k_{AB} = \sqrt{\frac{I_{AB}}{A}}$$

and so on.

The relation between radius of gyration and moment of inertia can be put in the form:

$$I = Ak^2 \qquad \textit{Eqn. } 5.12$$

From the above relation a geometric meaning can be assigned to the term 'radius of gyration.' We can consider k as the distance at which the complete area is squeezed and kept as a strip of negligible width (Fig. 5.20) such that there is no change in the moment of inertia.

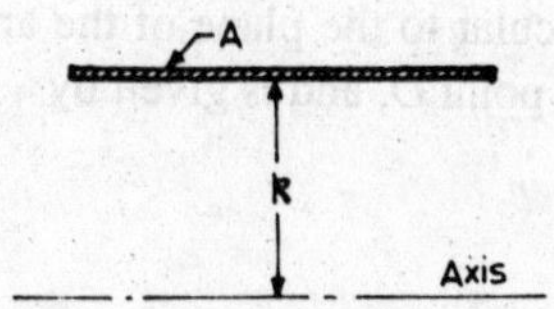

Fig. 5.20

5.11 THEOREMS OF MOMENTS OF INERTIA

There are two theorems of moment of inertia:
(1) Perpendicular axis theorem, and
(2) Parallel axis theorem.
These are explained and proved below.

Perpendicular Axis Theorem

The moment of inertia of an area about an axis perpendicular to its plane (polar moment of inertia) at any point O is equal to the sum of moments of inertia about

any two mutually perpendicular axis through the same point O and lying in the plane of the area.

Referring to Fig. 5.21, if $z - z$ is the axis normal to the plane of paper passing through point O, as per this theorem,

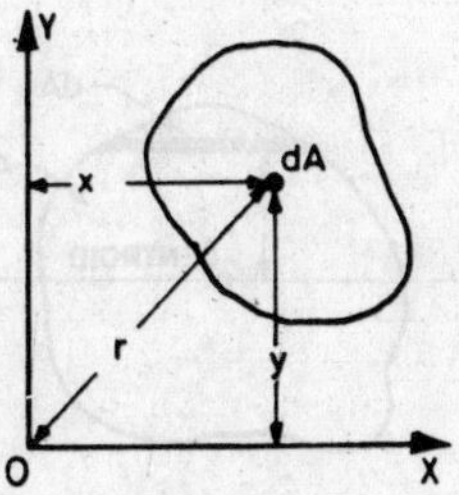

Fig. 5.21

$$I_{zz} = I_{xx} + I_{yy} \qquad \textit{Eqn. 5.13}$$

The above theorem can be easily proved. Let us consider an elemental area dA at a distance r from O. Let the coordinates of dA be x and y. Then from definition:

$$I_{zz} = \Sigma r^2 dA$$

$$= \Sigma(x^2 + y^2)dA$$

$$= \Sigma x^2 dA + \Sigma y^2 dA$$

$$I_{zz} = I_{xx} + I_{yy}$$

Parallel Axis Theorem

Moment of inertia about any axis in the plane of an area is equal to the sum of moment of inertia about a parallel centroidal axis and the product of area and square of the distance between the two parallel axes. Referring to Fig. 5.22, the above theorem means:

$$I_{AB} = I_{GG} + A\, y_c^2 \qquad \textit{Eqn. 5.14}$$

where

I_{AB} – moment of inertia about axis AB

I_{GG} – moment of inertia about centroidal axis GG parallel to AB.

A – The area of the plane figure given and

y_c – the distance between the axis AB and the parallel centroidal axis GG.

Proof: Consider an elemental parallel strip dA at a distance y from the centroidal axis (Fig. 5.22).

Then,

$$I_{AB} = \Sigma(y + y_c)^2 dA$$

$$= \Sigma(y^2 + 2y\,y_c + y_c^2)dA$$

$$= \Sigma y^2 dA + \Sigma 2y\,y_c dA + \Sigma y_c^2 dA$$

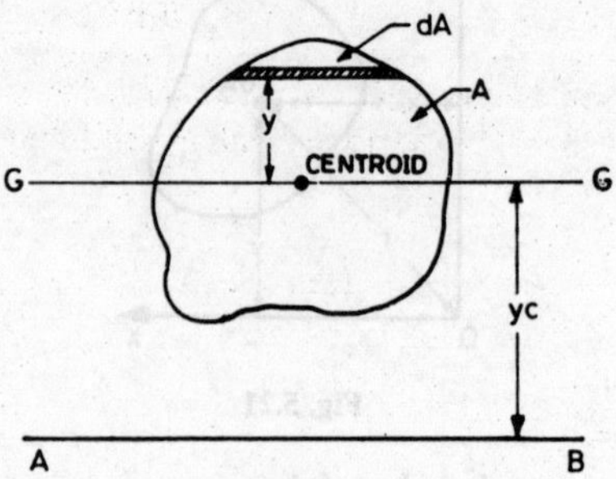

Fig. 5.22

Now,

$$\Sigma y^2\, dA = \text{Moment of inertia about the axis } GG$$

$$= I_{GG}$$

$$\Sigma 2\, yy_c\, dA = 2\, y_c\, \Sigma y\, dA$$

$$= 2 y_c A \frac{\Sigma\, ydA}{A}$$

In the above term $2y_c\, A$ is constant and $\frac{\Sigma\, ydA}{A}$ is the distance of centroid from the reference axis *GG*. Since *GG* is passing through the centroid itself $\frac{ydA}{A}$ is zero and hence the term $\Sigma\, 2\, y\, y_c\, dA$ is zero.

Now, the third term,

$$\Sigma\, y_c^2 dA = y_c^2 \Sigma\, dA$$

$$= A\, y_c^2$$

$$\therefore \quad I_{AB} = I_{GG} + A\, y_c^2$$

Note:

The above equation cannot be applied to any two parallel axes. One of the axes (*GG*) must be centroidal axis only.

5.12 MOMENT OF INERTIA FROM FIRST PRINCIPLES

For simple figures, moment of inertia can be obtained by writing down the general expression for an element and then carrying out integration so as to cover the entire area. This procedure is illustrated with the following three cases:

(1) Moment of inertia of rectangle about the centroidal axis
(2) Moment of inertia of a triangle about the base
(3) Moment of inertia of a circle about a diametral axis.

Moment of Inertia of a Rectangle about the Centroidal Axis – Consider a rectangle of width b and depth d (Fig. 5.23). Moment of inertia about the centroidal axis $x - x$ parallel to the short side is required.

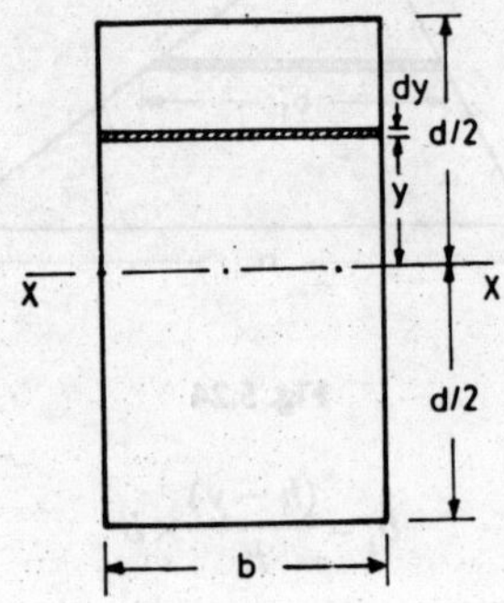

Fig. 5.23

Consider an elemental strip of width dy at a distance y from the axis. Moment of inertia of the elemental strip about the centroidal axis xx is:

$$= y^2\, dA$$

$$= y^2\, b\, dy$$

$$\therefore \qquad I_{xx} = \int_{-\frac{d}{2}}^{\frac{d}{2}} y^2\, b\, dy$$

$$= b\left[\frac{y^3}{3}\right]_{-\frac{d}{2}}^{\frac{d}{2}}$$

$$= b\left[\frac{d^3}{24} + \frac{d^3}{24}\right]$$

$$I_{xx} = \frac{bd^3}{12}$$

Moment of Intertia of a Triangle About its Base – Moment of inertia of a triangle with base width b and height h is to be determined about the base AB (Fig. 5.24)

Consider an elemental strip at a distance y from the base AB. Let dy be the thickness of the strip and dA its area. Width of this strip is given by:

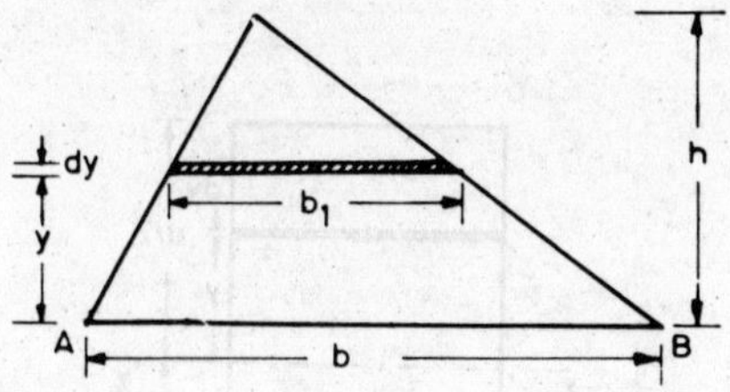

Fig. 5.24

$$b_1 = \frac{(h-y)}{h} \times b$$

Moment of inertia of this strip about AB

$$= y^2\, dA$$

$$= y^2\, b_1\, dy$$

$$= y^2 \frac{(h-y)}{h} \times b \times dy$$

$\therefore$ Moment of inertia of the triangle about AB,

$$I_{AB} = \int_0^h \frac{y^2 (h-y)\, bdy}{h}$$

$$= \int_0^h \left(y^2 - \frac{y^3}{h} \right) bdy$$

$$= \left[\frac{y^3}{3} - \frac{y^4}{4h} \right]_0^h$$

$$= b\left[\frac{h^3}{3} - \frac{h^4}{4h} \right]$$

$$I_{AB} = \frac{bh^3}{12}$$

Moment of Inertia of Circle about its Diametral Axis – Moment of inertia of a circle of radius R is required about it's diametral axis as shown in Fig. 5.25.

Consider an element of sides $rd\,\theta$ and dr as shown in the figure. Its moment of inertia about the diamentral axis $x - x$:

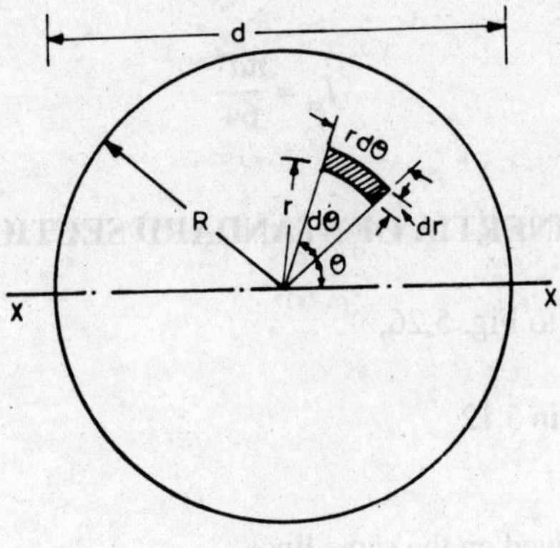

Fig. 5.25

$$= y^2\, dA$$

$$= (r \sin\theta)^2\, r\, d\theta\, dr$$

$$= r^3 \sin^2\theta\, d\theta\, dr$$

∴ Moment of inertia of the circle about $x - x$ is given by

$$I_{xx} = \int_0^R \int_0^{2\pi} r^3 \sin^2\theta\, d\theta\, dr$$

$$= \int_0^R \int_0^{2\pi} r^3 \frac{(1-\cos 2\theta)}{2}\, d\theta\, dr$$

$$= \int_0^R \frac{r^3}{2}\left[\theta - \frac{\sin 2\theta}{2}\right]_0^{2\pi} dr$$

$$= \left[\frac{r^4}{8}\right]_0^R [2\pi - 0 + 0 - 0]$$

$$= \frac{2\pi}{8} R^4$$

$$I_{xx} = \frac{\pi R^4}{4}$$

If d is the diameter of the circle, then

$$R = \frac{d}{2}$$

$$\therefore \quad I_{xx} = \frac{\pi}{4}\left(\frac{d}{2}\right)^4$$

$$I_{xx} = \frac{\pi d^4}{64}$$

5.13 MOMENT OF INERTIA OF STANDARD SECTIONS

Rectangle – Referring to Fig. 5.26,

(a) $I_{xx} = \frac{bd^3}{12}$ as derived in 5.12

(b) $I_{yy} = \frac{db^3}{12}$ can be derived on the same lines.

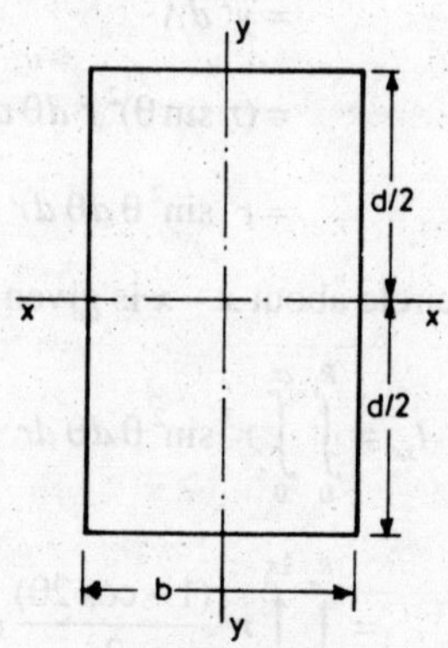

Fig. 5.26

(c) *About the base AB*, from parallel axis theorem,

$$I_{AB} = I_{xx} + Ay_c^2$$

$$= \frac{bd^3}{12} + bd\left(\frac{d}{2}\right)^2, \quad \text{since } y_c = \frac{d}{2}$$

$$= \frac{bd^3}{12} + \frac{bd^3}{4}$$

$$= \frac{bd^3}{3}$$

Hollow Rectangular Section – Referring to Fig. 5.27, Moment of inertia I_{xx} = Moment of inertia of larger rectangle – Moment of inertia of hollow portion. That is,

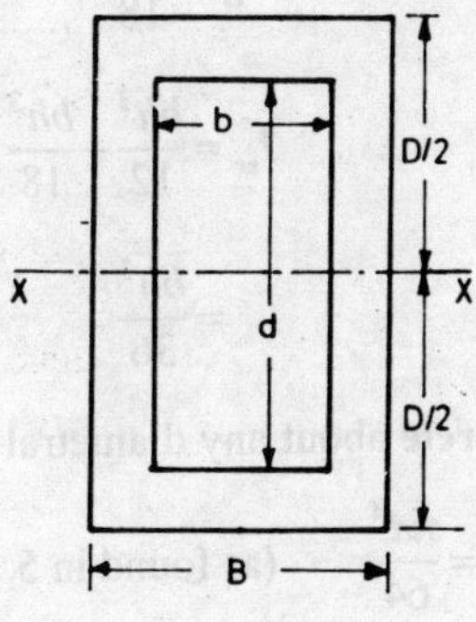

Fig. 5.27

$$= \frac{BD^3}{12} - \frac{bd^3}{12}$$

$$= \frac{1}{12}(BD^3 - bd^3)$$

Triangle – Referring to Fig. 5.28,

(a) About the base:
As found in 5.12,

$$I_{AB} = \frac{bh^3}{12}$$

(b) About centroidal axis, $x - x$ parallel to base:
From parallel axis theorem,

$$I_{AB} = I_{xx} + A\,y_c^2$$

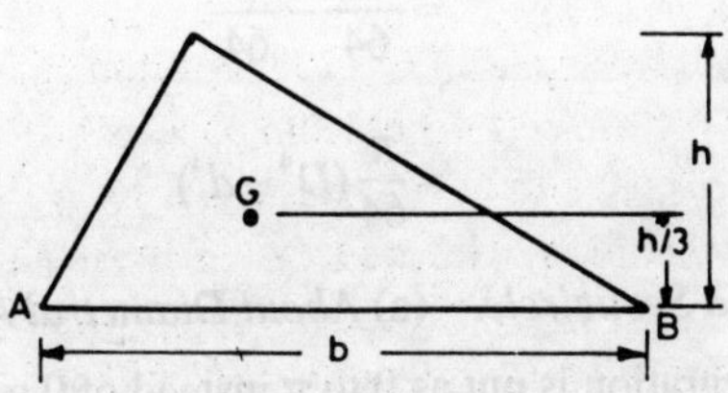

Fig. 5.28

Now, y_c the distance between the non-centroidal axis AB and centroidal axis $x - x$ is equal to $\frac{h}{3}$

$$\therefore \quad \frac{bh^3}{12} = I_{xx} + \frac{1}{2}bh\left(\frac{h}{3}\right)^2$$

$$= I_{xx} + \frac{bh^3}{18}$$

$$\therefore \quad I_{xx} = \frac{bh^3}{12} - \frac{bh^3}{18}$$

$$= \frac{bh^3}{36}$$

Moment of Inertia of a Circle about any diametral axis

$$= \frac{\pi d^4}{64} \quad \text{(as found in 5.12)}$$

Moment of Inertia of a Hollow Circle – Referring to Fig. 5.29,

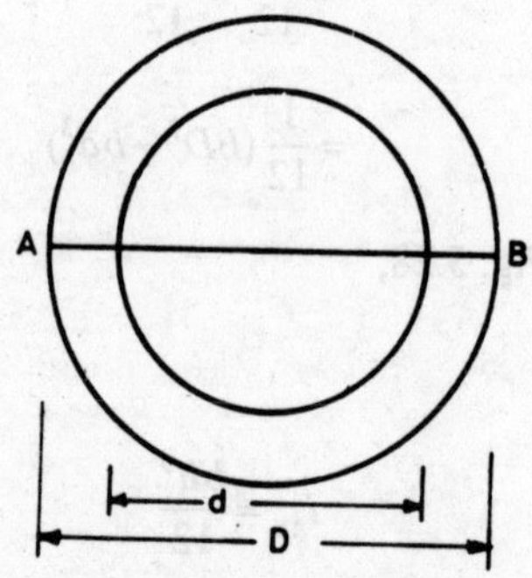

Fig. 5.29

I_{AB} = Moment of inertia of solid circle of diameter D about AB *minus* Moment of inertia of circle of diameter d about AB. That is,

$$= \frac{\pi D^4}{64} - \frac{\pi d^4}{64}$$

$$= \frac{\pi}{64}(D^4 - d^4)$$

Moment of Inertia of a Semicircle – (a) *About Diametral Axis:*

If the limit of integration is put as 0 to π instead of 0 to 2π in the derivation for the moment of inertia of a circle about diametral axis (Ref. 5.12), the moment of inertia of a semicircle is obtained. It can be observed that the moment of inertia of a semicircle (Fig. 5.30) about the diametral axis AB:

$$= \frac{1}{2} \times \frac{\pi d^4}{64} = \frac{\pi d^4}{128}$$

(b) *About centroidal axis x – x*: Now, the distance of centroidal axis y_c from the diametral axis is given by:

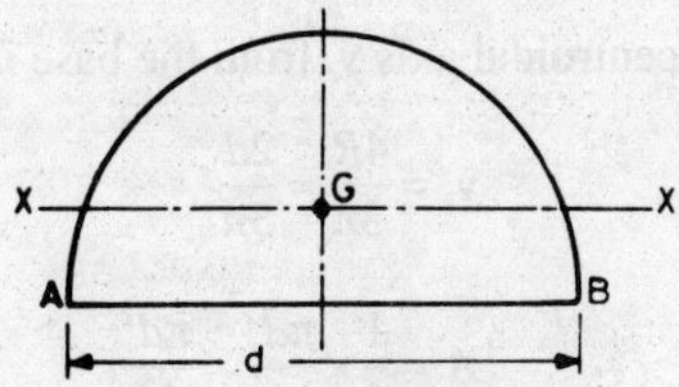

Fig. 5.30

$$y_c = \frac{4R}{3\pi} = \frac{2d}{3\pi}$$

and, Area $A = \frac{1}{2} \times \frac{\pi d^2}{4} = \frac{\pi d^2}{8}$

From parallel axes theorem,

$$I_{AB} = I_{xx} + A\, y_c^2$$

$$\frac{\pi d^4}{128} = I_{xx} + \frac{\pi d^2}{8} \times \left(\frac{2d}{3\pi}\right)^2$$

$$I_{xx} = \frac{\pi d^4}{128} - \frac{d^4}{18\pi}$$

$$= 0.0068598\, d^4$$

Moment of Inertia of a Quarter of a Circle – (a) *About the base*:

If the limit of integration is put as 0 to $\frac{\pi}{2}$ instead of 0 to 2 π in the derivation for moment of inertia of a circle (Ref. 5.12), the moment of inertia of a quarter of a circle is obtained. It can be observed that moment of inertia of the quarter of a circle about the base *AB*:

$$= \frac{1}{4} \times \frac{\pi d^4}{64} = \frac{\pi d^4}{256}$$

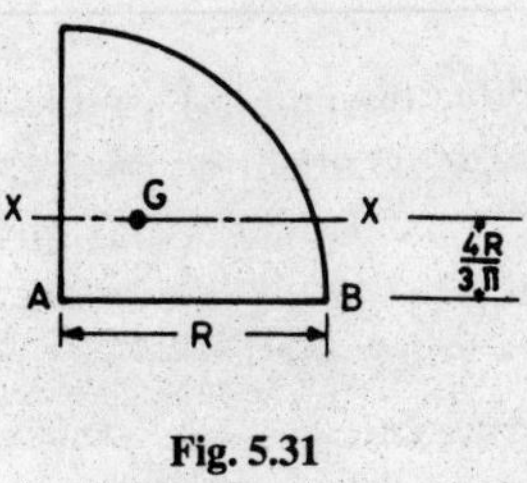

Fig. 5.31

(b) *About centroidal axis x – x*

Now, the distance of centroidal axis y_c from the base is given by:

$$y_c = \frac{4R}{3\pi} = \frac{2d}{3\pi}$$

and the area
$$A = \frac{1}{4} \times \frac{\pi d^2}{4} = \frac{\pi d^2}{16}$$

From parallel axes theorem,

$$I_{AB} = I_{xx} + A\,y_c^2$$

$$\frac{\pi d^4}{256} = I_{xx} + \frac{\pi d^2}{16}\left(\frac{2d}{3\pi}\right)^2$$

$$I_{xx} = \frac{\pi d^4}{256} - \frac{d^4}{36\pi}$$

$$= 0.00343 d^4$$

The moment of inertia of common standard sections are presented in Table 5.2.

Table 5.2 Moment of Inertia of Standard Sections

Shape	*Axis*	*Moment of Inertia*
Rectangle	(a) Centroidal axis $x - x$	$I_{xx} = \frac{bd^3}{12}$
	(b) Centroidal axis $y - y$	$I_{yy} = \frac{db^3}{12}$
	(c) $A - B$	$I_{AB} = \frac{bd^3}{3}$

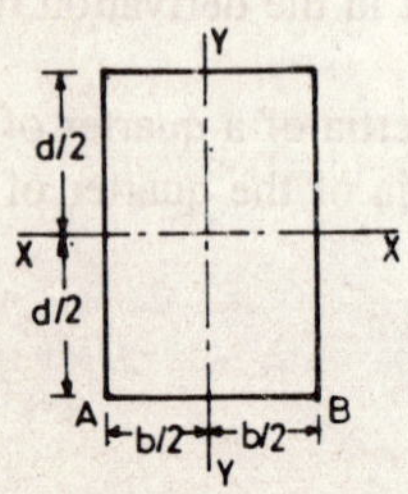

Fig.

(Contd...)

Shape	*Axis*	*Moment of Inertia*
Hollow Rectangle	Centroidal axis $x - x$	$I_{xx} = \dfrac{BD^3 - bd^3}{12}$

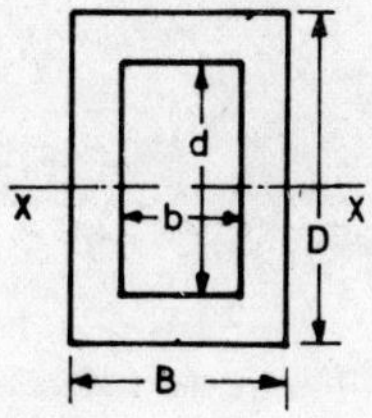

Fig.

Triangle	(a) Centroidal axis $x - x$	$I_{xx} = \dfrac{bh^3}{36}$
	(b) Base AB	$I_{AB} = \dfrac{bh^3}{12}$

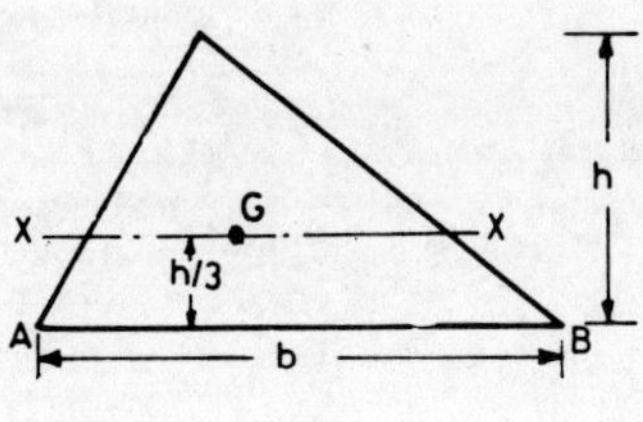

Fig.

Circle	Diametral axis	$I = \dfrac{\pi d^4}{64}$

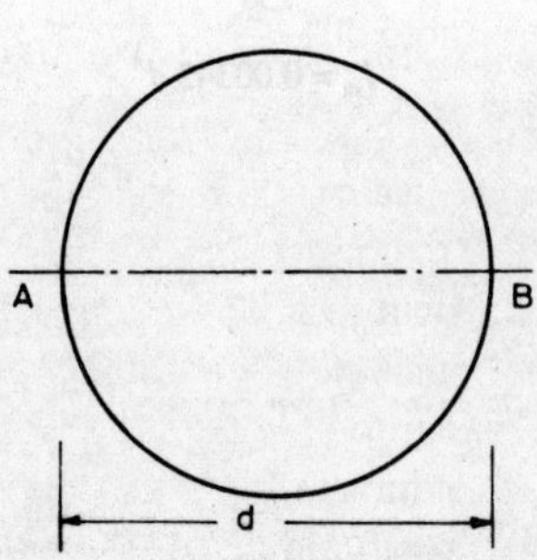

Fig.

(Contd...)

Shape	*Axis*	*Moment of Inertia*
Hollow Circle	Diametral axis	$I = \frac{\pi}{64}(D^4 - d^4)$
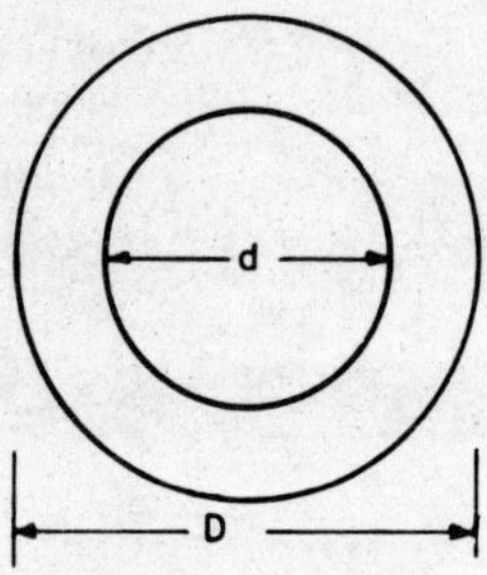 **Fig.**		
Semi Circle	(a) $A - B$	$I_{AB} = \frac{\pi d^4}{128}$
	(b) Centroidal axis	$I_{xx} = 0.0068598\ d^4$
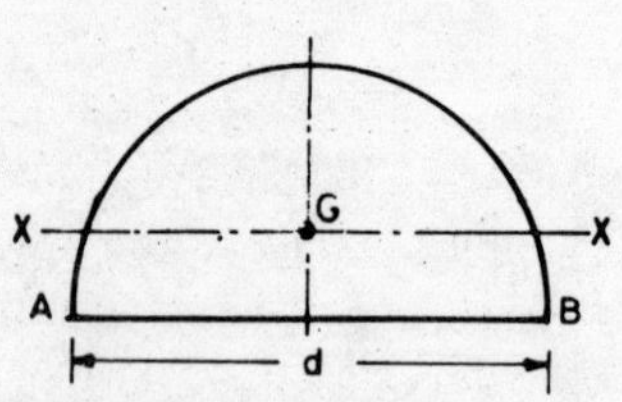 **Fig.**		
Quarter of a Circle	(a) $A - B$	$I_{AB} = \frac{\pi d^4}{256}$
	(b) Centroidal axis $x - x$	$I_{xx} = 0.00343\ d^4$
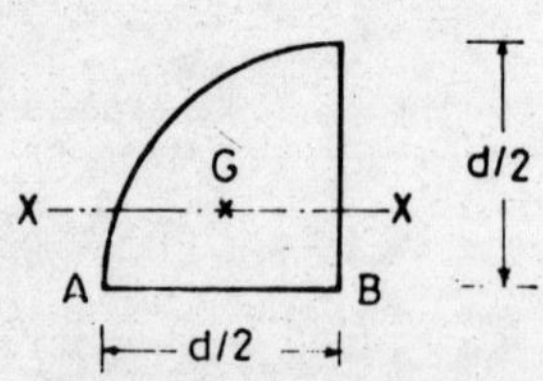 **Fig.**		

5.14 MOMENT OF INERTIA OF COMPOSITE SECTIONS

Beams and columns having composite sections are commonly used in structures. Moment of inertia of these sections about an axis can be found by the following steps:

1. Divide the given figure into a number of simple figures.
2. Locate the centroid of each simple figure by inspection or using standard expressions.
3. Find the moment of inertia of each simple figure about its centroidal axis. Add the term Ay^2 where A is the area of the simple figure and y is the distance of the centroid of the simple figure from the reference axis. This gives moment of inertia of the simple figure about the reference axis.
4. Sum up moments of inertia of all simple figures to get the moment of inertia of the composite section.

The procedure given above is illustrated below. Referring to the Fig. 5.32, it is required to find out the moment of inertia of the section about axis $A - B$.

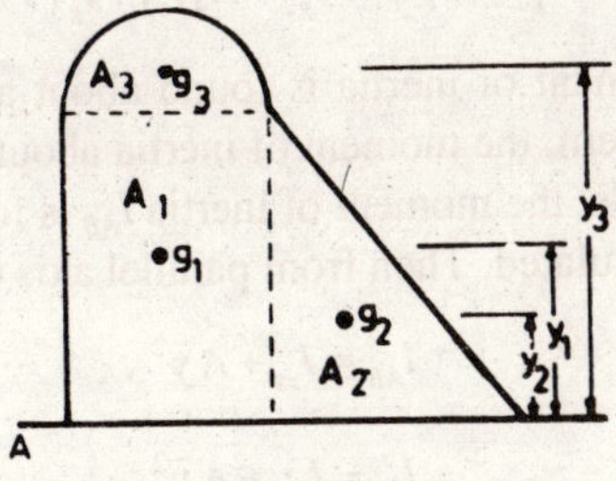

Fig. 5.32

1. The section in the figure is divided into a rectangle, a triangle and a semicircle. The areas of the simple figures A_1, A_2 and A_3 are calculated.
2. The centroids of the rectangle (g_1), triangle (g_2) and semicircle (g_3) are located. The distances y_1, y_2 and y_3 are found from the axis AB.
3. The moment of inertia of the rectangle about its centroid (I_{g1}) is calculated using standard expression. To this, the term $A_1 \, y_1^{\,2}$ is added to get the moment of inertia about the axis AB as:

$$I_1 = I_{g1} + A_1 \, y_1^{\,2}$$

Similarly, the moment of inertia of the triangle ($I_2 = I_{g2} + A_2 \, y_2^{\,2}$) and of semicircle ($I_3 = I_{g3} + A_3 \, y_3^{\,2}$) about axis AB are calculated.

(4) Moment of inertia of the composite section about AB is given by:

$$I_{AB} = I_1 + I_2 + I_3$$

$$= I_{g1} + A_1 y_1^2 + I_{g2} + A_2 y_2^2 + I_{g3} + A_3 y_3^2 \qquad \textit{Eqn.5.15}$$

In most engineering problems, moment of inertia about the centroidal axis is required. In such cases, first locate the centroidal axis as discussed in 5.7 and then find the moment of inertia about this axis.

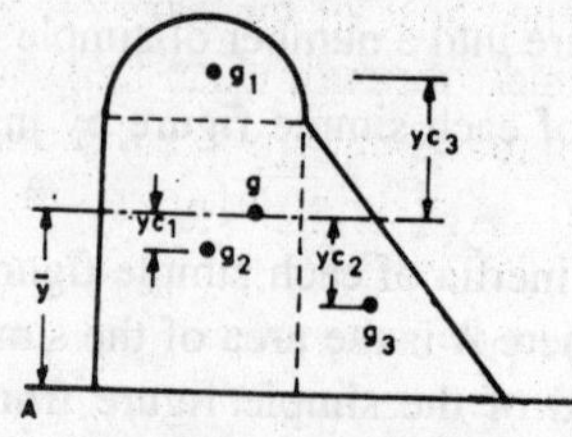

Fig. 5.33

Referring to Fig. 5.33, first the moment of area about any reference axis, say *AB*, is taken and is divided by the total area of the section to locate centroidal axis $x - x$. Then the distances of centroid of individual figures y_{c1}, y_{c2} and y_{c3} from the axis $x - x$ are determined. The moment of inertia of the composite section about the centroidal axis $x - x$ is calculated using the expression:

$$I_{xx} = I_{g1} + A_1 y_{c1}^2 + I_{g2} + A_2 y_{c2}^2 + I_{g3} + A_3 y_{c3}^2 \qquad Eqn.5.16$$

Sometimes the moment of inertia is found about a convenient axis and then using parallel axis theorem, the moment of inertia about centroidal axis is found.

In the above example, the moment of inertia I_{AB} is found and $\bar{y}$, the distance of *CG* from axis *AB* is calculated. Then from parallel axis theorem,

$$I_{AB} = I_{xx} + A\bar{y}^2$$

$$I_{xx} = I_{AB} - A\bar{y}^2$$

where *A* is the area of composite section.

Ex. 5.9: Determine the moment of inertia of the section shown in Fig. 5.34 about an axis passing through the centroid and parallel to the top most fibre of the section.

Also determine moment of inertia about the axis of symmetry. Hence find out radii of gyration.

The given composite section can be divided into two rectangles as follows:

Area $A_1 = 150 \times 10 = 1500 \text{ mm}^2$

Area $A_2 = 140 \times 10 = 1400 \text{ mm}^2$

Total Area $A = A_1 + A_2 = 2900 \text{ mm}^2$

Due to symmetry, centroid lies on the symmetric axis $y - y$.

The distance of the centroid from the top most fibre is given by:

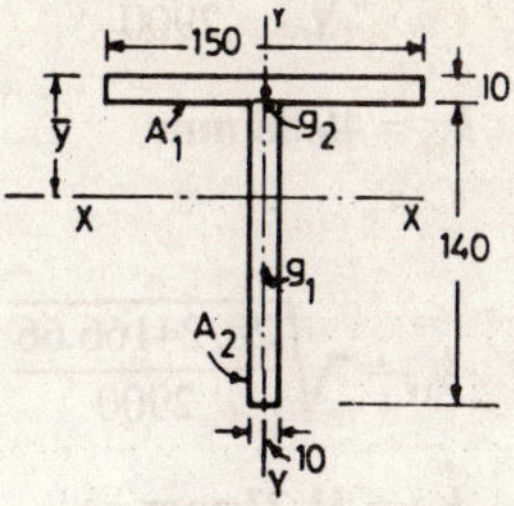

Fig. 5.34

$$\bar{y} = \frac{\text{Sum of moment of the areas about the top most fibre}}{\text{Total area}}$$

$$= \frac{1500 \times 5 + 1400(10 + 70)}{2900}$$

$$= 41.21 \text{ mm}$$

Referring to the centroidal axes $x - x$ and $y - y$, the centroid of A_1 is g_1 (0.0, 36.21) and that of A_2 is g_2 (0.0, 38.79).

Moment of inertia of the section about $x - x$ axis I_{xx} = Moment of inertia of A_1 about $x - x$ axis + moment of inertia of A_2 about $x - x$ axis.

$$\therefore \quad I_{xx} = \frac{150 \times 10^3}{12} + 1500\,(36.21)^2 + \frac{10 \times 140^3}{12} + 1400\,(38.79)^2$$

i.e. $\mathbf{I_{xx} = 63{,}72442.5\ mm^4}$ *Ans.*

Similarly,

$$I_{yy} = \frac{10 \times 150^3}{12} + \frac{140 \times 10^3}{12} = \mathbf{28{,}24166.7\ mm^4} \qquad \textit{Ans.}$$

Hence, the moment of inertia of the section about an axis passing through the centroid and parallel to the top most fibre is 63,72442.5 mm^4 and moment of inertia of the section about the axis of symmetry is 28,24166.66 mm^4.

The radius of gyration is given by:

$$k = \sqrt{\frac{I}{A}}$$

$$\therefore \quad k_{xx} = \sqrt{\frac{I_{xx}}{A}}$$

$$= \sqrt{\frac{63,72442.5}{2900}}$$

$$\mathbf{k_{xx} = 46.88\ mm} \qquad \textit{Ans.}$$

Similarly,

$$k_{yy} = \sqrt{\frac{28,24166.66}{2900}}$$

$$\mathbf{k_{yy} = 31.21\ mm} \qquad \textit{Ans.}$$

Ex. 5.10: Determine the moment of inertia of the L-section shown in the Fig. 5.33 about its centroidal axes parallel to the legs. Also find out the polar moment of inertia.

The given section is divided into two rectangles A_1 and A_2.

Area $A_1 = 125 \times 10 = 1250\ \text{mm}^2$

Area $A_2 = 75 \times 10 = 750\ \text{mm}^2$

Total Area $= 2000\ \text{mm}^2$

First, the centroid of the given section is to be located.

Two reference axes (1)–(1) and (2)–(2) are chosen as shown in Fig. 5.35.

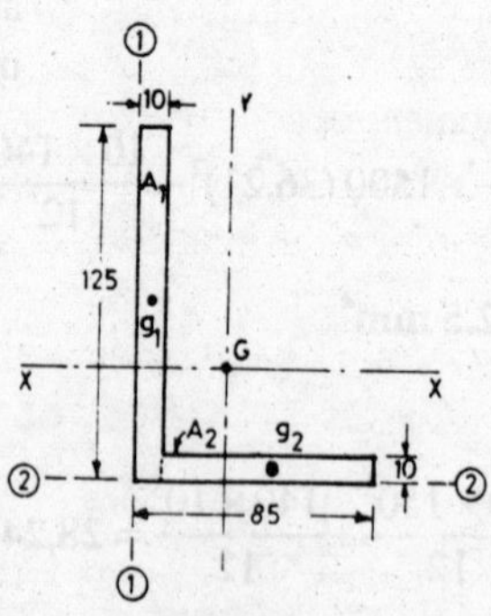

Fig. 5.35

The distance of centroid from the axis (1) – (1)

$$= \frac{\text{sum of moment of areas } A_1 \text{ and } A_2 \text{ about(1)} - (1)}{\text{Total area}}$$

$$\text{i.e. } \bar{x} = \frac{1250 \times 5 + 750\left(10 + \frac{75}{2}\right)}{2000}$$

$$= 20.94\ \text{mm}$$

Similarly,

the distance of the centroid from the axis (2) – (2)

$$= \bar{y} = \frac{1250 \times \frac{125}{2} + 750 \times 5}{2000}$$

$$= 40.94 \text{ mm}$$

With respect to the centroidal axes $x - x$ and $y - y$, the centroid of A_1 is g_1 (15.94, 21.56) and that of A_2 is g_2 (26.56, 35.94).

$\therefore$ I_{xx} = Moment of inertia of A_1 about $x - x$ axis + Moment of inertia of A_2 about $x - x$ axis

$$\therefore \quad I_{xx} = \frac{10 \times 125^3}{12} + 1250 \times 21.56^2 + \frac{75 \times 10^3}{12} + 750 \times 39.94^2$$

i.e $\quad$ $\mathbf{I_{xx} = 34{,}11298.9 mm^4}$ *Ans.*

Similarly,

$$I_{yy} = \frac{125 \times 10^3}{12} + 1250 \times 15.94^2 + \frac{10 \times 75^3}{12} + 750 \times 26.56^2$$

i.e. $\quad$ $\mathbf{I_{yy} = 12{,}08658.9 \text{ mm}^4}$ *Ans.*

Polar moment of inertia = $I_{xx} + I_{yy}$

$= 34{,}11298.9 + 12{,}08658.9$

$\mathbf{I_{zz} = 46{,}19956.8 \text{ mm}^4}$ *Ans.*

Ex. 5.11: Determine the moment of inertia of the symmetric I-section shown in Fig. 5.36 about its centroidal axes $x - x$ and $y - y$.

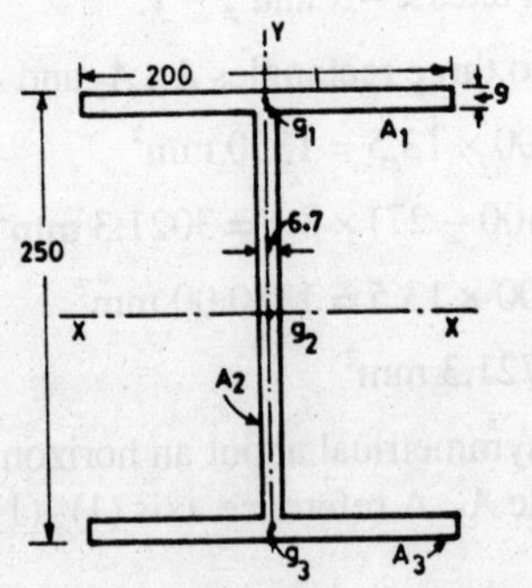

Fig. 5.36

Also, determine moment of inertia of the section about a centroidal axis perpendicular to $x - x$ axis and $y - y$ axis.

The section is divided into three rectangles A_1, A_2, A_3.

Area $\quad A_1 = 200 \times 9 = 1800 \text{ mm}^2$

Area $\quad A_2 = (250 - 9 \times 2) \times 6.7 = 1554.4 \text{ mm}^2$

Area $A_3 = 200 \times 9 = 1800 \text{ mm}^2$

Total Area $A = 5154.4 \text{ mm}^2$

The section is symmetrical about both $x - x$ and $y - y$ axes. Therefore, its centroid will coincide with the centroid of rectangle A_2.

With respect to the centroidal axes $x - x$ and $y - y$, the centroid of rectangle A_1 is g_1 (0.0, 120.5), that of A_2 is g_2 (0.0, 0.0) and that of A_3 is g_3 (0.0, 120.5).

I_{xx} = Moment of inertia of A_1 + Moment of inertia of A_2 + Moment of inertia of A_3 about $x - x$ axis

$$I_{xx} = \frac{200 \times 9^3}{12} + 1800 \times 120.5^2 + \frac{6.7 \times 232^3}{12} + 0 + \frac{200 \times 9^3}{12} + 1800(120.5)^2$$

$$\boldsymbol{I_{xx} = 5{,}92{,}69202 \text{ mm}^4}$$ *Ans.*

Similarly,

$$I_{yy} = \frac{9 \times 200^3}{12} + \frac{232 \times 6.7^3}{12} + \frac{9 \times 200^3}{12}$$

$$\boldsymbol{I_{yy} = 1{,}20{,}05815 \text{ mm}^4}$$ *Ans.*

Moment of inertia of the section about a centroidal axis perpendicular to $x - x$ and $y - y$ axis is nothing but polar moment of inertia, and is given by:

$$I_{zz} = I_{xx} + I_{yy}$$

$$= 5{,}92{,}69202 + 1{,}20{,}05815$$

$$\boldsymbol{I_{zz} = 7{,}12{,}75017 \text{ mm}^4}$$ *Ans.*

Ex. 5.12: Compute the second moment of area of the channel section shown in Fig. 5.38 about centroidal axes $x - x$ and $y - y$.

The section is divided into three rectangles A_1, A_2 and A_3.

Area $A_1 = 100 \times 13.5 = 1350 \text{ mm}^2$

Area $A_2 = (400 - 27) \times 8.1 = 3021.3 \text{ mm}^2$

Area $A_3 = 100 \times 13.5 = 1350.00 \text{ mm}^2$

Total Area $A = 5721.3 \text{ mm}^2$

The given section is symmetrical about an horizontal axis passing through the centroid g_2 of the rectangle A_2. A reference axis (1)–(1) is chosen as shown in Fig. 5.36.

The distance of the centroid of the section from (1) – (1)

$$= \frac{1350 \times 50 + 3021.3 \times \frac{8.1}{2} + 1350 \times 50}{5721.3}$$

$$= 25.73 \text{ mm}$$

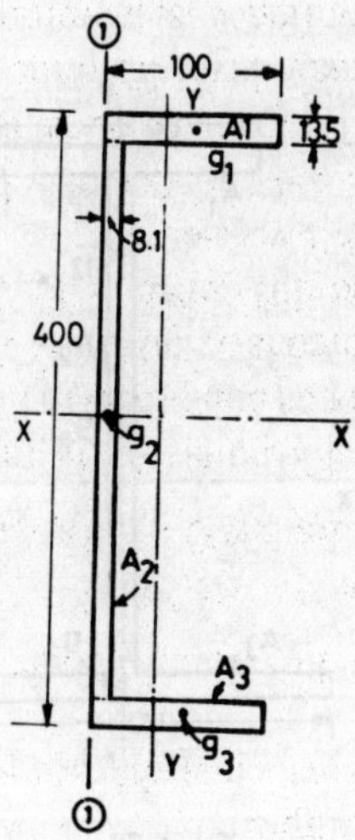

Fig. 5.37

With reference to the centroidal axes $x - x$ and $y - y$, the centroid of the rectangle A_1 is g_1 (24.27, 193.25), that of A_2 is g_2 (21.68, 0.0) and that of A_3 is g_3 (24.27, 193.25).

$\therefore$ I_{xx} = Moment of inertia of A_1, A_2 and A_3 about $x - x$

$$= \frac{100 \times 13.5^3}{12} + 1350 \times 193.25^2 + \frac{8.1 \times 373^3}{12} + \frac{100 \times 13.5^3}{12} + 1350 \times 193.25^2$$

$$\boldsymbol{I_{xx} = 1.359 \times 10^8 \text{ mm}^4} \qquad \textit{Ans.}$$

Similarly,

$$I_{yy} = \frac{13.5 \times 100^3}{12} + 1350 \times 24.27^2 + \frac{273 \times 8.1^3}{12} + 3021.3 \times 21.68^2 + \frac{13.5 \times 100^3}{12} + 1350 \times 24.27^2$$

$$\boldsymbol{I_{yy} = 52{,}72557.6 \text{ mm}^4} \qquad \textit{Ans.}$$

Ex. 5.13: Determine the polar moment of inertia of the I-section shown in the Fig. 5.38. Also determine the radii of gyration with respect to $x - x$ axis and $y - y$ axis.

The section is divided into three rectangles as shown in Fig. 5.38.

Area A_1 = $80 \times 12 = 960 \text{ mm}^2$

Area A_2 = $(150 - 22) \times 12 = 1536 \text{ mm}^2$

Area A_3 = $120 \times 10 = 1200 \text{ mm}^2$

Total area A = 3696 mm^2

Due to symmetry, centroid lies on axis $y - y$. The bottom fibre (1)–(1) is chosen as reference axis to locate the centroid.

The distance of the centroid from (1)–(1)

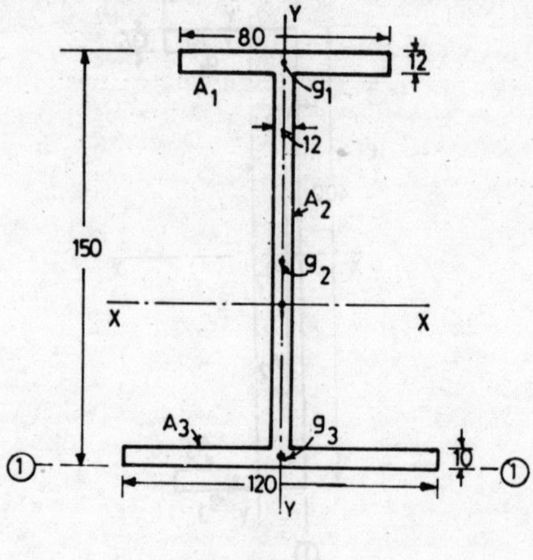

Fig. 5.38

$$= \frac{\text{Sum of moments of the areas of the rectangles about (1) – (1)}}{\text{Total area of sections}}$$

$$= \frac{960 \times (150 - 6) + 1536 \times \left(\frac{128}{2} + 10\right) + 1200 \times 5}{3696}$$

$$= 69.78 \text{ mm}$$

With reference to the centroidal axes $x - x$ and $y - y$, the centroid of the rectangles A_1 is g_1 (0.0, 74.22), that of A_2 is g_2 (0.0, 4.22) and that A_3 is g_3 (0.0, 64.78).

$$I_{xx} = \frac{80 \times 12^3}{12} + 960 \times 74.22^2 + \frac{12 \times 128^3}{12} + 1536 \times 4.22^2 + \frac{120 \times 10^3}{12} + 1200 \times 64.78^2$$

$$I_{xx} = 1,24,70028 \text{ mm}^4$$

$$I_{yy} = \frac{12 \times 80^3}{12} + \frac{128 \times 12^3}{12} + \frac{10 \times 120^3}{12}$$

$$= 19,70432 \text{mm}^4$$

Polar Moment of Inertia $= I_{xx} + I_{yy}$

$= 1,24,70027 + 19,70432$

$=$ **1,44,40459 mm^4** *Ans.*

$$\therefore \quad k_{xx} = \sqrt{\frac{I_{xx}}{A}} = \sqrt{\frac{1,24,70027}{3696}}$$

$=$ **58.09 mm** *Ans.*

$$k_{yy} = \sqrt{\frac{I_{yy}}{A}} = \sqrt{\frac{19,70432}{3696}}$$

$$= \mathbf{23.09 \text{ mm}} \qquad \textit{Ans.}$$

Ex. 5.14: Determine the moment of inertia of the built-up section shown in Fig. 5.39 about its centroidal axes $x - x$ and $y - y$.

The given composite section may be divided into simple rectangles and triangles as shown in the Fig. 5.39.

Area $A_1 = 100 \times 30 = 3000 \text{ mm}^2$

Area $A_2 = 100 \times 25 = 2500 \text{ mm}^2$

Area $A_3 = 200 \times 20 = 4000 \text{ mm}^2$

Area $A_4 = \frac{1}{2} \times 87.5 \times 20 = 875 \text{ mm}^2$

Area $A_5 = \frac{1}{2} \times 87.5 \times 20 = 875 \text{ mm}^2$

Total area $A = 11250 \text{ mm}^2$

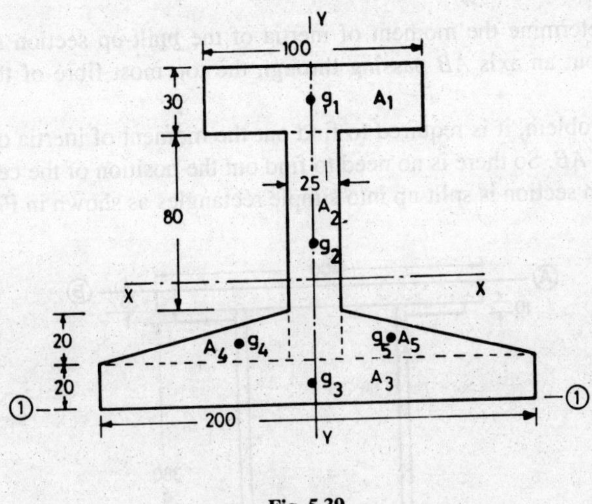

Fig. 5.39

Due to symmetry, centroid lies on the axis $y - y$.

A reference axis (1)–(1) is choosen as shown in the figure.

The distance of the centroidal axis from (1)–(1)

$$= \frac{\text{Sum of moments of areas about (1)} - (1)}{\text{Total area}}$$

$$\bar{y} = \frac{3000 \times 135 + 2500 \times 70 + 4000 \times 10 + 875\left(\frac{1}{3} \times 20 + 20\right) \times 2}{11250}$$

$$= 59.26\text{mm}$$

With reference to the centroidal axes x–x and y–y, the centroid of the rectangle A_1 is g_1 (0.0, 75.74), that of A_2 is g_2 (0.0, 10.74), that of A_3 is g_3 (0.0, 49.26), the centroid of triangle A_4 is g_4 (41.66, 32.59) and that of A_5 is g_5 (41.66, 32.59).

$$I_{xx} = \frac{100 \times 30^3}{12} + 3000 \times 75.74^2 + \frac{25 \times 100^3}{12} + 2500 \times 10.74^2 + \frac{200 \times 20^3}{12}$$

$$+ 4000 \times 49.26^2 + \frac{87.5 \times 20^3}{36} + 875 \times 32.59^2 + \frac{87.5 \times 20^3}{36} + 875 \times 32.59^2$$

$$I_{xx} = \mathbf{3,15,43447\ mm^4}$$ *Ans.*

$$I_{yy} = \frac{30 \times 100^3}{12} + \frac{100 \times 25^3}{12} + \frac{20 \times 200^3}{12} + \frac{20 \times 87.5^3}{36} + 875 \times 41.66^2$$

$$+ \frac{20 \times 87.5^3}{36} + 875 \times 41.66^2$$

$$I_{yy} = \mathbf{1,97,45122\ mm^4}$$ *Ans.*

Ex. 5.15: Determine the moment of inertia of the built-up section shown in the Fig. 5.40 about an axis *AB* passing through the top most fibre of the section as shown.

In this problem, it is required to find out the moment of inertia of the section about an axis *AB*. So there is no need to find out the position of the centroid.

The given section is split up into simple rectangles as shown in Fig. 5.40.

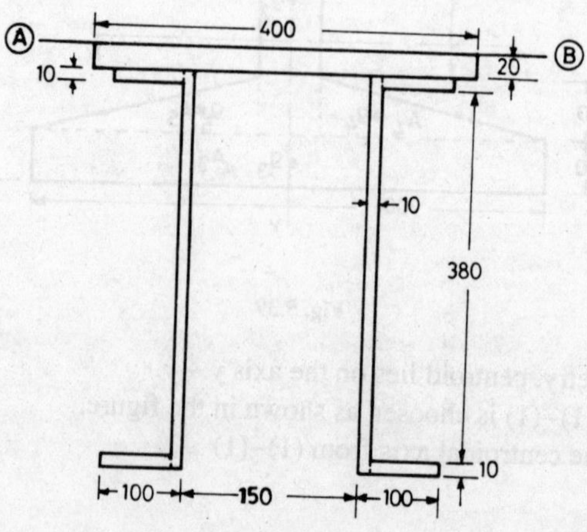

Fig. 5.40

Now,

Moment of Inertia about AB = Sum of moments of inertia of the rectangles about AB

$$= \frac{400 \times 20^3}{12} + 400 \times 20 \times 10^2 + \left[\frac{100 \times 10^3}{12} + 100 \times 10 \times (20+5)^2\right] \times 2$$

$$+ \left[\frac{10 \times 380^3}{12} + 10 \times 380 \times (30+190)^2\right] \times 2 + \left[\frac{100 \times 10^3}{12} + 100 \times 10 \times (20+10+380+5)^2\right] \times 2$$

$$\mathbf{I_{AB} = 8.06093 \times 10^8 \ mm^4}$$ *Ans.*

Ex. 5.16: Calculate the moment of inertia of the built-up section shown in Fig. 5.41 about a centroidal axis parallel to AB. All members are 10 mm thick.

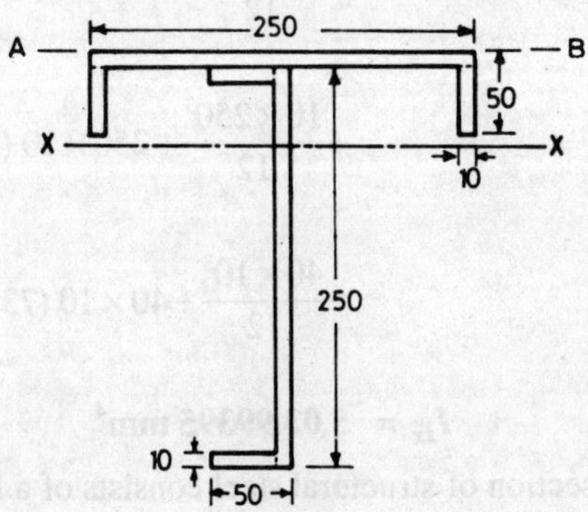

Fig. 5.41

The built-up section is divided into six simple rectangles as shown in the figure.

The distance of centroidal axis from AB

$$= \frac{\text{Sum of the moment of areas about } AB}{\text{Total Area}}$$

$$= \frac{\Sigma A_i \, y_i}{A}$$

Now,

$$\Sigma A_i \, y_i = 250 \times 10 \times 5 + 2 \times 40 \times 10 \times (10 + 20) + 40 \times 10 \times (10 + 5) + 40 \times 10 \times 255 + 250 \times 10 \times (10 + 125)$$

$$= 4{,}82000 \text{ mm}^3$$

$$A = 2 \times 250 \times 10 + 40 \times 10 \times 4$$

$$= 6600 \text{ mm}^2$$

$\therefore$

$$\bar{y} = \frac{\Sigma A_i \, y_i}{A} = \frac{4{,}82000}{6600}$$

$$= 73.03 \text{ mm}$$

Now,

Moment of inertia about the centroidal axis $\Big\} = \Big\{$ Sum of the moment of inertia of the individual rectangles

$$= \frac{250 \times 10^3}{12} + 250 \times 10 \times (73.03 - 5)^2$$

$$+ \left[\frac{10 \times 40^3}{12} + 40 \times 10\,(73.03 - 30)^2 \right] \times 2$$

$$+ \frac{40 \times 10^3}{12} + 40 \times 10\,(73.03 - 15)^2$$

$$+ \frac{10 \times 250^3}{12} + 250 \times 10\,(73.03 - 135)^2$$

$$+ \frac{40 \times 10^3}{12} + 40 \times 10\,(73.03 - 255)^2$$

$$\boldsymbol{I_{xx} = 5{,}03{,}99395 \text{ mm}^4}$$ *Ans.*

Ex. 5.17: A built-up section of structural steel consists of a flange plate 400 mm × 20 mm, a web plate 600 mm × 15 mm and two angles 150 mm × 150 mm × 10 mm assembled to form a section as shown in Fig. 5.42. Determine the moment of inertia of the section about the horizontal centroidal axis.

Each angle is divided into two rectangles as shown in Fig. 5.42.

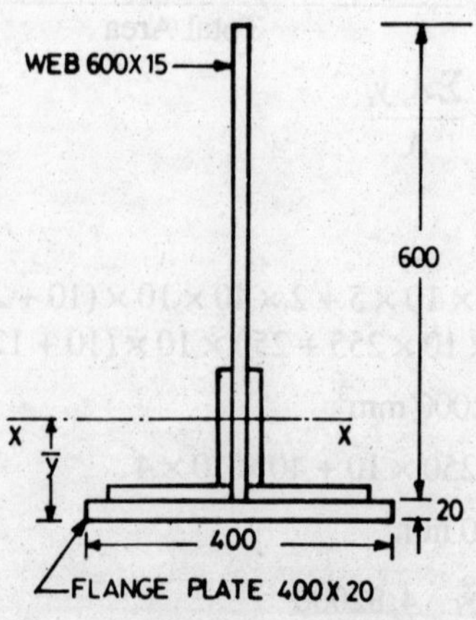

Fig. 5.42

The distance of the centroidal axis from the bottom fibres of section

$$= \frac{\text{Sum of the moment of the areas about bottom fibres}}{\text{Total area of the section}}$$

$$= \frac{\Sigma A_i\, y_i}{A}$$

Now,

$$\Sigma A_i\, y_i = 600 \times 15 \times \left(\frac{600}{2} + 20\right) + 140 \times 10 \times (70 + 30) \times 2$$

$$+ 150 \times 10 \times (5 + 20) \times 2 + 400 \times 20 \times 10$$

$$= 3315000\,\text{mm}^3$$

$$A = 600 \times 15 + 140 \times 10 \times 2 + 150 \times 10 \times 2 + 400 \times 20$$

$$= 22,800\,\text{mm}^2$$

$$\therefore \quad \bar{y} = \frac{\Sigma A_i\, y_i}{A} = \frac{19,65000}{22,800}$$

$$= 145.39\text{ mm}$$

$$\left.\begin{array}{l}\text{Moment of inertia of section about}\\ \text{centroidal axis}\end{array}\right\} = \left\{\begin{array}{l}\text{Sum of the moment of inertia of the}\\ \text{all simple figures about centroidal}\\ \text{axis}\end{array}\right.$$

$$= \frac{15 \times 600^3}{12} + 600 \times 15(145.39 - 320)^2$$

$$+ \left[\frac{10 \times 140^3}{12} + 1400(145.39 - 100)^2\right] \times 2$$

$$+ \left[\frac{150 \times 10^3}{12} + 1500 \times (145.39 - 15)^2\right] \times 2$$

$$+ \frac{400 \times 20^3}{12} + 400 \times 20 \times (145.39 - 10)^2$$

$$I_{xx} = 7.45156 \times 10^8\ \text{mm}^4 \qquad \textit{Ans.}$$

Ex. 5.18: Compute the moment of inertia of the 100 mm × 150 mm rectangle shown in Fig. 5.43 about $x - x$ axis to which it is inclined at an angle $\theta = \sin^{-1}\left(\frac{4}{5}\right)$

The rectangle is divided into four triangles as shown in the figure. [The dividing line between triangles A_1 and A_2 is parallel to $x - x$ axis].

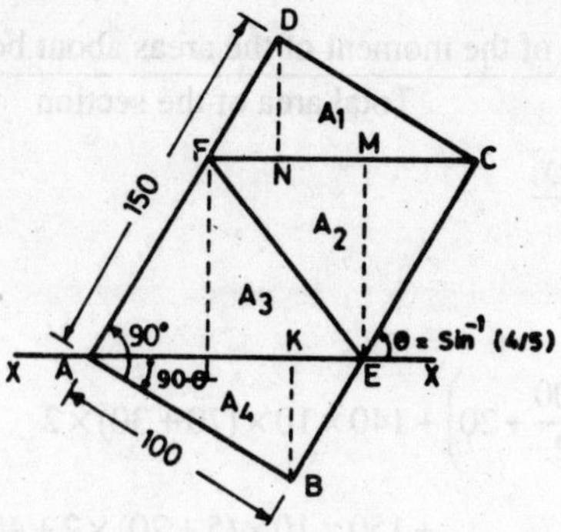

Fig. 5.43

Now $= \sin^{-1}\left(\frac{4}{5}\right) = 53.13°$

From the geometry of the Fig. 5.43,

$$BK = AB \sin (90° - \theta)$$
$$= 100 \sin (90° - 53.13°)$$
$$= 60 \text{ mm}$$
$$ND = BK = 60 \text{ mm}$$

$\therefore$ $$FD = \frac{60}{\sin\theta} = \frac{60}{\sin 53.13} = 75 \text{ mm}$$

$\therefore$ $$AF = 150 - FD = 75 \text{ mm}$$

Hence $$FL = ME = 75 \sin\theta = 60 \text{ mm}$$

$$AE = FC = \frac{AB}{\cos(90-\theta)} = \frac{100}{0.8} = 125 \text{ mm}$$

Moment of inertia of the section about $x - x$ axis = Sum of the moments of inertia of individual triangular areas about $x - x$ axis.

$$= I_{DFC} + I_{FCE} + I_{FEA} + I_{AEB}$$

$$= \frac{125 \times 60^3}{36} + \frac{1}{2} \times 125 \times 60 \times \left(60 + \frac{1}{3} \times 60\right)^2$$

$$+ \frac{125 \times 60^3}{36} + \frac{1}{2} \times 125 \times 60 \times \left(\frac{2}{3} \times 60\right)^2$$

$$+ \frac{125 \times 60^3}{36} + \frac{1}{2} \times 125 \times 60 \times \left(\frac{1}{3} \times 60\right)^2$$

$$+\frac{125\times 60^3}{36}+\frac{1}{2}\times 125\times 60\times\left(\frac{1}{3}\times 60\right)^2$$

$$I_{xx} = \mathbf{3{,}60{,}00000\ mm^4} \qquad \textit{Ans.}$$

Ex. 5.19: Find moment of inertia of the shaded area shown in the Fig. 5.44 about the axis *AB*.

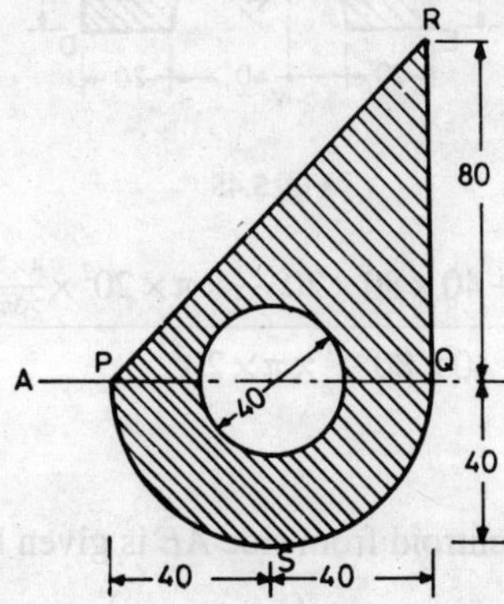

Fig. 5.44

The section is divided into a triangle *PQR*, a semicircle *PSQ* having base on axis *AB* and a circle having its centre on axis *AB*.

Now,

Moment of inertia of the section about axis *AB* = Moment of inertia of triangle *PQR* about *AB* + Moment of inertia of semicircle *PSQ* about *AB* – moment of inertia of circle about *AB*

$$=\frac{80\times 80^3}{12}+\frac{\pi}{128}\times 80^4-\frac{\pi}{64}\times 40^4$$

$$I_{AB} = \mathbf{42{,}92979\ mm^4} \qquad \textit{Ans.}$$

Ex. 5.20: Find the second moment of the shaded portion shown in the Fig. 5.45 about its centroidal axis.

The section is divided into three simple figures viz., a triangle *ABC*, a rectangle *ACDE* and a semicircle.

The distance of centroid of the section from base *ED*

$$=\frac{\text{Moment of area of triangle } ABC + \text{Moment of area of rectangle } ACDE - \text{Moment of semicircle}}{\text{Area of triangle ABC + Area of rectangle ACDE - Area of semicircle}}$$

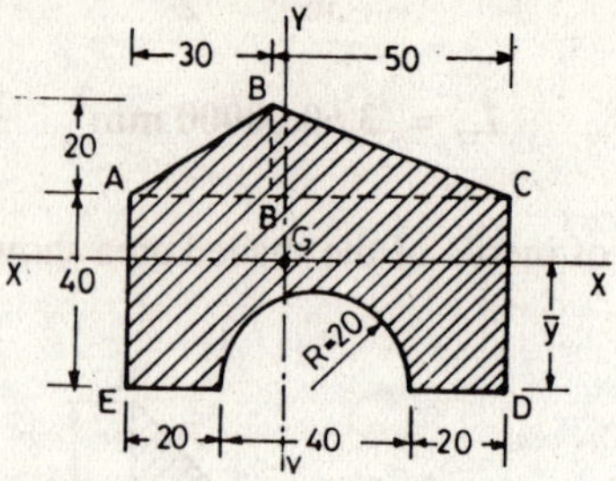

Fig. 5.45

$$\bar{y} = \frac{\frac{1}{2}\times 80\times 20\left(\frac{1}{3}\times 20+40\right)+40\times 80\times 20-\frac{1}{2}\times\pi\times 20^2\times\frac{4\times 20}{3\pi}}{\frac{1}{2}\times 80\times 20+40\times 80-\frac{1}{2}\times\pi\times 20^2}$$

$= 28.47$ mm

Similarly, the distance of centroid from side *AE* is given by:

$$\bar{x} = \frac{\frac{1}{2}\times 30\times 20\times\frac{2}{3}\times 30+\frac{1}{2}\times 50\times 20\times\left(\frac{1}{3}\times 50+30\right)+40\times 80\times 40-\frac{1}{2}\times\pi\times 20^2\times 40}{\frac{1}{2}\times 80\times 20+40\times 80-\frac{1}{2}\times\pi\times 20^2}$$

$= 39.21$ mm

Moment of inertia about centroidal $x - x$ axis = Moment of inertia of triangle *ABC* about $x - x$ axis + Moment of inertia of rectangle about $x - x$ axis − moment of semicircle about $x - x$ axis

$$\therefore\quad I_{xx} = \frac{80\times 20^3}{36}+\frac{1}{2}\times 80\times 20\left(60-\frac{2}{3}\times 20-28.47\right)^2+\frac{80\times 40^3}{12}+80\times 40\times(28.47-20)^2$$

$$-\left[0.0068598\times 20^4+\frac{1}{2}\pi\times 20^2\left(28.47-\frac{4\times 20}{3\pi}\right)^2\right]$$

I_{xx} = **6,86944 mm⁴** *Ans.*

Similarly,

$$I_{yy} = \frac{20\times 30^3}{36}+\frac{1}{2}\times 20\times 30\left(39.21-\frac{2}{3}\times 30\right)^2+\frac{20\times 50^3}{36}+\frac{1}{2}\times 20\times 50$$

$$\times\left[36.8364-\left(30+\frac{1}{3}\times 50\right)\right]^2+\frac{40+80^3}{12}+40\times 80(39.21-40)^2$$

$$-\frac{1}{2}\times\frac{\pi}{64}\times 20^4-\frac{1}{2}\times\pi\times 20^2(40-39.21)^2$$

I_{yy} = **19,27297.3 mm⁴** *Ans.*

EXERCISE

5.1 Determine the centroid of the built-up section in Fig. 5.46. Express the coordinates of centroid with respect to *x* and *y* axes shown.

[*Ans.* $\bar{x}$ = 48.91 mm; $\bar{y}$ = 61.30 mm]

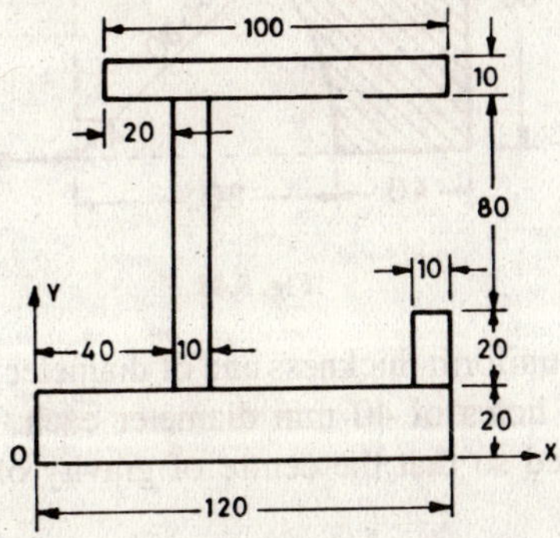

Fig. 5.46

5.2 Determine the centroid of the reinforced concrete retaining wall section shown in Fig. 5.47.

[*Ans.* $\bar{x}$ = 1.848 m; $\bar{y}$ = 1.825 m]

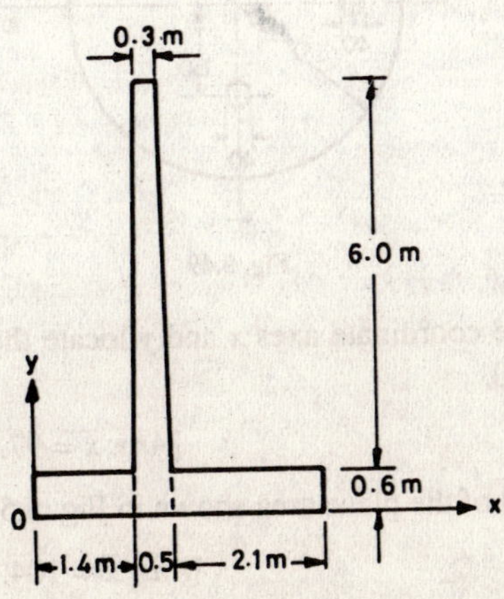

Fig. 5.47

5.3 Find the coordinates of the centroid of the shaded area with respect to the axes shown in Fig. 5.48.

[Ans. $\bar{x}$ = 43.98 mm; $\bar{y}$ = 70.15 mm]

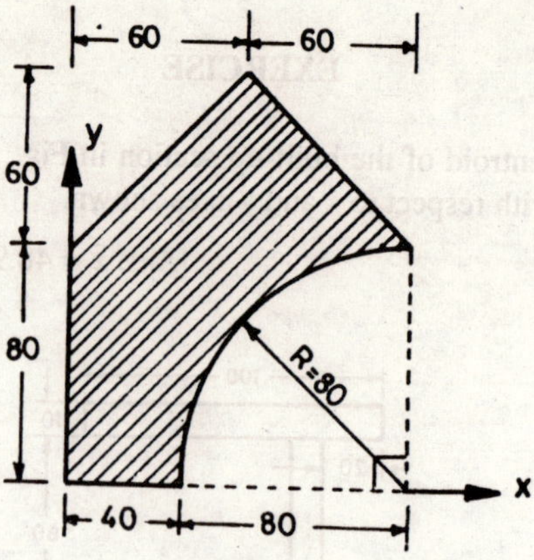

Fig. 5.48

5.4 A circular plate of uniform thickness and of diameter 500 mm as shown in Fig. 5.49 has two circular holes of 40 mm diameter each. Where should a 80 mm diameter hole be drilled so that the centre of gravity of the plate will be at the geometric centre.

[Ans. $x = 50$ mm; $y = 37.5$ mm]

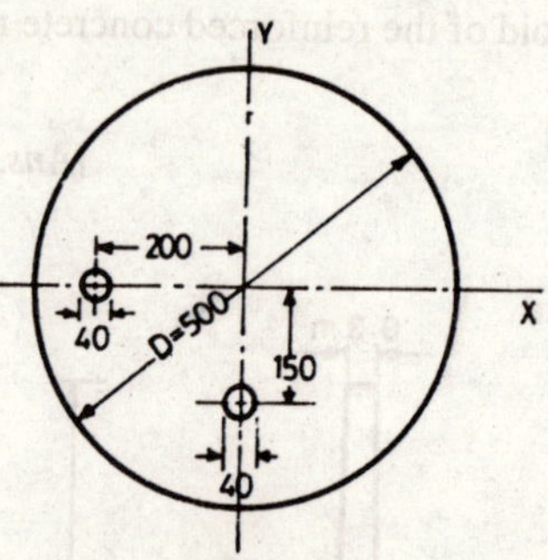

Fig. 5.49

5.5 With respect to the coordinate axes x and y locate the centroid of the shaded area shown in Fig. 5.50.

[Ans. $\bar{x} = 97.47$ mm; $\bar{y} = 70.69$ mm]

5.5 Locate the centroid of the plane area shown in Fig. 5.51.

[Ans. $\bar{x} = 104.10$ mm; $\bar{y} = 44.30$ mm]

5.7 Determine the coordinates of the centroid of shaded are shown in Fig. 5.52 with respect to the corner point O. Take $x = 40$ mm

[Ans. $\bar{x} = 235.45$ mm; $\bar{y} = 90.87$ mm]

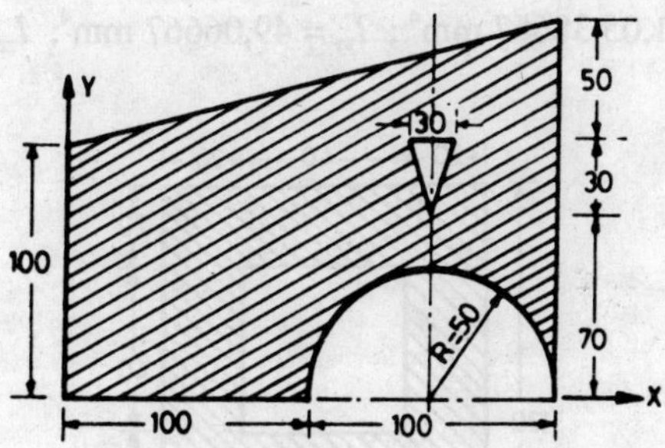

Fig. 5.50

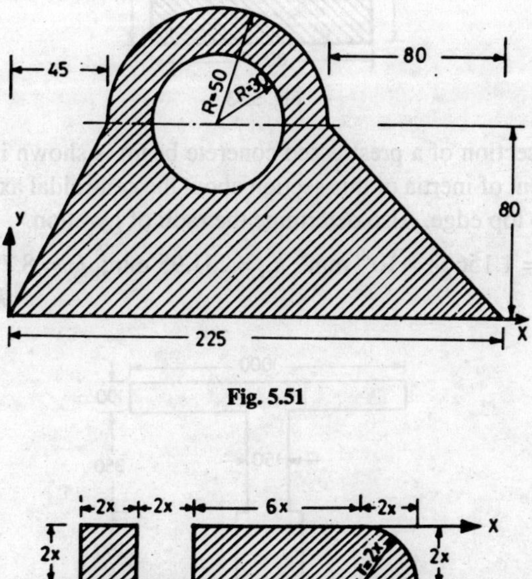

Fig. 5.51

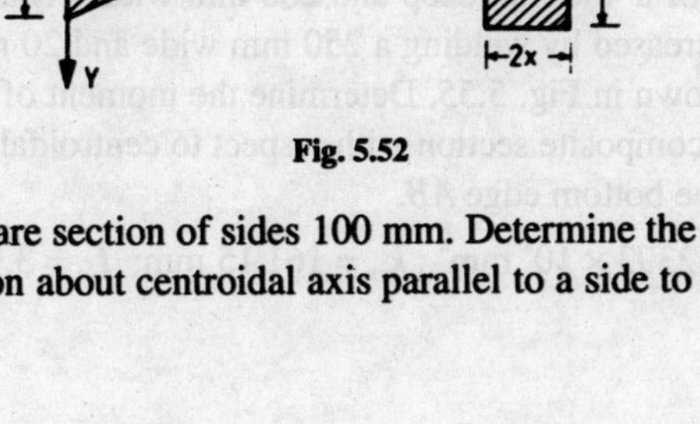

Fig. 5.52

5.8 ***ABCD*** **is a square section of sides 100 mm. Determine the ratio of moment of inertia of the section about centroidal axis parallel to a side to that about diagonal** ***AC*****.**

[*Ans.* 1]

5.9 The cross-section of a rectangular hollow beam is as shown in Fig. 5.53. Determine the polar moment of inertia of the section about centroidal axis.

[*Ans.* I_{xx} = 1,05,38667 mm^4 ; I_{yy} = 49,06667 mm^4; I_{zz} = 1,54,45334 mm^4]

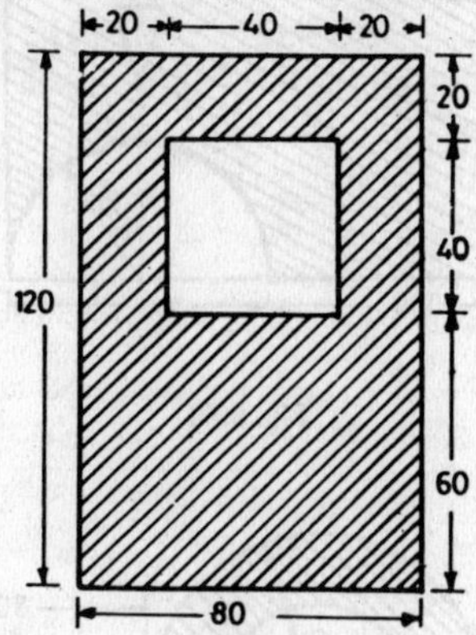

Fig. 5.53

5.10 The cross-section of a prestressed concrete beam is shown in Fig. 5.54. Calculate the moment of inertia of this section about the centroidal axis parallel to and perpendicular to top edge. Also determine the radii of gyration.

[*Ans.* I_{xx} = 1.15668 × 10^{10} mm^4; k_{xx} = 231.95 mm; I_{yy} = 8.75729 × 10^9 mm^4; k_{yy} = 201.82 mm]

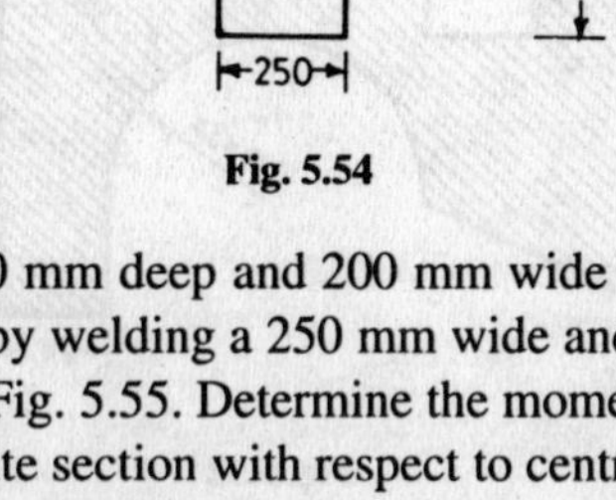

Fig. 5.54

5.11 The strength of a 400 mm deep and 200 mm wide I-beam of uniform thickness 10 mm, is increased by welding a 250 mm wide and 20 mm thick plate to its upper flange as shown in Fig. 5.55. Determine the moment of inertia and the radii of gyration of the composite section with respect to centroidal axes parallel to and perpendicular to the bottom edge *AB*.

[*Ans.* I_{xx} = 3.32393 × 10^8 mm^4 ; k_{xx} = 161.15 mm; I_{yy} = 3,94,06667 mm^4; k_{yy} = 55.49 mm]

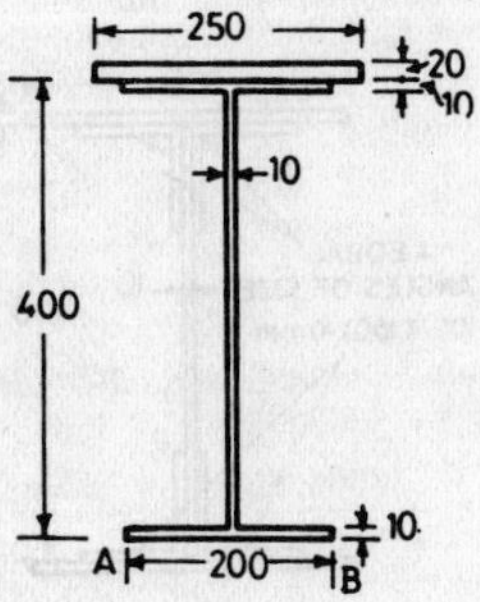

Fig. 5.55

5.12 The cross-section of a gantry girder is as shown in Fig. 5.56. It is made up of an I-section of depth 450 mm, flange width 200 mm and a channel of size 400 mm × 150 mm. Thickness of all members is 10 mm. Find the moment of inertia of the section about the horizontal centroid axis.

[*Ans.* $I_{xx} = 4.2198 \times 10^8$ mm^4]

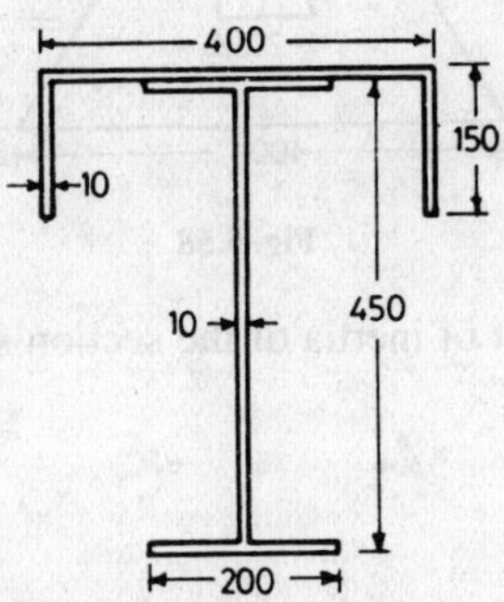

Fig. 5.56

5.13 A plate girder is made up of a web plate of size 400 mm × 10 mm, four angles of size 100 mm × 100 mm × 10 mm and cover plates of size 300 mm × 10 mm as shown in Fig. 5.57. Determine the moment of inertia about horizontal and vertical centroidal axes.

[*Ans.* $I_{xx} = 5.35786 \times 10^8$ mm^4; I_{yy} = 6,08,50,667 mm^4]

5.14 Determine the moment of inertia and radii of gyration of the area shown in Fig. 5.58 about the base *A* – *B* and the centroidal axis parallel to *AB*.

[*Ans.* I_{AB} = 48,15000 mm^4; I_{xx} = 18,24231 mm^4]

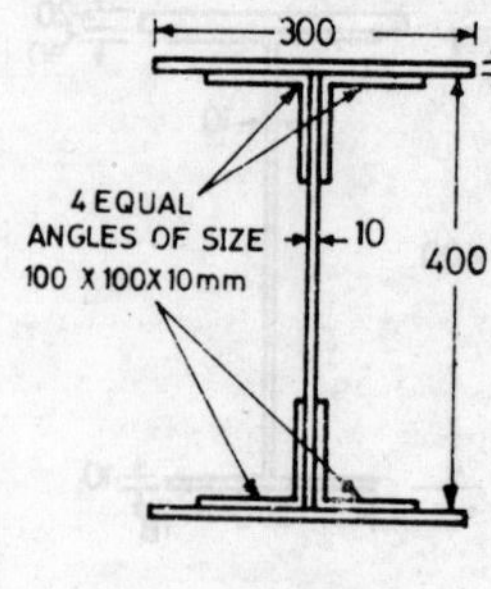

Fig. 5.57

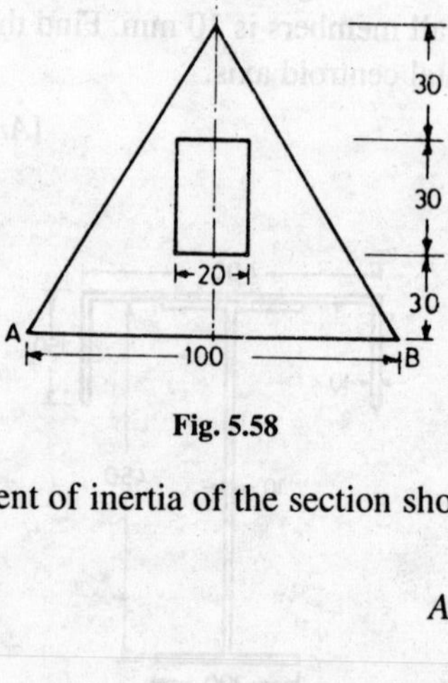

Fig. 5.58

5.15 Determine the moment of inertia of the section shown in Fig. 5.59 about the vertical centroidal axis.

Ans. I_{yy} = 5,03,82857 mm^4]

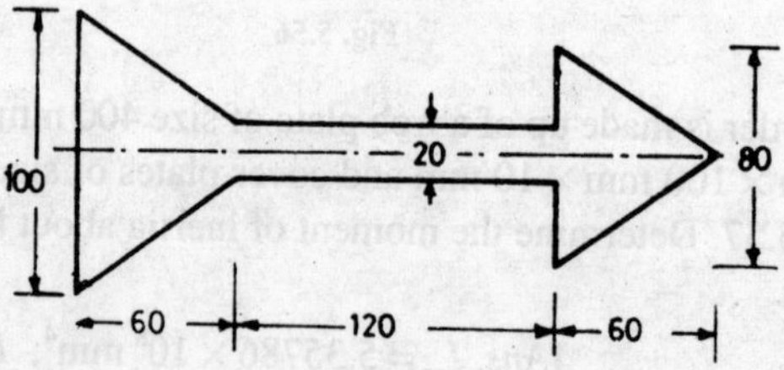

Fig. 5.59

5.16 A semi-circular cut is made in a rectangular wooden beam as shown in Fig. 5.60. Determine the polar moment of inertia of the section about the centroidal axis.

[*Ans.* I_{xx} = 33581456 mm^4; I_{yy} = 1,00,45631 mm^4; I_{zz} = 22098980 mm^4]

5.17 Determine the moment of inertia of the section shown in the Fig. 5.61 about

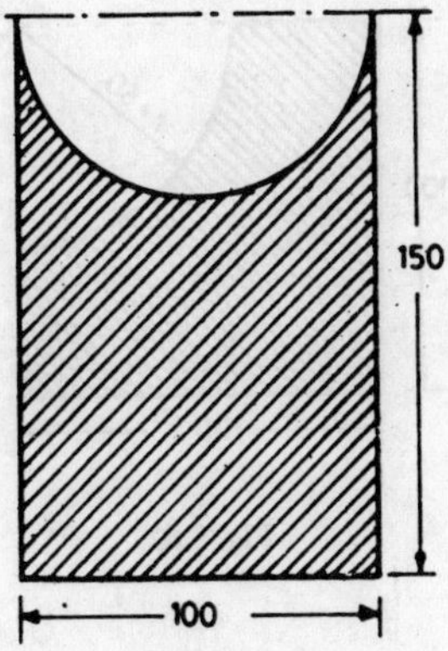

Fig. 5.60

the horizontal centroidal axis. Also find the moment of inertia of the section about the symmetrical axis. Hence find the polar moment of inertia.

[*Ans.* I_{xx} = 5409046 mm^4; I_{yy} = 1455310 mm^4; I_{zz} = 6864356 mm^4]

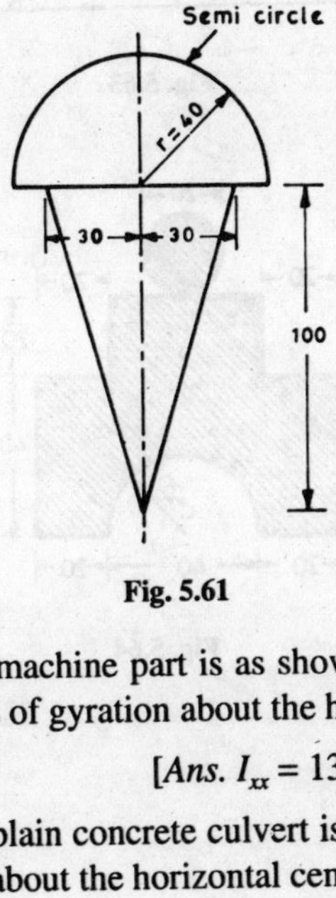

Fig. 5.61

5.18 The cross-section of a machine part is as shown in Fig. 5.62. Determine its moment of inertia and radius of gyration about the horizontal centroidal axis.

[*Ans.* I_{xx} = 13287754 mm^4; k_{xx} = 43.04 mm

5.19 The cross-section of a plain concrete culvert is as shown in Fig. 5.63. Determine the moment of inertia about the horizontal centroidal axes.

[*Ans.* $I_{xx} = 5.45865 \times 10^{10}$ mm^4]

5.20 Determine the centroid of the built-up section shown in Fig. 5.64 and find the moment of inertia and radius of gyration about the horizontal centroidal axis.

[*Ans.* I_{xx} = 1267942 mm^4; k_{xx} = 18.55 mm]

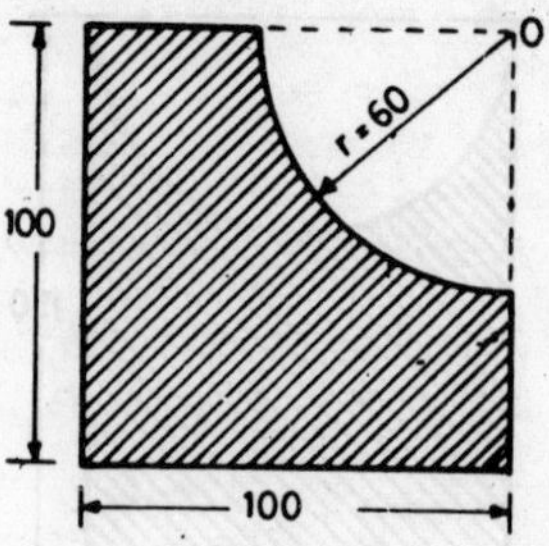

Fig. 5.62

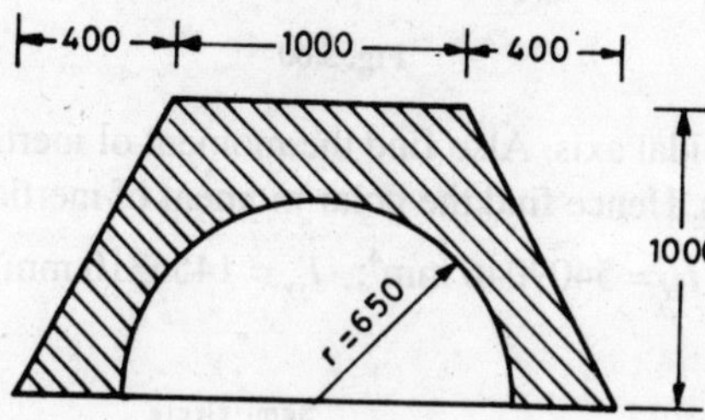

Fig. 5.63

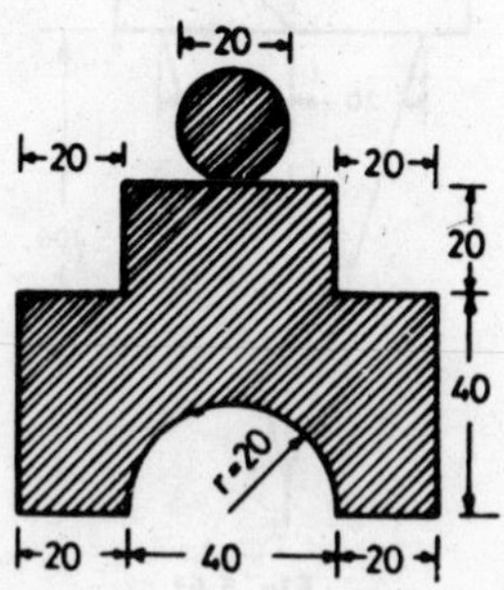

Fig. 5.64